Statistical Implications of Turing's Formula

Statistical Implications of Turing's Formula

Zhiyi Zhang

Department of Mathematics and Statistics, University of North Carolina, Charlotte, NC, US

Published by John Wiley & Sons, Inc., Hoboken, New Jersey

Published simultaneously in Canada

Library of Congress Cataloging-in-Publication Data:

Names: Zhang, Zhiyi, 1960-
Title: Statistical implications of Turing's formula / Zhiyi Zhang.
Description: Hoboken, New Jersey : John Wiley & Sons, Inc., [2017] | Includes bibliographical references and index.
Identifiers: LCCN 2016027830| ISBN 9781119237068 (cloth) | ISBN 9781119237099 (epub)
Subjects: LCSH: Mathematical statistics–Textbooks. | Probabilities–Textbooks.
Classification: LCC QA273 .Z43 2017 | DDC 519.5–dc23 LC record available at https://lccn.loc.gov/2016027830

Cover image courtesy Author Zhiyi Zhang

Set in 10/12pt, WarnockPro by SPi Global, Chennai, India.

Printed in the United States of America

10 9 8 7 6 5 4 3 2 1

To my family and all my teachers

Contents

Preface

This book introduces readers to Turing's formula and then re-examines several core statistical issues of modern data science from Turing's perspective. Turing's formula was a remarkable invention of Alan Turing during World War II in an early attempt to decode the German enigmas. The formula looks at the world of randomness through a unique and powerful binary perspective – unmistakably of Turing. However, Turing's formula was not well understood for many years. Research amassed during the last decade has brought to light profound and new statistical implications of the formula that were previously not known. Recently, and only recently, a relatively clear and systematic description of Turing's formula, with its statistical properties and implications, has become possible. Hence this book.

Turing's formula is often perceived as having a mystical quality. I was awestruck when I first learned of the formula 10 years ago. Its anti-intuitive implication was simply beyond my immediate grasp. However, I was not along in this regard. After turning it over in my mind for a while, I mentioned to two of my colleagues, both seasoned mathematicians, that there might be a way to give a nonparametric characterization to tail probability of a random variable beyond data range. To that, their immediate reaction was, "tell us more when you have figured it out." Some years later, a former doctoral student of mine said to me, "I used to refuse to think about anti-intuitive mathematical statements, but after Turing's formula, I would think about a statement at least twice however anti-intuitive it may sound." Still another colleague of mine recently said to me, "I read everything you wrote on the subject, including details of the proofs. But I still cannot see intuitively why the formula works." To that, I responded with the following two points:

1) Our intuition is a bounded mental box within which we conduct intellectual exercises with relative ease and comfort, but we must admit that this box also reflects the limitations of our experience, knowledge, and ability to reason.
2) If a fact known to be true does not fit into one's current box of intuition, is it not time to expand the boundary of the box to accommodate the true fact?

My personal journey in learning about Turing's formula has proved to be a rewarding one. The experience of observing Turing's formula totally outside of my box of intuition initially and then having it gradually snuggled well within the boundary of my new box of intuition is one I wish to share.

Turing's formula itself, while extraordinary in many ways, is not the only reason for this book. Statistical science, since R.A. Fisher, has come a long way and continues to evolve. In fact, the frontier of Statistics has largely moved on to the realm of nonparametrics. The last few decades have witnessed great advances in the theory and practice of nonparametric statistics. However in this realm, a seemingly impenetrable wall exists: how could one possibly make inference about the tail of a distribution beyond data range? In front of this wall, many, if not most, are discouraged by their intuition from exploring further. Yet it is often said in Statistics that "it is all in the tail." Statistics needs a trail to get to the other side of the wall. Turing's formula blazes a trail, and this book attempts to mark that trail.

Turing's formula is relevant to many key issues in modern data sciences, for example, Big Data. Big Data, though as of yet not a field of study with a clearly defined boundary, unambiguously points to a data space that is a quantum leap away from what is imaginable in the realm of classical statistics in terms of data volume, data structure, and data complexity. Big Data, however defined, issues fundamental challenges to Statistics. To begin, the task of retrieving and analyzing data in a vastly complex data space must be in large part delegated to a machine (or software), hence the term Machine Learning. How does a machine learn and make judgment? At the very core, it all boils down to a general measure of association between two observable random elements (not necessarily random variables). At least two fundamental issues immediately present themselves:

1) *High Dimensionality.* The complexity of the data space suggests that a data observation can only be appropriately registered in a very high-dimensional space, so much so that the dimensionality could be essentially infinite. Quickly, the usual statistical methodologies run into fundamental conceptual problems.
2) *Discrete and Non-ordinal Nature.* The generality of the data space suggests that possible data values may not have a natural order among themselves: different gene types in the human genome, different words in text, and different species in an ecological population are all examples of general data spaces without a natural "neighborhood" concept.

Such issues would force a fundamental transition from the platform of random variables (on the real line) to the platform of random elements (on a general set or an alphabet). On such an alphabet, many familiar and fundamental concepts of Statistics and Probability no longer exist, for example, moments, correlation, tail, and so on. It would seem that Statistics is in need of a rebirth to tackle these issues.

The rebirth has been taking place in Information Theory. Its founding father, Claude Shannon, defined two conceptual building blocks: entropy (in place of moments) and mutual information (in place of correlation) in his landmark paper (Shannon, 1948). Just as important as estimating moments and coefficient of correlation for random variables, entropy and mutual information must be estimated for random elements in practice. However, estimation of entropy and estimation of mutual information are technically difficult problems due to the curse of "*High Dimensionality*" and "*Discrete and Non-ordinal Nature.*" For about 50 years since (Shannon, 1948), advances in this arena have been slow to come. In recent years however, research interest, propelled by the rapidly increasing level of data complexity, has been reinvigorated and, at the same time, has been splintered into many different perspectives. One in particular is Turing's perspective, which has brought about significant and qualitative improvement to these difficult problems. This book presents an overview of the key results and updates the frontier in this research space.

The powerful utility of Turing's perspective can also be seen in many other areas. One increasingly important modern concept is Diversity. The topics of what it is and how to estimate it are rapidly moving into rigorous mathematical treatment. Scientists have passionately argued about them for years but largely without consensus. Turing's perspective gives some very interesting answers to these questions. This book gives a unified discussion of diversity indices, hence making good reading for those who are interested in diversity indices and their estimation. The final two chapters of the book speak to the issues of tail classification and, if classified, how to perform a refined analysis for a parametric tail model via Turing's perspective. These issues are scientifically relevant in many fields of study.

I intend this book to serve two groups of readers:

1) *Textbook* for graduate students. The material is suitable for a topic course at the graduate level for students in Mathematics, Probability, Statistics, Computer Science (Artificial Intelligence, Machine Learning, Big Data), and Information Theory.
2) *Reference book* for researchers and practitioners. This book offers an informative presentation of many of the critical statistical issues of modern data science and with updated new results. Both researchers and practitioners will find this book a good learning resource and enjoy the many relevant methodologies and formulas given and explained under one cover.

For a better flow of the presentation, some of the lengthy but instructive proofs are placed at the end of each chapter.

The seven chapters of this book may be naturally organized into three groups. Group 1 includes Chapters 1 and 2. Chapter 1 gives an introduction to Turing's formula; and Chapter 2 translates Turing's formula into a particular perspective (referred to as Turing's perspective) as embodied in a class of indices (referred to as Generalized Simpson's Indices). Group 1 may be considered as the theoretical foundation of the whole book. Group 2 includes Chapters 3–5. Chapter 3 takes Turing's perspective into entropy estimation, Chapter 4 takes it into diversity estimation, and Chapter 5 takes it into estimation of various information indices. Group 2 may be thought of as consisting of applications of Turing's perspective. Chapters 6 and 7 make up Group 3. Chapter 6 discusses the notion of tail on alphabets and offers a classification of probability distributions. Chapter 7 offers an application of Turing's formula in estimating parametric tails of random variables. Group 3 may be considered as a pathway to further research.

The material in this book is relatively new. In writing the book, I have made an effort to let the book, as well as its chapters, be self-contained. On the one hand, I wanted the material of the book to flow in a linearly coherent manner for students learning it for the first time. In this regard, readers may experience a certain degree of repetitiveness in notation definitions, lemmas, and even proofs across chapters. On the other hand, I wanted the book to go beyond merely stating established results and referring to proofs published elsewhere. Many of the mathematical results in the book have instructive value, and their proofs indicate the depth of the results. For this reason, I have included many proofs that might be judged overly lengthy and technical in a conventional textbook, mostly in the appendices.

It is important to note that this book, as the title suggests, is essentially a monograph on Turing's formula. It is not meant to be a comprehensive learning resource on topics such as estimation of entropy, estimation of diversity, or estimation of information. Consequently, many worthy methodologies in these topics have not been included. By no means do I suggest that the methodologies discussed in this book are the only ones with scientific merit. Far from it, there are many wonderful ideas proposed in the existing literature but not mentioned among the pages of this book, and assuredly many more are yet to come.

I wish to extend my heartfelt gratitude to those who have so kindly allowed me to bend their ears over the years. In particular, I wish to thank Hongwei Huang, Stas Molchanov, and Michael Grabchak for countless discussions on Turing's formula and related topics; my students, Chen Chen, Li Liu, Ann Stewart, and Jialin Zhang for picking out numerous errors in an earlier draft of the book; and The University of North Carolina at Charlotte for granting me a sabbatical leave in Spring 2015, which allowed me to bring this book to a complete draft. Most importantly, I wish to thank my family, wife Carol, daughter Katherine, and son Derek, without whose love and unwavering support this book would not have been possible.

Zhiyi Zhang (张稚逸)
May 2016 *Charlotte, North Carolina*

1

Turing's Formula

Consider the population of all birds in the world along with all its different species, say $\{s_k; k \geq 1\}$, and denote the corresponding proportion distribution by $\{p_k; k \geq 1\}$ where p_k is the proportion of the kth bird species in the population. Suppose a random sample of $n = 2000$ is to be taken from the population, and let the bird counts for the different species be denoted by $\{Y_k; k \geq 1\}$. If it is of interest to estimate p_1 the proportion of birds of species s_1 in the population, then $\hat{p}_1 = Y_1/n$ is an excellent estimator; and similarly so is $\hat{p}_k = Y_k/n$ for p_k for every particular k.

To illustrate, consider a hypothetical sample of size $n = 2000$ with bird counts given in Table 1.1 or a version rearranged in decreasing order of the observed frequencies as in Table 1.2.

With this sample, one would likely estimate p_1 by $\hat{p}_1 = 300/2000 = 0.15$ and p_2 by $\hat{p}_2 = 200/2000 = 0.10$, and so on.

The total number of bird species observed in this sample is 30. Yet it is clear that the bird population must have more than just 30 different species. A natural follow-up question would then be as follows:

> *What is the total population proportion of birds belonging to species other than those observed in the sample?*

The follow-up question implies a statistical problem of estimation with a target, or estimand, being the collective proportion of birds of the species not represented in the sample. For convenience, let this target be denoted as π_0. It is important to note that π_0 is a random quantity depending on the sample and therefore is not an estimand in the usual statistical sense. In the statistics

Statistical Implications of Turing's Formula, First Edition. Zhiyi Zhang.

Table 1.1 Bird sample

s_k	1	2	3	4	5	6	7	8	9	10
y_k	300	200	300	200	100	100	100	100	0	100
s_k	11	12	13	14	15	16	17	18	19	20
y_k	100	80	70	0	30	50	6	1	2	1
s_k	21	22	23	24	25	26	27	28	29	30
y_k	1	1	0	0	1	1	1	1	1	1
s_k	31	32	33	34	35	36	37	38	39	…
y_k	50	100	1	1	0	0	0	0	0	…

Table 1.2 Rearranged bird sample

s_k	1	3	2	4	5	6	7	8	10	11
y_k	300	300	200	200	100	100	100	100	100	100
s_k	32	12	13	16	31	15	17	19	18	20
y_k	100	80	70	50	50	30	6	2	1	1
s_k	21	22	25	26	27	28	29	30	33	34
y_k	1	1	1	1	1	1	1	1	1	1
s_k	9	14	23	24	35	36	37	38	39	…
y_k	0	0	0	0	0	0	0	0	0	…

literature, $1 - \pi_0$ is often referred to as the *sample coverage of the population*, or in short *sample coverage* , or just *coverage*. Naturally π_0 may be referred to as the *noncoverage*.

The noncoverage π_0 defined with a random sample of size n is an interesting quantity. It is sometimes interpreted as the "probability" of discovering a new species, because, in a loose sense, the chance that the "next" bird is of a new, or previously unobserved, species is π_0. This interpretation is however somewhat misleading. The main issue of such an interpretation is the lack of clarification of the underlying experiment (and its sample space). Words such as "probability" and "next" can only have meaning in a well-specified experiment. While it is quite remarkable that π_0 could be reasonably and nonparametrically estimated by Turing's formula, π_0 is not a probability associated with the sample space of the experiment where the sample of size n is drawn. Further discussion of this point is given in Section 1.5.

Turing's formula, also sometimes known as the Good–Turing formula, is an estimator of π_0 introduced by Good (1953) but largely credited to Alan Turing. Let N_1 denote the number of species each of which is represented by exactly one observation in a random sample of size n. Turing's formula is given by $T_1 = N_1/n$. For the bird example given in Table 1.2, $N_1 = 12, n = 2000$, and therefore $T_1 = 0.006$ or 0.6%.

1.1 Turing's Formula

Let $\mathscr{X} = \{\ell_k; k \geq 1\}$ be a countable alphabet with letters $\ell_1, \ldots$; and let $\mathbf{p} = \{p_k; k \geq 1\}$ be a probability distribution on $\mathscr{X}$. Let $\mathbf{Y} = \{Y_k; k \geq 1\}$ be the observed frequencies of the letters in an *identically and independently distributed* (*iid*) random sample of size n. For any integer r, $1 \leq r \leq n$, let the number of letters in the alphabet that are each represented exactly r times in the sample be denoted by

$$N_r = \sum_{k\geq 1} 1[Y_k = r] \tag{1.1}$$

where $1[\cdot]$ is the indicator function. Let the total probability associated with letters that are represented exactly $r - 1$ times in the sample be denoted by

$$\pi_{r-1} = \sum_{k\geq 1} p_k 1[Y_k = r - 1]. \tag{1.2}$$

Of special interest is the case of $r = 1$ when

$$N_1 = \sum_{k\geq 1} 1[Y_k = 1]$$

and

$$\pi_0 = \sum_{k\geq 1} p_k 1[Y_k = 0] \tag{1.3}$$

representing, respectively,

N_1: the number of letters of $\mathscr{X}$ that each appears exactly once; and
π_0: the total probability associated with the unobserved letters of $\mathscr{X}$ in an *iid* sample of size n.

The following expression is known as Turing's formula:

$$T_1 = \frac{N_1}{n}. \tag{1.4}$$

Turing's formula has been extensively studied during the many years following its introduction by Good (1953) and has been demonstrated, mostly through numerical simulations, to provide a satisfactory estimate of π_0 for a wide range of distributions. These studies put forth a remarkable yet puzzling implication: the total probability on the unobserved subset of $\mathscr{X}$ may be well estimated nonparametrically. No satisfactory interpretation was given in the literature until Robbins (1968).

Robbins' Claim: Let π_0 be defined with an *iid* random sample of size n as in (1.3). Let Turing's formula be defined with an augmented *iid* sample of size $n+1$ by adding one new *iid* observation to the original sample of size n; and let the resulting estimator be denoted by T_1^+. Then T_1^+ is an unbiased estimator of π_0, in the sense of $\mathrm{E}\,(T_1^+ - \pi_0) = 0$.

Robbins' claim is easily verified. Let N_1^+ be the number of letters represented exactly once in the augmented sample of size $n+1$, with observed frequencies $\{Y_k^+; k \geq 1\}$.

$$\mathrm{E}\,(N_1^+) = \sum_{k\geq 1} \mathrm{E}\,(1[Y_k^+ = 1]) = \sum_{k\geq 1} (n+1)p_k(1-p_k)^n \text{ and}$$
$$\mathrm{E}\,(T_1^+) = \mathrm{E}(N_1^+/(n+1)) = \sum_{k\geq 1} p_k(1-p_k)^n.$$

On the other hand, with the sample of size n,

$$\mathrm{E}\,(\pi_0) = \sum_{k\geq 1} p_k \mathrm{E}\,(1[Y_k = 0]) = \sum_{k\geq 1} p_k(1-p_k)^n.$$

Hence, $\mathrm{E}\,(T_1^+ - \pi_0) = 0$.

Robbins' claim provides an intuitive interpretation in the sense that

1) T_1^+ is an unbiased and, therefore, a good estimator of π_0;
2) the difference between T_1 and T_1^+ should be small; and therefore
3) T_1 should be a good estimator of π_0.

However, Robbins' claim still leaves much to be desired. Suppose for a moment that $\mathrm{E}\,(T_1^+ - \pi_0) = 0$ is an acceptable notion of a good estimator for T_1^+, it does not directly address the performance of T_1 as an estimator of π_0, not even the bias $\mathrm{E}\,(T_1 - \pi_0)$, let alone other important statistical properties of $T_1 - \pi_0$. In short, the question whether Turing's formula works at all, or if it does how well, is not entirely resolved by Robbins' claim.

In addition to the unusual characteristic that the estimand π_0 is a random variable, it is to be noted that both the estimator T_1 and the estimand π_0 converge to zero in probability (see Exercise 6). Therefore, it is not sufficiently adequate to characterize the performance of T_1 by the bias $\mathrm{E}\,(T_1 - \pi_0)$ alone. Any reasonable characterization of its performance must take into consideration the vanishing rate of the estimand π_0.

To put the discussion in perspective, a notion of performance is needed however minimal it may be. Let $\mathbf{X}$ denote a data sample. Let $\hat{\theta}_n(\mathbf{X})$ be an estimator of a random estimand $\theta_n(\mathbf{X})$. Let

$$\frac{\mathrm{E}\,(\hat{\theta}_n(\mathbf{X}) - \theta_n(\mathbf{X}))}{\mathrm{E}\,(\theta_n(\mathbf{X}))} \tag{1.5}$$

whenever $\mathrm{E}\,(\theta_n(\mathbf{X})) \neq 0$ be referred to as the *relative bias* of $\hat{\theta}_n(\mathbf{X})$ estimating $\theta_n(\mathbf{X})$.

Definition 1.1 *$\hat{\theta}_n(\mathbf{X})$ is said to be an asymptotically relatively unbiased estimator of $\theta_n(\mathbf{X})$ if*

$$\frac{\mathrm{E}\,(\hat{\theta}_n(\mathbf{X}) - \theta_n(\mathbf{X}))}{\mathrm{E}\,(\theta_n(\mathbf{X}))} \to 0$$

as $n \to \infty$, provided that $\mathrm{E}\,(\theta_n(\mathbf{X})) \neq 0$.

When $\theta_n(\mathbf{X}) = \theta \neq 0$ does not depend on sample data $\mathbf{X}$, an asymptotically relatively unbiased estimator is an asymptotically unbiased estimator in the usual statistical sense.

Let $K = \sum_{k\geq 1} 1[p_k > 0]$ be referred to as the *effective cardinality* of $\mathscr{X}$, or just simply *cardinality* whenever there is no risk of ambiguity. $K < \infty$ implies that there are only finite letters in $\mathscr{X}$ with positive probabilities.

Example 1.1 *For any probability distribution $\{p_k\}$ such that $1 < K < \infty$, Turing's formula T_1 is not an asymptotically relatively unbiased estimator of π_0. This is so because, letting $p_\wedge = \min\{p_k > 0; k \geq 1\}$,*

$$\begin{aligned}
0 &< \frac{\mathrm{E}\,(T_1 - \pi_0)}{\mathrm{E}\,(\pi_0)} \\
&= \frac{\sum_{k\geq 1} p_k(1-p_k)^{n-1} - \sum_{k\geq 1} p_k(1-p_k)^n}{\sum_{k\geq 1} p_k(1-p_k)^n} \\
&= \frac{\sum_{k\geq 1} p_k^2(1-p_k)^{n-1}}{\sum_{k\geq 1} p_k(1-p_k)^n} \\
&= \frac{\sum_{k\geq 1} p_k^2[(1-p_k)/(1-p_\wedge)]^{n-1}}{\sum_{k\geq 1} p_k(1-p_k)[(1-p_k)/(1-p_\wedge)]^{n-1}} \\
&\to \frac{p_\wedge^2}{p_\wedge(1-p_\wedge)} = \frac{p_\wedge}{(1-p_\wedge)} \neq 0
\end{aligned}$$

as $n \to \infty$.

Example 1.1 demonstrates that there exist distributions on $\mathscr{X}$ under which Turing's formula is not an asymptotically relatively unbiased estimator. In retrospect, this is not so surprising. On a finite alphabet, the probabilities associated with letters not covered by a large sample, π_0, should rapidly approach zero, at least on an intuitive level, as manifested by the fact that

$\mathrm{E}(\pi_0) = \mathcal{O}((1-p_\wedge)^n)$ (see Exercise 7). On the other hand, the number of singletons, that is, letters that are each observed exactly once in the sample $(\sum_{k\geq 1} 1[Y_k = 1])$, should also approach zero rapidly, as manifested by the fact that $\mathrm{E}(\sum_{k\geq 1} 1[Y_k = 1)]) = \mathcal{O}\,(n(1-p_\wedge))^n$. The fact that T_1 and π_0 are vanishing at the same rate leads to the result of Example 1.1. This example also suggests that perhaps Turing's formula may not always be a good estimator of its target π_0, at least in its current form. However, Turing's formula is not an asymptotically relatively unbiased estimator only when $K < \infty$.

Theorem 1.1 *Turing's formula T_1 is an asymptotically relatively unbiased estimator of π_0 if and only if $K = \infty$.*

To prove Theorem 1.1, the following three lemmas are needed.

Lemma 1.1 *Suppose $\{p_k\}$ is a strictly decreasing infinite probability sequence with $p_k > 0$ for every k, $k \geq 1$. Then, as $n \to \infty$,*

$$r_n = \frac{\sum_{k=1}^{\infty} p_k^2(1-p_k)^n}{\sum_{k=1}^{\infty} p_k(1-p_k)^n} \to 0. \tag{1.6}$$

Proof: Let k_0 be any specific index. r_n may be re-expressed as follows.

$$\begin{aligned}
0 < r_n &= \frac{\sum_{k=1}^{\infty} p_k^2(1-p_k)^n}{\sum_{k=1}^{\infty} p_k(1-p_k)^n} \\
&= \frac{\sum_{k=1}^{k_0-1} p_k^2(1-p_k)^n + \sum_{k=k_0}^{\infty} p_k^2(1-p_k)^n}{\sum_{k=1}^{\infty} p_k(1-p_k)^n} \\
&\leq \frac{\sum_{k=1}^{k_0-1} p_k^2(1-p_k)^n + p_{k_0} \sum_{k=k_0}^{\infty} p_k(1-p_k)^n}{\sum_{k=1}^{\infty} p_k(1-p_k)^n} \\
&= \frac{\sum_{k=1}^{k_0-1} p_k^2(1-p_k)^n}{\sum_{k=1}^{\infty} p_k(1-p_k)^n} + \frac{p_{k_0} \sum_{k=k_0}^{\infty} p_k(1-p_k)^n}{\sum_{k=1}^{\infty} p_k(1-p_k)^n} \\
&=: r_{n,1} + r_{n,2}
\end{aligned}$$

where the symbol ":=" stands for the defining equality, for example, "$a := b$" and "$a =: b$" represent "a is defined by b" and "a defines b" respectively.

For any given $\varepsilon > 0$, let $k_0 = \min\{k : p_k < \varepsilon/2\}$. First it is observed that

$$0 < r_{n,2} < \frac{p_{k_0} \sum_{k=k_0}^{\infty} p_k(1-p_k)^n}{\sum_{k=k_0}^{\infty} p_k(1-p_k)^n} = p_{k_0} < \varepsilon/2.$$

Next it is observed that, letting $q_k = 1 - p_k$,

$$0 < r_{n,1} = \frac{\sum_{k=1}^{k_0-1} p_k^2(1-p_k)^n}{\sum_{k=1}^{\infty} p_k(1-p_k)^n} < \frac{\sum_{k=1}^{k_0-1} p_k^2(1-p_k)^n}{p_{k_0}(1-p_{k_0})^n} = \frac{\sum_{k=1}^{k_0-1} p_k^2(q_k/q_{k_0})^n}{p_{k_0}}.$$

In the numerator of the above-mentioned last expression, $0 < q_k/q_{k_0} < 1$ for every $k = 1, \dots, k_0 - 1$, therefore $(q_k/q_{k_0})^n \to 0$ as $n \to \infty$. In addition, since there are finite terms in the summation, $r_{n,1} \to 0$. That is to say that, for any $\varepsilon > 0$ as given earlier, there exists an integer N_ε such that for every $n > N_\varepsilon$, $r_{n,1} < \varepsilon/2$, or $r_n \le r_{n,1} + r_{n,2} < \varepsilon/2 + \varepsilon/2 = \varepsilon$. □

Lemma 1.2 *Suppose $\{p_k\}$ is a strictly decreasing infinite probability sequence with $p_k > 0$ for every k, $k \ge 1$. Then, as $n \to \infty$,*

$$r_n^* = \frac{\sum_{k=1}^{\infty} p_k^2 (1-p_k)^{n-1}}{\sum_{k=1}^{\infty} p_k (1-p_k)^n} \to 0. \tag{1.7}$$

Proof: By Lemma 1.1,

$$r_n = r_n^* - \frac{\sum_{k=1}^{\infty} p_k^3 (1-p_k)^{n-1}}{\sum_{k=1}^{\infty} p_k (1-p_k)^n} \to 0,$$

and therefore it suffices to show that

$$\frac{\sum_{k=1}^{\infty} p_k^3 (1-p_k)^{n-1}}{\sum_{k=1}^{\infty} p_k (1-p_k)^n} \to 0.$$

Noting $p/(1-p)$ is an increasing function for $p \in (0,1)$, for each p_k of any given distribution $\{p_k; k \ge 1\}$, $p_k/(1-p_k) \le p_\vee/(1-p_\vee)$ where $p_\vee = \max\{p_k; k \ge 1\}$. Therefore,

$$\begin{aligned}
\frac{\sum_{k=1}^{\infty} p_k^3 (1-p_k)^{n-1}}{\sum_{k=1}^{\infty} p_k (1-p_k)^n} &= \frac{\sum_{k=1}^{\infty} p_k^2 (1-p_k)^n \left(\frac{p_k}{1-p_k}\right)}{\sum_{k=1}^{\infty} p_k (1-p_k)^n} \\
&\le \left(\frac{p_\vee}{1-p_\vee}\right) \frac{\sum_{k=1}^{\infty} p_k^2 (1-p_k)^n}{\sum_{k=1}^{\infty} p_k (1-p_k)^n} \\
&= \left(\frac{p_\vee}{1-p_\vee}\right) r_n \to 0.
\end{aligned}$$

□

With a little adjustment, the proof for Lemma 1.1 goes through with any infinite probability sequence that is not necessarily strictly decreasing.

Lemma 1.3 *If $\{p_k\}$ is a probability sequence with positive mass on each letter in an infinite subset of $\mathscr{X}$, then $r_n \to 0$ and $r_n^* \to 0$ as $n \to \infty$ where r_n and r_n^* are as in (1.6) and (1.7).*

The proof of Lemma 1.3 is left as an exercise (see Exercise 8).

Proof of Theorem 1.1: The sufficiency part of the theorem immediately follows Lemma 1.3. The necessity part of the theorem follows the fact that if $K < \infty$, then Example 1.1 provides a counterargument. □

Both random quantities, T_1 and π_0, converge to zero in probability. One would naturally wonder if a naive estimator, say $\hat{\pi}_0 = 1/n$, could perform reasonably well in estimating π_0. In view of (1.5), the relative bias of $\hat{\pi}_0 = 1/n$, when $K < \infty$ as in Example 1.1, is

$$\frac{\mathrm{E}\,(\hat{\pi}_0 - \pi_0)}{\mathrm{E}\,(\pi_0)} = \frac{1/n}{\sum_{k\geq 1} p_k^2(1-p_k)^{n-1}} - 1 > \frac{1/n}{(1-p_\wedge)^{n-1}} - 1 \to \infty$$

indicating $\hat{\pi}_0$ as a very inadequate estimator. In fact, since the decreasing rate of $\mathrm{E}(\pi_0) \to 0$ could vary over a wide range of different distributions on $\mathscr{X}$, for any naive estimator of the form $\hat{\pi}_0 = g(n) \to 0$, there exist arbitrarily many distributions under which the relative bias of such a $\hat{\pi}_0$ diverges to infinity. This fact will become clearer in Chapter 6.

Consider next the variance of $T_1 - \pi_0$.

$$\begin{aligned}
\sigma_n^2 &= \mathrm{Var}\,(T_1 - \pi_0) = \mathrm{E}\,((T_1-\pi_0)^2) - (\mathrm{E}\,(T_1-\pi_0))^2 \\
&= \mathrm{E}\,(T_1^2 - 2T_1\,\pi_0 + \pi_0^2) - \left[\sum_{k\geq 1} p_k^2(1-p_k)^{n-1}\right]^2 \\
&= \mathrm{E}\left(n^{-2}\sum_{i\geq 1}\sum_{j\geq 1} 1[Y_i = 1]1[Y_j = 1]\right) \\
&\quad - 2\mathrm{E}\left(n^{-1}\sum_{i\geq 1}\sum_{j\geq 1} p_j 1[Y_i = 1]1[Y_j = 0]\right) \\
&\quad + \mathrm{E}\left(\sum_{i\geq 1}\sum_{j\geq 1} p_i\,p_j 1[Y_i = 0]1[Y_j = 0]\right) - \left[\sum_{k\geq 1} p_k^2(1-p_k)^{n-1}\right]^2 \\
&= n^{-2}\left(n\sum_{i\geq 1} p_i(1-p_i)^{n-1} + n(n-1)\sum_{i\neq j, i\geq 1, j\geq 1} p_i\,p_j(1-p_i-p_j)^{n-2}\right) \\
&\quad - 2\left(\sum_{i\neq j, i\geq 1, j\geq 1} p_i\,p_j(1-p_i-p_j)^{n-1}\right) \\
&\quad + \left(\sum_{i\geq 1} p_i^2(1-p_i)^n + \sum_{i\neq j, i\geq 1, j\geq 1} p_i\,p_j(1-p_i-p_j)^n\right) \\
&\quad - \left[\sum_{k\geq 1} p_k^2(1-p_k)^{n-1}\right]^2 \\
&= \frac{1}{n}\left(\sum_{k\geq 1} p_k(1-p_k)^{n-1} - \sum_{i\neq j, i\geq 1, j\geq 1} p_i\,p_j(1-p_i-p_j)^{n-2}\right) \\
&\quad + \sum_{i\neq j, i\geq 1, j\geq 1} p_i\,p_j(1-p_i-p_j)^{n-2}(p_i+p_j)^2 + \sum_{i\geq 1} p_k^2(1-p_k)^{n-1}
\end{aligned}$$

$$
\begin{aligned}
&- \sum_{k\geq 1} p_k^3(1-p_k)^{n-1} - \left[\sum_{k\geq 1} p_k^2(1-p_k)^{n-1}\right]^2 \\
&= \frac{1}{n}\sum_{k\geq 1} p_k(1-p_k)^{n-1} \\
&\quad - \frac{1}{n}\sum_{i\geq 1}\sum_{j\geq 1} p_i\, p_j(1-p_i-p_j)^{n-2} + \frac{1}{n}\sum_{i\geq 1} p_i^2(1-2p_i)^{n-2} \\
&\quad + \sum_{i\geq 1}\sum_{j\geq 1} p_i\, p_j(1-p_i-p_j)^{n-2}(p_i+p_j)^2 - 4\sum_{k\geq 1} p_k^4(1-2p_k)^{n-2} \\
&\quad + \sum_{k\geq 1} p_k^2(1-p_k)^{n} - \left[\sum_{k\geq 1} p_k^2(1-p_k)^{n-1}\right]^2 .
\end{aligned} \tag{1.8}
$$

It can be verified that $\lim_{n\to\infty} \sigma_n^2 = 0$ (see Exercise 10).

Denote $\zeta_{2,n-1} = \sum_{k\geq 1} p_k^2(1-p_k)^{n-1}$. By Chebyshev's theorem, for any $m > 0$,

$$
\mathrm{P}\left(|(T_1 - \pi_0) - \zeta_{2,n-1}| \leq m\sigma_n\right) \geq 1 - \frac{1}{m^2}
$$

or

$$
\mathrm{P}\left(T_1 - \zeta_{2,n-1} - m\sigma_n \leq \pi_0 \leq T_1 - \zeta_{2,n-1} + m\sigma_n\right) \geq 1 - \frac{1}{m^2}. \tag{1.9}
$$

Replacing the $\{p_k\}$ in $\zeta_{2,n-1}$ and σ_n as in (1.8) by $\{\hat{p}_k\}$ based on a sample and denoting the resulting values as $\hat{\zeta}_{2,n-1}$, and $\hat{\sigma}_n$ respectively, one may choose to consider

$$
\left(T_1 - \hat{\zeta}_{2,n-1}\right) \pm m\hat{\sigma}_n. \tag{1.10}
$$

as a conservative and approximate confidence interval for π_0 at a confidence level of $1 - 1/m^2$.

Example 1.2 *Use the data in Table 1.2 to find a 95% confidence interval for π_0. For $1 - 1/m^2 = 0.95$, $m = \sqrt{20} = 4.4721$. Using $n = 2000$, $T_1 = 0.006$, $\hat{\zeta}_{2,n-1} = 1.2613 \times 10^{-6}$ and $\hat{\sigma}_n = 0.0485$, a conservative and approximate 95% confidence interval according to (1.10) is $(-0.2109, 0.2229)$.*

Remark 1.1 *While the statement of (1.9) is precisely correct under any distribution $\{p_k\}$ on $\mathscr{X}$, (1.10) is a very weakly justified confidence interval for π_0. In addition to the fact that the confidence level associated with (1.10) is a conservative lower bound, $\hat{\zeta}_{2,n-1}$ and $\hat{\sigma}_n$ are merely crude approximations to $\zeta_{2,n-1}$ and σ_n, respectively. One must exercise caution when (1.10) is used.*

Remark 1.2 *By Definition 1.1, it is clear that a good estimator is one with a bias, namely $\mathrm{E}(T_1 - \pi_0)$, that vanishes faster than the expected value of the target, namely $\mathrm{E}(\pi_0)$, as n increases indefinitely. Given that the bias is $\zeta_{2,n-1} = \sum_{k\geq 1} p_k^2(1-p_k)^{n-1}$, one could possibly entertain the idea of a modification to*

Turing's formula T_1 to reduce the bias, possibly to the extent such that, in view of Example 1.1, a modified Turing's formula may become asymptotically and relative unbiased even for distributions with $K < \infty$. As it turns out, there are many such modifications, one of which is presented in Section 1.4.

1.2 Univariate Normal Laws

To better understand several important normal laws concerning the asymptotic behavior of $T_1 - \pi_0$, consider first a slightly more general probability model on the alphabet $\mathscr{X}$, namely $\{p_{k,n}; k \geq 1\}$, instead of $\{p_k; k \geq 1\}$. The subindex n in $p_{k,n}$ suggests that the probability distribution could, but not necessarily, be dynamically changing as n changes. In this sense, $\{p_k; k \geq 1\}$ is a special case of $\{p_{k,n}; k \geq 1\}$.

Thirty years after Good (1953) introduced Turing's formula, Esty (1983) established an asymptotic normal law of $T_1 - \pi_0$ as stated in Theorem 1.2.

Theorem 1.2 *Let the probability distribution $\{p_{k,n}; k \geq 1\}$ be such that*

1) $\mathrm{E}(N_1/n) \to c_1$ *where* $0 < c_1 < 1$, *and*
2) $\mathrm{E}(N_2/n) \to c_2 \geq 0$.

Then

$$\frac{n(T_1 - \pi_0)}{\sqrt{N_1 + 2N_2 - N_1^2/n}} \xrightarrow{L} N(0, 1) \tag{1.11}$$

where N_1 and N_2 are given by (1.1), π_0 is given by (1.2), and T_1 is given by (1.4).

Example 1.3 *Let $\{p_{k,n}; k = 1, \dots, n\}$ be such that $p_{k,n} = 1/n$ for every k.*

$$\begin{aligned}
\mathrm{E}(N_1/n) &= \sum_{k\geq 1} \mathrm{E}(1[Y_k = 1])/n = \sum_{k\geq 1} \mathrm{P}(Y_k = 1)/n \\
&= \sum_{k\geq 1} np_{k,n}(1 - p_{k,n})^{n-1}/n = (1 - 1/n)^{n-1} \to e^{-1}; \\
\mathrm{E}(N_2/n) &= \sum_{k\geq 1} \mathrm{E}(1[Y_k = 2])/n = \sum_{k\geq 1} \mathrm{P}(Y_k = 2)/n \\
&= \sum_{k\geq 1} [n(n-1)/2!]\, p_{k,n}^2(1 - p_{k,n})^{n-2}/n \\
&= (1 - 1/n)^{n-1}/2 \to \frac{1}{2}e^{-1}.
\end{aligned}$$

The conditions of Theorem 1.2 are satisfied and therefore (1.11) holds.

Example 1.3 has a subtle but comforting implication in applications: there exist probability distributions on finite alphabets of cardinality K_n (finite at least at the time a sample of size n is taken) such that the normality of

Theorem 1.2 holds. In this sense, Theorem 1.2 provides a theoretical justification for approximate inferential procedures derived from (1.11) to be used in applications where the cardinalities of alphabets are often more appropriately assumed to be finite.

The asymptotic normality of Turing's formula given in (1.11) was the first of its kind and the proof of Theorem 1.2 was a technical trailblazer. However, the sufficient condition in Theorem 1.2 is overly restrictive and consequently no fixed distribution $\{p_k; k \geq 1\}$ satisfies the sufficient condition. The last fact is demonstrated by Lemma 1.4.

Lemma 1.4 *Let $\{p_k; k \geq 1\}$ be any probability distribution on alphabet $\mathcal{X}$. Then*

a) $\lim_{n\to\infty} \mathrm{E}(N_1/n) = 0$, *and*
b) $\lim_{n\to\infty} \mathrm{E}(N_2/n) = 0$.

Proof: For Part (a), noting first $\mathrm{E}(N_1/n) = \sum_{k\geq 1} p_k(1-p_k)^{n-1}$ and then $p_k(1-p_k)^{n-1} \leq p_k$ and $\sum_{k\geq 1} p_k = 1$, by the dominated convergence theorem,

$$\lim_{n\to\infty} \mathrm{E}(N_1/n) = \sum_{k\geq 1} p_k \lim_{n\to\infty} (1-p_k)^{n-1} = 0.$$

For Part (b), noting first $\mathrm{E}(N_2/n) = (1/2)\sum_{k\geq 1}(n-1)p_k^2(1-p_k)^{n-2}$ and then that $(n-1)p(1-p)^{n-2}$ attains its maximum at $p = 1/(n-1)$, one has

$$(n-1)p_k(1-p_k)^{n-2} \leq [(n-2)/(n-1)]^{n-2} < 1.$$

Again by the dominated convergence theorem,

$$\lim_{n\to\infty} \mathrm{E}(N_2/n) = (1/2)\sum_{k\geq 1} \lim_{n\to\infty} [(n-1)p_k^2(1-p_k)^{n-2}] = 0.$$

□

Zhang and Huang (2008) modified Esty's sufficient condition in Theorem 1.2 and established Theorems 1.3 and 1.4 for fixed probability distributions $\{p_k; k \geq 1\}$.

Condition 1.1 *There exists a $\delta \in (0, 1/4)$ such that, as $n \to \infty$,*

1) $n^{1-4\delta}\, \mathrm{E}(N_1/n) \to c_1 \geq 0$,
2) $n^{1-4\delta}\, \mathrm{E}(N_2/n) \to c_2/2 \geq 0$, *and*
3) $c_1 + c_2 > 0$.

Theorem 1.3 *If a given probability distribution $\{p_k; k \geq 1\}$ satisfies Condition 1.1, then*

$$n^{1-2\delta}\,(T_1 - \pi_0) \xrightarrow{L} N(0, c_1 + c_2). \qquad (1.12)$$

The proof of Theorem 1.3 is lengthy and therefore is omitted here. Interested readers may refer to Zhang and Huang (2008) for details. The rate of the weak convergence of (1.12) plays an important role throughout this book and therefore is assigned a special notation for convenience

$$g(n, \delta) = n^{1-2\delta}. \tag{1.13}$$

Let

$$\hat{c}_1 = \frac{g^2(n,\delta)}{n^2} N_1 \quad \text{and} \quad \hat{c}_2 = \frac{2g^2(n,\delta)}{n^2} N_2. \tag{1.14}$$

Lemma 1.5 *Under Condition 1.1,*

$$\hat{c}_1 \xrightarrow{p} c_1 \quad \textit{and} \quad \hat{c}_2 \xrightarrow{p} c_2.$$

The proof of Lemma 1.5 is left as an exercise (see Exercise 11).

Theorem 1.4 is a corollary of Theorem 1.3.

Theorem 1.4 *If a given probability distribution $\{p_k; k \geq 1\}$ satisfies Condition 1.1, then*

1)

$$\frac{n\,(T_1 - \pi_0)}{\sqrt{\mathrm{E}(N_1) + 2\mathrm{E}(N_2)}} \xrightarrow{L} N(0,1), \quad \textit{and} \tag{1.15}$$

2)

$$\frac{n\,(T_1 - \pi_0)}{\sqrt{N_1 + 2N_2}} \xrightarrow{L} N(0,1). \tag{1.16}$$

Proof: For Part 1,

$$\begin{aligned}
\frac{n\,(T_1 - \pi_0)}{\sqrt{\mathrm{E}(N_1) + 2\mathrm{E}(N_2)}} &= \frac{n}{n^{1-2\delta}} \frac{\sqrt{c_1 + c_2}}{\sqrt{\mathrm{E}(N_1) + 2\mathrm{E}(N_2)}} \frac{g(n,\delta)\,(T_1 - \pi_0)}{\sqrt{c_1 + c_2}} \\
&= \left(\frac{\sqrt{c_1 + c_2}}{\sqrt{n^{-4\delta}\mathrm{E}(N_1) + 2n^{-4\delta}\mathrm{E}(N_2)}} \right) \left[\frac{g(n,\delta)\,(T_1 - \pi_0)}{\sqrt{c_1 + c_2}} \right].
\end{aligned}$$

In the above-mentioned last expression, the first factor converges to 1 by Condition 1.1 and the second factor converges weakly to the standard normal distribution by Theorem 1.3. The result of (1.15) follows Slutsky's theorem.

For Part 2,

$$\begin{aligned}
\frac{n\,(T_1 - \pi_0)}{\sqrt{N_1 + 2N_2}} &= \frac{n}{n^{1-2\delta}} \frac{\sqrt{c_1 + c_2}}{\sqrt{N_1 + 2N_2}} \frac{g(n,\delta)\,(T_1 - \pi_0)}{\sqrt{c_1 + c_2}} \\
&= \left(\frac{\sqrt{c_1 + c_2}}{\sqrt{n^{-4\delta}N_1 + 2n^{-4\delta}N_2}} \right) \left[\frac{g(n,\delta)\,(T_1 - \pi_0)}{\sqrt{c_1 + c_2}} \right].
\end{aligned}$$

In the above-mentioned last expression, the first factor converges to 1 by Lemma 1.5 and the second factor converges weakly to the standard normal distribution by Theorem 1.3. The result of (1.16) follows Slutskys theorem. □

An interesting feature of Theorem 1.4 is that, although Condition 1.1 calls for the existence of a $\delta \in (0, 1/4)$, the results of Theorem 1.4 do not depend on such a δ, as evidenced by its total absence in (1.15) and (1.16). Also to be noted is the fact that the conditions of Theorems 1.2 and 1.3 do not overlap. The condition of Theorem 1.2 essentially corresponds to $\delta = 1/4$ in the form of the condition of Theorem 1.4, which is on the boundary of the interval $(0, 1/4)$ but not in the interior of the interval.

Remark 1.3 *The value of δ represents a tail property of the underlying distribution. The fact that (1.15) does not depend on δ may be considered as a consequence of a domain of attraction of certain types of distributions sharing the same asymptotic normality via Turing's formula. Discussion of domains of attraction on alphabets is given in Chapter 6.*

The following example shows that the class of probability distributions satisfying the condition of Theorem 1.4 is in fact nonempty. It also gives a description of the bulk of distributions in the family that supports the normal laws in (1.15) and (1.16).

Example 1.4 *If $p_k = c_\lambda\ k^{-\lambda}$ for all $k \geq 1$ where $\lambda > 1$ and c_λ is such that $\sum_{k\geq 1} p_k = 1$, then $\{p_k; k \geq 1\}$ satisfies the conditions of Theorem 1.4.*

To see that the claim of Example 1.4 is true, the following two lemmas are needed.

Lemma 1.6 *(Euler–Maclaurin) For each n, $n = 1, 2, \ldots$, let $f_n(x)$ be a continuous function of x on $[x_0, \infty)$ where x_0 is a positive integer. Suppose there exists a sequence of real values, x_n satisfying $x_n \geq x_0$, such that $f_n(x)$ is increasing on $[x_0, x_n]$ and decreasing on $[x_n, \infty)$. If $f_n(x_0) \to 0$ and $f_n(x_n) \to 0$, then*

$$\lim_{n\to\infty} \sum_{k\geq x_0} f_n(k) = \lim_{n\to\infty} \int_{x_0}^{\infty} f_n(x)\ dx.$$

Proof: It can be verified that

$$\sum_{x_0\leq k\leq x_n} f_n(k) - f_n(x_n) \leq \int_{x_0}^{x_n} f_n(x)\ dx \leq \sum_{x_0+1\leq k<x_n} f_n(k) + f_n(x_n) \quad \text{and}$$

$$\sum_{k>x_n} f_n(k) - f_n(x_n) \leq \int_{x_n}^{\infty} f_n(x)\ dx \leq \sum_{k\geq x_n} f_n(k) + f_n(x_n).$$

Adding the corresponding parts of the above-mentioned two expressions and taking limits give

$$\lim_{n\to\infty}\sum_{k=x_0}^{\infty} f_n(k) - 2\lim_{n\to\infty} f_n(x_n) \le \lim_{n\to\infty}\int_{x_0}^{\infty} f_n(x)\, dx$$

$$\le \lim_{n\to\infty}\sum_{k=x_0}^{\infty} f_n(k) - \lim_{n\to\infty} f_n(x_0) + 2\lim_{n\to\infty} f_n(x_n).$$

The desired result follows from the conditions of the lemma. □

Condition 1.2 *There exists a* $\delta \in (0, 1/4)$ *such that, as* $n \to \infty$,

1) $n^{1-4\delta}\sum_{k\ge1} p_k\ e^{-np_k} \to c_1 \ge 0$,
2) $n^{2-4\delta}\sum_{k\ge1} p_k^2 e^{-np_k} \to c_2 \ge 0$, *and*
3) $c_1 + c_2 > 0$.

Lemma 1.7 *Conditions 1.1 and 1.2 are equivalent.*

The proof of Lemma 1.7 is nontrivial and is found in Zhang (2013a), where a more general statement is proved in the appendix.

In Example 1.4, let $\delta = (4\lambda)^{-1}$ and therefore $g(n,\delta) = n^{1-2\delta} = n^{1-\frac{1}{2\lambda}}$. By Lemma 1.7, it suffices to check Condition 1.2 instead of Condition 1.1:

$$\frac{g^2(n,\delta)}{n}\sum_{k\ge1} p_k\ e^{-np_k} = n^{1-\frac{1}{\lambda}}\sum_{k=k_0}^{\infty} c_\lambda\ k^{-\lambda} e^{-nc_\lambda\ k^{-\lambda}} = \sum_{k=k_0}^{\infty} f_n(k)$$

where

$$f_n(x) = n^{1-\frac{1}{\lambda}} c_\lambda\ x^{-\lambda} e^{-nc_\lambda\ x^{-\lambda}}.$$

Since it is easily verified that

$$f_n'(x) = -\lambda c_\lambda\ n^{1-\frac{1}{\lambda}} x^{-(\lambda+1)}\ (1 - nc_\lambda\ x^{-\lambda})\ e^{-nc_\lambda\ x^{-\lambda}},$$

$f_n(x)$ increases and decreases, respectively, over the following two sets

$$[1, (nc_\lambda)^{1/\lambda}] \quad \text{and} \quad [(nc_\lambda)^{1/\lambda}, \infty)\ .$$

Let $x_0 = 1$ and $x_n = (nc_\lambda)^{1/\lambda}$. It is clear that $f_n(x_0) \to 0$ and

$$\begin{aligned} f_n(x_n) &= n^{1-\frac{1}{\lambda}} c_\lambda (nc_\lambda)^{-1} e^{-nc_\lambda (nc_\lambda)^{-1}} \\ &= n^{1-\frac{1}{\lambda}} c_\lambda (nc_\lambda)^{-1} e^{-1} \\ &= e^{-1} n^{-1/\lambda} \to 0. \end{aligned}$$

Invoking the Euler–Maclaurin lemma, with changes of variable $t = x^{-\lambda}$ and then $s = nc_\lambda t$,

$$n^{1-\frac{1}{\lambda}}\sum_{k=1}^{\infty} p_k\ e^{-np_k} \simeq \int_{x_0}^{\infty} n^{1-\frac{1}{\lambda}} c_\lambda\ x^{-\lambda} e^{-nc_\lambda\ x^{-\lambda}}\ dx$$

$$
\begin{aligned}
&= \frac{c_\lambda}{\lambda}\int_0^{x_0^{-\lambda}} n^{1-\frac{1}{\lambda}} t^{-\frac{1}{\lambda}} e^{-nc_\lambda t} dt \\
&= \frac{c_\lambda}{\lambda} n^{1-\frac{1}{\lambda}} \int_0^{x_0^{-\lambda}} (nc_\lambda t)^{-\frac{1}{\lambda}} (nc_\lambda)^{-1+\frac{1}{\lambda}} e^{-nc_\lambda t} d(nc_\lambda t) \\
&= \frac{c_\lambda}{\lambda} n^{1-\frac{1}{\lambda}} (nc)^{-1+\frac{1}{\lambda}} \int_0^{nc_\lambda\ x_0^{-\lambda}} s^{-\frac{1}{\lambda}} e^{-s} ds \\
&= \frac{(c_\lambda)^{\frac{1}{\lambda}}}{\lambda} n^0 \int_0^{nc_\lambda\ x_0^{-\lambda}} s^{-\frac{1}{\lambda}} e^{-s} ds \\
&= \frac{(c_\lambda)^{\frac{1}{\lambda}}}{\lambda} \int_0^{nc_\lambda\ x_0^{-\lambda}} s^{\left(1-\frac{1}{\lambda}\right)-1} e^{-s} ds \\
&= \frac{(c_\lambda)^{\frac{1}{\lambda}}}{\lambda} \Gamma\left(1-\frac{1}{\lambda}\right) \left[\frac{1}{\Gamma\left(1-\frac{1}{\lambda}\right)} \int_0^{nc_\lambda\ x_0^{-\lambda}} s^{\left(1-\frac{1}{\lambda}\right)-1} e^{-s} ds\right] \\
&\to \frac{(c_\lambda)^{\frac{1}{\lambda}}}{\lambda} \Gamma\left(1-\lambda^{-1}\right) > 0
\end{aligned}
$$

where "$\simeq$" indicates equality in limit as $n \to \infty$.

Thus, the sufficient condition of Theorem 1.4 is verified and therefore the claim in Example 1.4 is true.

It is important to note that Condition 1.1 (and therefore Condition 1.2) is in fact a tail property of the underlying distribution, that is, the values of p_k for $k < k_0$ for any arbitrarily fixed positive integer k_0 are completely irrelevant to whether these conditions hold. To see this fact, it suffices to note, for example,

$$
\mathrm{E}\,(N_1/n) = \sum_{k\geq 1} p_k(1-p_k)^{n-1} = \sum_{k<k_0} p_k(1-p_k)^{n-1} + \sum_{k\geq k_0} p_k(1-p_k)^{n-1}.
$$

The first of the two terms in the above-mentioned last expression converges to zero exponentially fast, and therefore the question whether $\mathrm{E}\,(N_1/n)$ converges to zero at a power decaying rate rests entirely with the second of the two terms. A similar argument applies to $\mathrm{E}\,(N_2/n)$.

Example 1.5 *Let $\{p_k\}$ be such that p_k is not specified for $k < k_0$ and $p_k = ck^{-\lambda}$ for $k \geq k_0$ for some integer k_0, where $\lambda > 1$ and c are such that $\sum_{k\geq 1} p_k = 1$, then $\{p_k; k \geq 1\}$ satisfies Condition 1.1.*

The proof of Example 1.5 is left as an exercise (see Exercise 12).

Using Theorem 1.4, an approximate $100(1-\alpha)\%$ confidence interval for π_0 can be devised:

$$
\frac{N_1}{n} \pm z_{\alpha/2} \frac{\sqrt{N_1 + 2N_2}}{n}. \tag{1.17}
$$

Example 1.6 *Use the data in Table 1.1 to construct a 95% confidence interval for π_0. With $n = 2000$, $N_1 = 12$, and $N_2 = 1$ by (1.17), a 95% confidence interval for π_0 is*

$$\frac{N_1}{n} \pm z_{\alpha/2}\frac{\sqrt{N_1 + 2N_2}}{n} = \frac{12}{2000} \pm 1.96\frac{\sqrt{12 + 2 \times 1}}{2000}$$
$$= 0.006 \pm 0.0037 = (0.0023, 0.0097).$$

While the sufficient conditions of asymptotic normality in Theorems 1.2 and 1.4 are often easy to check, they are both superseded by a necessary and sufficient condition given by Zhang and Zhang (2009). Let

$$s_n^2 = n\sum_{k\geq 1} p_{k,n}\, e^{-np_{k,n}} + n^2\sum_{k\geq 1} p_{k,n}^2 e^{-np_{k,n}},$$

where $\{p_{k,n}\}$ is a probability distribution on the alphabet $\mathscr{X}$. Consider Condition 1.3 below.

Condition 1.3 *As $n \to \infty$,*

1) $\mathrm{E}(N_1) + \mathrm{E}(N_2) \to \infty$, *and*

2) $\left(\frac{n}{s_n}\right)^2 \sum_{k\geq 1} p_{k,n}^2 e^{-np_{k,n}} 1[np_{k,n} > \varepsilon s_n] \to 0$, *for any* $\varepsilon > 0$.

The following theorem is due to Zhang and Zhang (2009).

Theorem 1.5 *A probability distribution $\{p_{k,n}\}$ on $\mathscr{X}$ satisfies Condition 1.3 if and only if*

$$\frac{n(T_1 - \pi_0)}{\sqrt{\mathrm{E}(N_1) + 2\mathrm{E}(N_2)}} \xrightarrow{L} N(0, 1). \tag{1.18}$$

Condition 1.3 is less restrictive than either of the conditions of Theorems 1.2 and 1.4. Condition 1.3 is quite subtle and can be difficult to verify for a particular distribution. However, it does identify some distributions that do not admit the asymptotic normality (1.18) for Turing's formula.

Example 1.7 *If $\{p_k\}$ is a probability distribution on a finite alphabet, then the normality of (1.18) does not hold.*

To see this, one only needs to check the first part of Condition 1.3. Let $p_\wedge = \min\{p_k > 0; k \geq 1\}$:

$$\mathrm{E}(N_1) = n\sum_{k\geq 1} p_k(1 - p_k)^{n-1} \leq n(1 - p_\wedge)^{n-1} \to 0,$$

$$E(N_2) = \frac{n(n-1)}{2} \sum_{k\geq 1} p_k^2(1-p_k)^{n-2} \leq n^2(1-p_\wedge)^{n-2} \to 0,$$

and therefore $E(N_1) + E(N_2) \to 0$.

Example 1.8 *If $\{p_k\}$ is such that $p_k = c_0\ e^{-k}$ where $c_0 = e - 1$ for all $k \geq 1$, then the normality of (1.18) does not hold.*

A verification of the statement in Example 1.8 will be a lengthy and tedious one and therefore is left as an exercise. (See Exercise 14. This is a difficult exercise and readers may wish to defer it until Chapter 6.)

Theorem 1.6 below is a corollary to Theorem 1.5.

Theorem 1.6 *If a probability distribution $\{p_{k,n}\}$ on $\mathscr{X}$ satisfies Condition 1.3, then*

$$\frac{n(T_1 - \pi_0)}{\sqrt{N_1 + 2N_2}} \xrightarrow{L} N(0,1) \tag{1.19}$$

Remark 1.4 *In view of Theorems 1.4 and 1.5 and of Examples 1.4, 1.7, and 1.8, one might be tempted to reach a conclusion that Turing's formula would admit asymptotic normality when the underlying distribution has a thick tail. As it turns out, the notion of a "thin tail" or a "thick tail" is not easily defined on an alphabet $\mathscr{X}$. The natural and intuitive notion of a tail of a distribution on the real line does not carry over well to a countable alphabet. The asymptotic behavior of quantities, such as $E(N_1)$, could be quite intricate and will be dealt with in all other subsequent chapters, in particular Chapter 6.*

Remark 1.5 *Theorem 1.5 is established (and is stated earlier) under a general distribution family $\{p_{k,n}\}$ on $\mathscr{X}$ whose members may dynamically change with the sample size n. The theorem remains valid if restricted to the subclass of fixed distributions $\{p_k\}$. For pedagogical and statistical simplicity, the distributions on $\mathscr{X}$ considered in the remainder of this book are all members of the fixed family, unless otherwise specified.*

In the same spirit of the relative bias given in Definition 1.1, one may also be interested in various asymptotic aspects of the following random variable

$$\frac{T_1 - \pi_0}{\pi_0} = \frac{T_1}{\pi_0} - 1.$$

Condition 1.4 *There exists a $\delta \in (0, 1/4)$ such that, as $n \to \infty$,*

1) $n^{1-4\delta} E(N_1/n) \to c_1 > 0$, and
2) $n^{1-4\delta} E(N_2/n) \to \frac{c_2}{2} \geq 0$.

Condition 1.4 is slightly more restrictive than Condition 1.1 in that c_1 in Part 1 of Condition 1.4 is required to be strictly greater than zero.

Theorem 1.7 *If* $\{p_k; k \geq 1\}$ *satisfies Condition 1.4, then*

$$n^{2\delta}\left(\frac{T_1-\pi_0}{\pi_0}\right) = n^{2\delta}\left(\frac{T_1}{\pi_0}-1\right) \xrightarrow{L} N(0,\sigma^2) \tag{1.20}$$

where $\sigma^2 = (c_1+c_2)/c_1^2$.

Proof: Note first

$$\begin{aligned} n^{2\delta}\left(\frac{T_1-\pi_0}{\pi_0}\right) &= \left(\frac{n^{2\delta}}{g(n,\delta)\pi_0}\right)[g(n,\delta)\,(T_1-\pi_0)\,] \\ &= \left(\frac{1}{n^{1-4\delta}\pi_0}\right)[g(n,\delta)\,(T_1-\pi_0)\,]\,. \end{aligned}$$

By Theorem 1.3, $g(n,\delta)(T_1-\pi_0) \xrightarrow{L} N(0, c_1+c_2)$, and therefore it suffices to show that $n^{1-4\delta}\pi_0 \xrightarrow{p} c_1 > 0$. Toward that end, consider first

$$\begin{aligned} \mathrm{E}\left(n^{1-4\delta}\pi_0\right) &= n^{1-4\delta}\sum_{k\geq 1} p_k(1-p_k)^n \\ &= \left(\frac{n}{n+1}\right)^{1-4\delta}(n+1)^{1-4\delta}\sum_{k\geq 1} p_k(1-p_k)^n \\ &= \left(\frac{n}{n+1}\right)^{1-4\delta}\left[\frac{g^2(n+1,\delta)}{n+1}\sum_{k\geq 1} p_k(1-p_k)^n\right]. \end{aligned}$$

In the above-mentioned last expression, the first factor converges to 1 and the second factor converges to c_1 due to Part 1 of Condition 1.4. Therefore, it is established that $\mathrm{E}\,(n^{1-4\delta}\pi_0) \to c_1 > 0$.

Consider next

$$\begin{aligned} \mathrm{Var}\,(n^{1-4\delta}\pi_0) &= n^{2-8\delta}\,[\mathrm{E}(\pi_0^2) - (\mathrm{E}(\pi_0))^2] \\ &= n^{2-8\delta}\left[\mathrm{E}\left(\left(\sum_{k\geq 1} p_k 1[Y_k=0]\right)^2\right) - \left(\sum_{k\geq 1} p_k(1-p_k)^n\right)^2\right] \\ &= n^{2-8\delta}\left[\sum_{k\geq 1} p_k^2(1-p_k)^n + \sum_{i\neq j} p_i\, p_j(1-p_i-p_j)^n - \left(\sum_{k\geq 1} p_k(1-p_k)^n\right)^2\right] \end{aligned}$$

$$= n^{2-8\delta}\left(\sum_{k\geq 1} p_k^2(1-p_k)^n - \sum_{k\geq 1} p_k^2(1-p_k)^{2n}\right)$$
$$+ n^{2-8\delta}\left(\sum_{i\neq j} p_i\, p_j(1-p_i-p_j)^n\right.$$
$$\left. - \sum_{i\neq j} p_i\, p_j(1-p_i)^n(1-p_j)^n\right)$$
$$\leq n^{2-8\delta}\sum_{k\geq 1} p_k^2(1-p_k)^n.$$

The last inequality holds since for every (p_i, p_j),

$$1-p_i-p_j \leq 1-p_i-p_j+p_i\, p_j = (1-p_i)(1-p_k).$$

However, by Condition 1.4,

$$n^{2-8\delta}\sum_{k\geq 1} p_k^2(1-p_k)^n = \left[\frac{n^{2-8\delta}}{(n+2)^{2-4\delta}}\right]\left[\frac{2(n+2)^2}{(n+2)(n+1)}\right]$$
$$\times\left[\frac{g^2(n+2,\delta)}{(n+2)^2}\sum_{k\geq 1}\binom{n+2}{2} p_k^2(1-p_k)^n\right] \to 0,$$

since the three factors in the last expression converge, respectively, to zero at the rate of $\mathcal{O}(n^{-4\delta})$, 2, and $c_2/2$. It follows that Var $(n^{1-4\delta}\pi_0) \to 0$, and therefore $n^{1-4\delta}\pi_0 \xrightarrow{p} c_1$. The result of the theorem immediately follows Slutsky's theorem. □

Remark 1.6 *The respective weak convergences of $g(n,\delta)(T_1-\pi_0)$ and $n^{2\delta}(T_1/\pi_0-1)$ provide two different perspectives in terms of inference regarding π_0. It may be interesting to note that, under Condition 1.4, the rates of weak convergence of $(T_1-\pi_0)$ and (T_1/π_0-1) are, respectively, $g(n,\delta)=n^{1-2\delta}$ and $n^{2\delta}$. Since $\delta\in(0,1/4)$, it follows that $2\delta\in(0,1/2)$ and that $1-2\delta\in[1/2,1)$, and therefore $n^{1-2\delta}$ always increases at a faster rate than $n^{2\delta}$. The faster convergence of $(T_1-\pi_0)$ suggests that the associated additive perspective of Turing's formula (as opposed to the multiplicative perspective associated with T_1/π_0) may perhaps be intrinsically more effective.*

Corollary 1.1 *If $\{p_k; k\geq 1\}$ satisfies Condition 1.4, then*

1)

$$\frac{\mathrm{E}(N_1)\,(T_1/\pi_0-1)}{\sqrt{\mathrm{E}(N_1)+2\mathrm{E}(N_2)}} \xrightarrow{L} N(0,1), \quad \textit{and} \tag{1.21}$$

2)

$$\frac{N_1\,(T_1/\pi_0-1)}{\sqrt{N_1+2N_2}} \xrightarrow{L} N(0,1). \tag{1.22}$$

Proof: For Part 1, by Theorem 1.7 and Condition 1.4,

$$\frac{n^{2\delta}\,(T_1/\pi_0 - 1)}{\sqrt{\frac{\mathrm{E}(N_1)/n^{4\delta}+2\mathrm{E}(N_2)/n^{4\delta}}{(\mathrm{E}(N_1)/n^{4\delta})^2}}} = \frac{(T_1/\pi_0 - 1)}{\sqrt{\frac{\mathrm{E}(N_1)+2\mathrm{E}(N_2)}{(\mathrm{E}(N_1))^2}}}$$

$$= \frac{\mathrm{E}(N_1)\,(T_1/\pi_0 - 1)}{\sqrt{\mathrm{E}(N_1) + 2\mathrm{E}(N_2)}} \xrightarrow{L} N(0, 1).$$

For Part 2, the result immediately follows the facts that $N_1/n^{4\delta} \xrightarrow{p} c_1$ and that $N_2/n^{4\delta} \xrightarrow{p} c_2/2$ of Lemma 1.5. □

Once again, while the convergence rate of $T_1/\pi_0 - 1$ depends on the underlying distribution, the left-hand side of Part 2 of Corollary 1.1 is a function of observable components except π_0, thus allowing statistical inference regarding π_0. Using (1.22), it can be verified that an approximate $100 \times (1 - \alpha)\%$ confidence interval for π_0 is

$$\frac{T_1}{1 + z_{\alpha/2}\frac{\sqrt{N_1+2N_2}}{N_1}} \le \pi_0 \le \frac{T_1}{1 - z_{\alpha/2}\frac{\sqrt{N_1+2N_2}}{N_1}} \tag{1.23}$$

(see Exercise 24).

Example 1.9 *Use (1.23) and the data in Table 1.1 to construct a 95% confidence intervalfor π_0. With $n = 2000$, $N_1 = 12$, $N_2 = 1$, and $T_1 = 0.006$,*

$$z_{\alpha/2}\frac{\sqrt{N_1 + 2N_2}}{N_1} = 1.96\frac{\sqrt{12+2}}{12} = 0.6111,$$

the 95% confidence interval is

$$\left(\frac{0.006}{1+0.6111}, \frac{0.006}{1-0.6111}\right) = (0.0037, 0.0150).$$

In comparing the two 95% confidence intervals of Examples 1.6 and 1.9, that is, (0.0023, 0.0097) and (0.0037, 0.0150), respectively, it may be interesting to note that one is larger than the other, and that the smaller interval is not entirely nested in the larger interval.

Lemma 1.8 *Let X_n be a sequence of random variables satisfying*

$$h(n)(X_n - \theta) \xrightarrow{L} N(0, \sigma^2)$$

where $h(\cdot)$ is a strictly increasing function satisfying $\lim_{t\to\infty} h(t) = \infty$, θ and $\sigma > 0$ are two fixed constants. Then, for any given $\varepsilon > 0$,

$$\lim_{n\to\infty} \mathrm{P}(|X_n - \theta| < \varepsilon) = 1.$$

Proof: Without loss of generality, assume that $\sigma = 1$:

$$\begin{aligned} P(|X_n - \theta| < \varepsilon) &= P(-\varepsilon < X_n - \theta < \varepsilon) \\ &= P(X_n - \theta < \varepsilon) - P(X_n - \theta \leq -\varepsilon) \\ &= P(h(n)(X_n - \theta) < h(n)\varepsilon) - P(h(n)(X_n - \theta) \leq -h(n)\varepsilon). \end{aligned}$$

It is desired to show that, for any given α, there exists a sufficiently large integer N such that for all $n > N$, $P(|X_n - \theta| < \varepsilon) > 1 - \alpha$.

Toward that end, let $z_{\alpha/2}$ be the $100(1-\alpha/2)$th percentile and $N_1 = [g^{-1}(z_{\alpha/2}/\varepsilon)]$ where $h^{-1}(\cdot)$ is the inverse function of $h(\cdot)$. For any $n > N_1$, noting $n > h^{-1}(z_{\alpha/2}/\varepsilon)$,

$$P(|X_n - \theta| < \varepsilon) \geq P(h(n)(X_n - \theta) < z_{\alpha/2}) - P(h(n)(X_n - \theta) \leq -z_{\alpha/2}).$$

On the other hand, by the given asymptotic normality, there exist sufficiently large integers N_2 and N_3 such that for all $n > N_2$

$$|P(h(n)(X_n - \theta) < z_{\alpha/2}) - \Phi(z_{\alpha/2})| < \alpha/2$$

and for all $n > N_3$

$$|P(h(n)(X_n - \theta) < -z_{\alpha/2}) - \Phi(-z_{\alpha/2})| < \alpha/2$$

where $\Phi(\cdot)$ is the *cdf* of the standard normal distribution. Consequently for all $n > \max\{N_2, N_3\}$,

$$P(h(n)(X_n - \theta) < z_{\alpha/2}) - P(h(n)(X_n - \theta) < -z_{\alpha/2}) > 1 - \alpha.$$

The proof is completed by defining $N = \max\{N_1, N_2, N_3\}$. □

Corollary 1.2 *If $\{p_k; k \geq 1\}$ satisfies Condition 1.4, then*

$$\frac{T_1}{\pi_0} \xrightarrow{p} 1. \tag{1.24}$$

Proof: The result immediately follows Theorem 1.7 and Lemma 1.8. □

Property (1.24) is sometimes referred to as the multiplicative consistency of Turing's formula. Interested readers may wish to see Ohannessian and Dahleh (2012) for a more detailed discussion on such consistency under slightly more general conditions.

Corollary 1.3 *If $\{p_k; k \geq 1\}$ satisfies Condition 1.4, then*

1)

$$\frac{E(N_1)\,(\ln T_1 - \ln \pi_0)}{\sqrt{E(N_1) + 2E(N_2)}} \xrightarrow{L} N(0,1) \ \textit{and} \tag{1.25}$$

2)
$$\frac{N_1\,(\ln T_1 - \ln \pi_0)}{\sqrt{N_1 + 2N_2}} \xrightarrow{L} N(0,1). \tag{1.26}$$

The proof of Corollary 1.3 is left as an exercise (see Exercise 23).

1.3 Multivariate Normal Laws

Turing's formula T_1 is only one member of a family introduced by Good (1953), albeit the most famous one. Recall the notations, for every integer r, $1 \le r \le n$, N_r, and π_{r-1} as defined in (1.1) and (1.2). The following may be referred to as the rth order Turing's formula.

$$T_r = \binom{n}{r-1}\binom{n}{r}^{-1} N_r = \frac{r}{n-r+1} N_r. \tag{1.27}$$

T_r may be thought of as an estimator of π_{r-1}. The main objective of this section is to give a multivariate normal law for the vector

$$(T_1, T_2, \dots, T_R)^\tau$$

where R is an arbitrarily fixed positive integer. Toward that end, a normal law for T_r, where r is any positive integer, is first given.

The following is a sufficient condition under which many of the subsequent results are established. Let $g(n,\delta) = n^{1-2\delta}$ as in (1.13).

Condition 1.5 *There exists a $\delta \in (0, 1/4)$ such that as $n \to \infty$,*

1) $\frac{g^2(n,\delta)}{n^2} \mathrm{E}(N_r) \to \frac{c_r}{r!} \ge 0,$
2) $\frac{g^2(n,\delta)}{n^2} \mathrm{E}(N_{r+1}) \to \frac{c_{r+1}}{(r+1)!} \ge 0,$ *and*
3) $c_r + c_{r+1} > 0.$

Theorem 1.8 *Under Condition 1.5,*

$$g(n,\delta)(T_r - \pi_{r-1}) \xrightarrow{L} N\left(0, \frac{c_{r+1} + rc_r}{(r-1)!}\right).$$

The proof of Theorem 1.8 is based on a direct evaluation of the characteristic function of $g(n,\delta)(T_r - \pi_{r-1})$, which can be shown to have a limit, as $n \to \infty$, of

$$\exp\left\{-\frac{t^2}{2}\left[\frac{c_{r+1}}{(r-1)!} + \frac{rc_r}{(r-1)!}\right]\right\}.$$

The details of the proof are found in Zhang (2013a).

Theorem 1.9 below is a restatement of Theorem 1.8.

Theorem 1.9 *Under Condition 1.5,*

$$\frac{n(T_r - \pi_{r-1})}{\sqrt{r^2 E(N_r) + (r+1) r E(N_{r+1})}} \xrightarrow{L} N(0,1).$$

The proof of Theorem 1.9 is straightforward and is left as an exercise (see Exercise 16).

Lemma 1.9 *Let*

$$\hat{c}_r = \frac{r! g^2(n,\delta)}{n^2} N_r.$$

Under Condition 1.5, $\hat{c}_r$ converges to c_r in probability.

By Condition 1.5 and Chebyshev's inequality, it suffices to show that

$$\mathrm{Var}(\hat{c}_r) \to 0.$$

A proof of that is found in Zhang (2013a).

Theorem 1.10 *Under Condition 1.5,*

$$\frac{n(T_r - \pi_{r-1})}{\sqrt{r^2 N_r + (r+1) r N_{r+1}}} \xrightarrow{L} N(0,1).$$

Theorem 1.10 is a direct consequence of Theorem 1.9 and Lemma 1.9. The proof is left as an exercise (see Exercise 17).

Similar to the asymptotic results in Theorems 1.3 and 1.4, it may be of interest to note that the results of Theorems 1.9 and 1.10 require no further knowledge of $g(n,\delta)$, that is, the knowledge of δ, other than its existence.

To establish the asymptotic bivariate distribution of (T_{r_1}, T_{r_2}) where r_1 and r_2 are two different positive integers, the following condition plays a central role.

Condition 1.6 *Let $g(n,\delta)$ be as in (1.13). There exists a $\delta \in (0, 1/4)$ such that as $n \to \infty$,*

1) $\frac{g^2(n,\delta)}{n^2} \mathrm{E}(N_{r_1}) \to \frac{c_{r_1}}{r_1!} \geq 0,$
2) $\frac{g^2(n,\delta)}{n^2} \mathrm{E}(N_{r_1+1}) \to \frac{c_{r_1+1}}{(r_1+1)!} \geq 0,$
3) $c_{r_1} + c_{r_1+1} > 0,$
4) $\frac{g^2(n,\delta)}{n^2} \mathrm{E}(N_{r_2}) \to \frac{c_{r_2}}{r_2!} \geq 0,$
5) $\frac{g^2(n,\delta)}{n^2} \mathrm{E}(N_{r_2+1}) \to \frac{c_{r_2+1}}{(r_2+1)!} \geq 0,$ *and*
6) $c_{r_2} + c_{r_2+1} > 0.$

Lemma 1.10 *For any two constants, a and b satisfying $a^2 + b^2 > 0$, assuming that $r_1 < r_2 - 1$ and that Condition 1.6 holds, then*

$$g(n,\delta)[a(T_{r_1} - \pi_{r_1-1}) + b(T_{r_2} - \pi_{r_2-1})] \xrightarrow{L} N(0,\sigma^2)$$

where

$$\sigma^2 = a^2\frac{c_{r_1+1} + r_1\ c_{r_1}}{(r_1-1)!} + b^2\frac{c_{r_2+1} + r_2\ c_{r_2}}{(r_2-1)!}.$$

Lemma 1.11 *For any two constants, a and b satisfying $a^2 + b^2 > 0$, assuming that $r_1 = r_2 - 1$ and that Condition 1.6 holds, then*

$$g(n,\delta)[a(T_{r_1} - \pi_{r_1-1}) + b(T_{r_2} - \pi_{r_2-1})] \xrightarrow{L} N(0,\sigma^2)$$

where

$$\sigma^2 = a^2\frac{c_{r_1+1} + r_1\ c_{r_1}}{(r_1-1)!} - 2ab\frac{c_{r_2}}{(r_1-1)!} + b^2\frac{c_{r_2+1} + r_2\ c_{r_2}}{(r_2-1)!}.$$

The proofs of both Lemmas 1.10 and 1.11 are based on a straightforward evaluation of the characteristic functions of the underlying statistic. They are lengthy and tedious and they are found in Zhang (2013a).

Let

$$\begin{aligned}
\sigma_r^2 &= r^2\mathrm{E}(N_r) + (r+1)r\mathrm{E}(N_{r+1}),\\
\rho_r(n) &= -r(r+1)\mathrm{E}(N_{r+1})/(\sigma_r\ \sigma_{r+1}),\\
\rho_r &= \lim_{n\to\infty}\rho_r(n),\\
\hat{\sigma}_r^2 &= r^2N_r + (r+1)rN_{r+1},\\
\hat{\rho}_r = \hat{\rho}_r(n) &= -r(r+1)\frac{N_{r+1}}{\sqrt{\hat{\sigma}_r^2\hat{\sigma}_{r+1}^2}}.
\end{aligned}$$

By the consistency of Lemma 1.9 and the definition of multivariate normality of a random vector, the following two corollaries and two theorems are immediate.

Corollary 1.4 *Assume that $r_1 < r_2 - 1$ and that Condition 1.6 holds, then*

$$n\left(\frac{T_{r_1} - \pi_{r_1-1}}{\sigma_{r_1}}, \frac{T_{r_2} - \pi_{r_2-1}}{\sigma_{r_2}}\right)^{\tau} \xrightarrow{L} MVN\ (\emptyset, I_{2\times 2})$$

where $\emptyset$ is a zero vector and $I_{2\times 2}$ is the two-dimensional identity matrix.

Corollary 1.5 *Assume that $r_1 = r_2 - 1$ and that Condition 1.6 holds, then*

$$n\left(\frac{T_{r_1} - \pi_{r_1-1}}{\sigma_{r_1}}, \frac{T_{r_2} - \pi_{r_2-1}}{\sigma_{r_2}}\right)^{\tau} \xrightarrow{L} MVN\ \left(\emptyset, \begin{pmatrix} 1 & \rho_{r_1} \\ \rho_{r_1} & 1 \end{pmatrix}\right).$$

Corollaries 1.4 and 1.5 suggest that, in the following series

$$\left\{ n \left(\frac{T_r - \pi_{r-1}}{\sigma_r} \right) ; r \geq 1 \right\},$$

any two entries are asymptotically independent unless they are immediate neighbors.

Theorem 1.11 *For any positive integer R, if Condition 1.6 holds for every r, $1 \leq r \leq R$, then*

$$n \left(\frac{T_1 - \pi_0}{\sigma_1}, \frac{T_2 - \pi_1}{\sigma_2}, \dots, \frac{T_R - \pi_{R-1}}{\sigma_R} \right)^{\tau} \xrightarrow{L} MVN(\emptyset, \Sigma)$$

where $\emptyset$ is a zero vector, $\Sigma = (a_{i,j})$ is an $R \times R$ covariance matrix with all the diagonal elements being $a_{r,r} = 1$ for $r = 1, \dots, R$, the super-diagonal and the sub-diagonal elements being $a_{r,r+1} = a_{r+1,r} = \rho_r$ for $r = 1, \dots, R-1$, and all the other off-diagonal elements being zeros.

Let $\hat{\Sigma}$ be the resulting matrix of Σ with ρ_r replaced by $\hat{\rho}_r(n)$ for all r. Let $\hat{\Sigma}^{-1}$ denote the inverse of $\hat{\Sigma}$ and $\hat{\Sigma}^{-1/2}$ denote any $R \times R$ matrix satisfying $\hat{\Sigma}^{-1} = (\hat{\Sigma}^{-1/2})^{\tau} \hat{\Sigma}^{-1/2}$.

Theorem 1.12 *For any positive integer R, if Condition 1.6 holds for every r, $1 \leq r \leq R$, then*

$$n\hat{\Sigma}^{-1/2} \left(\frac{T_1 - \pi_0}{\hat{\sigma}_1}, \frac{T_2 - \pi_1}{\hat{\sigma}_2}, \dots, \frac{T_R - \pi_{R-1}}{\hat{\sigma}_R} \right)^{\tau} \xrightarrow{L} MVN(\emptyset, I_{R\times R}).$$

Theorem 1.12 has a profound statistical implication that can be explored by considering the following three points:

1) The alphabet $\mathscr{X} = \{\ell_k; k \geq 1\}$ is dynamically partitioned by a sample of size n into $n+1$ groups

$$\{\ell_k : Y_k = 0\}, \{\ell_k : Y_k = 1\}, \dots, \{\ell_k : Y_k = r\}, \dots, \{\ell_k : Y_k = n\}.$$

   The probability sequence $\{p_k; k \geq 1\}$ is also dynamically sorted accordingly into the $n+1$ groups with respect to the partition above, that is,

$$\{p_k : Y_k = 0\}, \{p_k : Y_k = 1\}, \dots, \{p_k : Y_k = r\}, \dots, \{p_k : Y_k = n\}.$$

2) For any chosen fixed integer $R \geq 1$, the respective group total probabilities for the first $R-1$ groups, the entries in the second row below,

$\{\ell_k : Y_k = 0\}$	$\{\ell_k : Y_k = 1\}$	$\dots$	$\{\ell_k : Y_k = R-1\}$
$\sum_{k:Y_k=0} p_k$	$\sum_{k:Y_k=1} p_k$	$\dots$	$\sum_{k:Y_k=R-1} p_k$

(1.28)

   may be statistically estimated according to Theorem 1.12.

3) The union of the first R partitions, that is,

$$\cup_{r=0}^{R-1}\{\ell_k : Y_k = r\},$$

dynamically covers a subset of $\mathcal{X}$ of letters with very low probabilities, or informally a "tail", for any n large or small.

These three points are perhaps better demonstrated via an example. Consider a special case of a discrete distribution with $\{p_k\}$ following a discrete power law, known as the Pareto law, in the tail, that is,

$$p_k = Ck^{-\lambda} \tag{1.29}$$

for all $k > k_0$ where $C > 0$ and $\lambda > 1$ are unknown parameters describing the tail of the probability distribution beyond an unknown positive integer k_0. For $k = 1, \dots, k_0$, the probabilities are nonnegative but are otherwise not specified. This partially parametric probability model is subsequently referred to as the "power tail model" or the "Pareto tail model," for which Condition 1.6 may be verified. Suppose it is of interest to estimate C and λ. An estimation procedure may be devised based on Theorem 1.12. This problem gets a full treatment in Chapter 7. However, the following gives an intuitive perspective to the problem, and the perspective is relevant in much of the development throughout the subsequent chapters.

Under the model (1.29), $\mathcal{X} = \{1, 2, \dots\}$ and the R groups in the partition by a sample with observed frequencies, $\{Y_1, Y_2, \dots\}$, are subsets of the positive integers on the real line. Two key issues are to be resolved in estimating C and λ.

1) The basis of a statistical estimation.
2) The issue of an unknown k_0.

The basis of a statistical argument is the asymptotic normality of Theorem 1.12 with each π_r represented under the semiparametric tail model, that is, for each $r = 1, \dots, R$

$$\pi_{r-1} = \sum_{k:Y_k=r-1} p_k = \sum_{k:Y_k=r-1} Ck^{-\lambda}. \tag{1.30}$$

Noting that T_r is observable and π_{r-1} is also an observable function of C and λ, the asymptotic distribution in Theorem 1.12 has an explicit likelihood function, which will be referred to as the *asymptotic likelihood function* and is a function of the data and the parameters C and λ. Let $(\hat{C}, \hat{\lambda})$ be the maximum likelihood estimator of (C, λ) based on the asymptotic distribution of Theorem 1.12, which is shown to exist uniquely in Chapter 7 and is referred to thereafter as the asymptotic maximum likelihood estimator or *AMLE*. The *AMLE* is also shown to be consistent.

Regarding the issue of the unknown k_0, since the semiparametric model is only valid for $k > k_0$, the second equality in (1.30) may not necessarily be true for some k. However for every r, as n increases indefinitely, the minimum value of k in the rth partition group, that is, $\min\{k : Y_k = r - 1\}$ rapidly increases in probability to pass the fixed albeit unknown k_0 (see Exercise 18). This argument essentially validates the second equality in (1.30) and, in turn, validates the above-mentioned asymptotic argument. The fact that the estimation procedure does not require a preset threshold for the unknown k_0, in contrast to the required abundance threshold in Hill's estimator (see Hill (1975)) for continuous random variables, is quite remarkable. In short, by Theorem 1.12, the vector $(T_1, \dots, T_R)^\tau$ may be viewed as a statistical "window on the tail" in capturing a parametric tail behavior of an underlying probability distribution, possibly beyond data range.

1.4 Turing's Formula Augmented

For $r = 1, \dots, n$, recall

$$N_r = \sum_{k\geq 1} 1[Y_k = r]$$

which is the total number of letters represented exactly r times in the sample. Chao, Lee, and Chen (1988) proposed the following augmented Turing's formula:

$$T^\sharp = \sum_{r=1}^{n} (-1)^{r+1} \binom{n}{r}^{-1} N_r. \tag{1.31}$$

Noting that both n and r are finite integers,

$$\begin{aligned}
\mathrm{E}\,(T^\sharp - \pi_0) &= \sum_{r=1}^{n} (-1)^{r+1} \binom{n}{r}^{-1} \mathrm{E}(N_r) - \mathrm{E}(\pi_0) \\
&= \sum_{r=1}^{n} (-1)^{r+1} \binom{n}{r}^{-1} \binom{n}{r} \sum_{k\geq 1} p_k^r (1-p_k)^{n-r} \\
&\quad - \sum_{k\geq 1} p_k (1-p_k)^n \\
&= \sum_{k\geq 1} p_k (1-p_k)^{n-1} - \sum_{k\geq 1} p_k (1-p_k)^n \\
&\quad + \sum_{r=2}^{n} (-1)^{r+1} \sum_{k\geq 1} p_k^r (1-p_k)^{n-r} \\
&= \sum_{k\geq 1} p_k^2 (1-p_k)^{n-1} + \sum_{r=2}^{n} (-1)^{r+1} \sum_{k\geq 1} p_k^r (1-p_k)^{n-r} \\
&= \sum_{k\geq 1} p_k^2 (1-p_k)^{n-1} - \sum_{k\geq 1} p_k^2 (1-p_k)^{n-2} \\
&\quad + \sum_{r=3}^{n} (-1)^{r+1} \sum_{k\geq 1} p_k^r (1-p_k)^{n-r}
\end{aligned}$$

$$= -\sum_{k\geq1} p_k^3(1-p_k)^{n-2} + \sum_{r=3}^{n}(-1)^{r+1}\sum_{k\geq1} p_k^r(1-p_k)^{n-r}$$

$$\vdots$$

$$= (-1)^{n+1}\sum_{k\geq1} p_k^{n+1}.$$

The bias of $T^\sharp$ is therefore, letting $p_\vee = \max\{p_k; k \geq 1\}$,

$$|\mathrm{E}\,(T^\sharp - \pi_0)\,| = \sum_{k\geq1} p_k^{n+1} \leq p_\vee^n. \tag{1.32}$$

It is clear that unless $p_\vee = 1$, that is, $\{p_k\}$ has all of its probability mass on a single letter, the bias of $T^\sharp$ decays at least exponentially fast in sample size n.

Theorem 1.13 *If the effective cardinality of $\mathcal{X}$ is greater than or equal to three, that is, $K \geq 3$, then the augmented Turing's formula $T^\sharp$ is asymptotically and relatively unbiased.*

Proof: To be instructive, consider first the case that $p_\vee < 1/2$, which implies that $p_\vee/(1-p_\vee) \in (0,1)$. By Definition 1.1, as $n \to \infty$,

$$\frac{|\mathrm{E}\,(T^\sharp - \pi_0)\,|}{\mathrm{E}(\pi_0)} = \frac{\sum_{k\geq1} p_k^{n+1}}{\sum_{k\geq1} p_k(1-p_k)^n} \leq \left(\frac{p_\vee}{1-p_\vee}\right)^n \to 0.$$

If $p_\vee \geq 1/2$, then since $K \geq 3$, there must exist a $k = k^\tau$ such that $0 < p_{k'} < 1 - p_\vee$ and hence $1 - p_{k'} > p_\vee$, that is, $p_\vee/(1-p_{k'}) < 1$. On the other hand,

$$\frac{|\mathrm{E}\,(T^\sharp - \pi_0)\,|}{\mathrm{E}(\pi_0)} = \frac{\sum_{k\geq1} p_k^{n+1}}{\sum_{k\geq1} p_k(1-p_k)^n} \leq \frac{p_\vee^n}{p_{k'}(1-p_{k'})^n}$$

$$= \frac{1}{p_{k'}}\left(\frac{p_\vee}{1-p_{k'}}\right)^n \to 0.$$

□

Theorem 1.13 suggests that the augmented Turing's formula is of some material improvement to Turing's formula, particularly in view of Example 1.1. In addition, the augmented Turing's formula $T^\sharp$ also admits asymptotic normality.

Theorem 1.14 *Let $T^\sharp$ be as in (1.31). If the probability distribution $\{p_k; k \geq 1\}$ satisfies Condition 1.1, then*

$$\frac{n\,(T^\sharp - \pi_0)}{\sqrt{E(N_1) + 2E(N_2)}} \xrightarrow{L} N(0,1). \tag{1.33}$$

To prove Theorem 1.14, the following two lemmas are useful.

Lemma 1.12 *For each integer $n > 0$, let X_n be a nonnegative random variable with finite mean $m_n = E(X_n)$. If $\lim_{n\to\infty} m_n = 0$, then $X_n \xrightarrow{p} 0$.*

The proof is left as an exercise (see Exercise 9).

Lemma 1.13 *For any constant $\delta \in (0,1)$ and a probability sequence $\{p_k; k \geq 1\}$, as $n \to \infty$,*

$$1.\ n^\delta \sum_{k\geq 1} p_k^2 (1-p_k)^n \to 0 \quad and \quad 2.\ n^\delta \sum_{k\in S} p_k^2 (1-p_k)^n \to 0$$

where S is any subset of $\{k; k \geq 1\}$.

Proof: For Part 1, since $\delta \in (0,1)$, there exists $\delta_1 \in (\delta, 1)$. Let

$$I = \{k : p_k < 1/n^{\delta_1}\} \quad \text{and} \quad I^c = \{k : p_k \geq 1/n^{\delta_1}\}\,.$$

$$\begin{aligned}
0 &\leq n^\delta \sum_{k\geq 1} p_k^2 (1-p_k)^n \\
&= n^\delta \sum_{k\in I^c} p_k^2 (1-p_k)^n + n^\delta \sum_{k\in I} p_k^2 (1-p_k)^n \\
&\leq n^\delta \sum_{k\in I^c} p_k \ \ (1-1/n^{\delta_1})^n + n^\delta \sum_{k\in I} p_k / n^{\delta_1} \\
&\leq n^\delta \ (1-1/n^{\delta_1})^{n^{\delta_1+(1-\delta_1)}} + n^\delta / n^{\delta_1} \\
&= n^\delta \ [\ (1-1/n^{\delta_1})^{n^{\delta_1}}\ ]^{n^{1-\delta_1}} + n^{-(\delta_1-\delta)} \to 0.
\end{aligned}$$

Part 1 implies Part 2. □

Proof of Theorem 1.14: Let $\sigma_n = \sqrt{E(N_1) + 2E(N_2)}$. By Slutsky's theorem, the objective is to show

$$\frac{n(T^\sharp - \pi_0)}{\sigma_n} - \frac{n(T_1 - \pi_0)}{\sigma_n} \xrightarrow{p} 0. \tag{1.34}$$

Note first that by Condition 1.1,

$$\frac{\sigma_n}{n^{2\delta}} = \sqrt{\frac{E(N_1) + 2E(N_2)}{n^{4\delta}}} \to c_1 + c_2 > 0.$$

Note second that

$$\begin{aligned}
\frac{n(T^\sharp - \pi_0)}{\sigma_n} - \frac{n(T_1 - \pi_0)}{\sigma_n} &= \frac{n}{\sigma_n}\ (T^\sharp - T_1) \\
&= \frac{n}{\sigma_n} \left[\sum_{r=2}^{n} (-1)^{r+1} \binom{n}{r}^{-1} N_r \right] \\
&= \left(\frac{n^{2\delta}}{\sigma_n}\right)\ n^{1-2\delta} \left[\sum_{r=2}^{n} (-1)^{r+1} \binom{n}{r}^{-1} N_r \right].
\end{aligned}$$

Therefore, (1.34) holds if and only if

$$n^{1-2\delta}\left[\sum_{r=2}^{n}(-1)^{r+1}\binom{n}{r}^{-1}N_r\right]\xrightarrow{p}0. \tag{1.35}$$

Separating the $\sum_{r=2}^{n}$ in (1.35) into two parts, $\sum_{odd}$ and $\sum_{even}$, where $\sum_{odd}$ sums over all odd values of r and $\sum_{even}$ sums over all even values of r in $\{2, \dots, n\}$, the left-hand side of (1.35) may be re-expressed as

$$a_n - b_n := n^{1-2\delta}\sum_{odd}\binom{n}{r}^{-1}N_r - n^{1-2\delta}\sum_{even}\binom{n}{r}^{-1}N_r.$$

Now the objective becomes to show, respectively,

$$a_n \xrightarrow{p} 0 \quad and \quad b_n \xrightarrow{p} 0.$$

However, since both a_n and b_n are nonnegative, it suffices to show that $a_n + b_n \xrightarrow{p} 0$, that is,

$$a_n + b_n = n^{1-2\delta}\sum_{r=2}^{n}\binom{n}{r}^{-1}N_r \xrightarrow{p} 0.$$

By Lemma 1.12, it suffices to show

$$\mathrm{E}(a_n + b_n) = n^{1-2\delta}\sum_{r=2}^{n}\sum_{k\ge1}p_k^r(1-p_k)^{n-r} \longrightarrow 0.$$

Let $\delta^* = 1 - 2\delta$,

$$I_1 = \{k\colon p_k = 1/2\},$$
$$I_2 = \{k\colon p_k > 1/2\},$$
$$I_3 = \{k\colon p_k < 1/2\}.$$

$$\begin{aligned}0 &\le n^{\delta^*}\sum_{r=2}^{n}\sum_{k\ge1}p_k^r(1-p_k)^{n-r}\\ &= n^{\delta^*}\sum_{r=2}^{n}\left(\sum_{k\in I_1}+\sum_{k\in I_2}+\sum_{k\in I_3}\right)p_k^r(1-p_k)^{n-r}\\ &=: d_1 + d_2 + d_3.\end{aligned}$$

In the following, it remains to show that $d_i \to 0$, for each $i \in \{1, 2, 3\}$, respectively.

If I_1 is empty, then $d_1 = 0$. If not, then for each index $k \in I_1$,

$$n^{\delta^*}\sum_{r=2}^{n}p_k^r(1-p_k)^{n-r} = n^{\delta^*}\sum_{r=2}^{n}(1/2)^n = n^{\delta^*}(n-1)(1/2)^n \to 0.$$

Since there are at most two indices in I_1, $d_1 \to 0$.

If I_2 is empty, then $d_2 = 0$. If not, then for each index $k \in I_2$,

$$
\begin{aligned}
n^{\delta^*} \sum_{r=2}^{n} p_k^r (1-p_k)^{n-r} &= n^{\delta^*} \left[\sum_{r=2}^{n} \left(\frac{p_k}{1-p_k} \right)^r \right] (1-p_k)^n \\
&= n^{\delta^*} \left\{ \left(\frac{p_k}{1-p_k} \right)^2 \left[\frac{1 - \left(\frac{p_k}{1-p_k} \right)^{n-1}}{1 - \frac{p_k}{1-p_k}} \right] \right\} \\
&\quad \times (1-p_k)^n \\
&= n^{\delta^*} \left[\frac{p_k^2 (1-p_k)^{n-1}}{1-2p_k} \right] \left[1 - \frac{p_k^{n-1}}{(1-p_k)^{n-1}} \right] \\
&= n^{\delta^*} \left[\frac{p_k^2 (1-p_k)^{n-1}}{1-2p_k} \right] \\
&\quad - n^{\delta^*} \left[\frac{p_k^2 (1-p_k)^{n-1}}{1-2p_k} \right] \frac{p_k^{n-1}}{(1-p_k)^{n-1}} \\
&= n^{\delta^*} \left[\frac{p_k^2 (1-p_k)^{n-1}}{1-2p_k} \right] - n^{\delta^*} \left(\frac{p_k^{n+1}}{1-2p_k} \right) \to 0.
\end{aligned}
$$

Since there is at most one index in I_2, $d_2 \to 0$.

For d_3, using the same argument as that for d_2 earlier,

$$
\begin{aligned}
d_3 &= n^{\delta^*} \sum_{k \in I_3} p_k^2 (1-p_k)^{n-1} (1-2p_k)^{-1} - n^{\delta^*} \sum_{k \in I_3} p_k^{n+1} (1-2p_k)^{-1} \\
&=: d_{31} - d_{32}.
\end{aligned}
$$

It suffices to show $d_{31} \to 0$ and $d_{32} \to 0$. Let $p_\vee = \max\{p_k; k \in I_3\}$. Then $p_\vee < 1/2$ and

$$
\begin{aligned}
d_{31} &= n^{\delta^*} \sum_{k \in I_3} p_k^2 (1-p_k)^{n-1} (1-2p_k)^{-1} \\
&\le n^{\delta^*} (1-p_\vee)^{-1} (1-2p_\vee)^{-1} \sum_{k \in I_3} p_k^2 (1-p_k)^n \to 0
\end{aligned}
$$

by Part 2 of Lemma 1.13. Finally,

$$
d_{32} = n^{\delta^*} \sum_{k \in I_3} p_k^{n+1} (1-2p_k)^{-1} \le n^{\delta^*} (1-2p_\vee)^{-1} (1/2)^n \to 0.
$$

□

The consistency of Lemma 1.5 immediately leads to the following practically useful statement.

Theorem 1.15 *Let $T^\sharp$ be as in (1.31). Under Condition 1.1,*

$$\frac{n\,(T^\sharp - \pi_0)}{\sqrt{N_1 + 2N_2}} \xrightarrow{L} N(0,1). \tag{1.36}$$

Example 1.10 *Construct a 95% confidence interval for π_0 using Theorem 1.15 and the bird data in Table 1.2.*

Solution: There are only a few values of r for which N_r is nonzero in the data set and they are as follows:

r	1	2	6	30	50	70	80	100	200	300
N_r	12	1	1	1	2	1	1	7	2	2

By the sheer magnitude of $\binom{n}{r}$, and hence the insignificance of $\binom{n}{r}^{-1}$, with large values of r, the additive terms in (1.31) do not register but in a few terms. Consequently, $T^\sharp = 0.00599$, and by Theorem 1.15 and Example 1.6, a 95% confidence interval for π_0 is

$$0.0059995 \pm 0.0037 = (0.0022995, 0.0096995).$$

One drawback of $T^\sharp$ is that it may theoretically assume a negative value though practically impossible with a sufficiently large n. One may wish to further augment $T^\sharp$ by means of, for example,

$$T^{\sharp\sharp} = \max\left\{T^\sharp, 0\right\}. \tag{1.37}$$

It can be verified that $T^{\sharp\sharp}$ also admits asymptotic normality of Theorems 1.14 and 1.15 (see Exercise 19).

In the same spirit of (1.31), the rth order Turing's formula may also be augmented similarly. Noting for each integer r, $1 \le r \le n-1$, $\pi_{r-1} = \sum_{k\ge1} p_k 1[Y_k = r-1]$,

$$\begin{aligned}
\mathrm{E}\left(\binom{n}{r-1}^{-1}\pi_{r-1}\right) &= \sum_{k\ge1} p_k^r(1-p_k)^{n-(r-1)} \\
&= \sum_{k\ge1} p_k^r(1-p_k)^{n-r} - \sum_{k\ge1} p_k^{r+1}(1-p_k)^{n-r} \\
&= \sum_{k\ge1} p_k^r(1-p_k)^{n-r} - \sum_{k\ge1} p_k^{r+1}(1-p_k)^{n-r-1} + \cdots \\
&\quad + (-1)^{n-r+1}\sum_{k\ge1} p_k^{n+1},
\end{aligned}$$

and

$$\mathrm{E}\left(\binom{n}{r}^{-1}N_r\right) = \sum_{k\ge1} p_k^r(1-p_k)^{n-r},$$

it may be verified that the augmented rth order Turing's formula,

$$T_r^{\sharp} = \binom{n}{r-1} \sum_{m=r}^{n} (-1)^{m-r} \binom{n}{m}^{-1} N_m \tag{1.38}$$

has bias

$$|E(T_r^{\sharp} - \pi_{r-1})| = \binom{n}{r-1} \sum_{k\geq 1} p_k^{n+1} \leq \binom{n}{r-1} p_{\vee}^n = \mathcal{O}\left(n^{r-1} p_{\vee}^n\right),$$

where $p_{\vee} = \max\{p_k; k \geq 1\}$, decaying rapidly when r is a small fixed integer.

1.5 Goodness-of-Fit by Counting Zeros

Turing's formula is mostly about sorting letters, observed in an *iid* sample of size n, into equal frequency groups and counting the different letters in each group, namely N_r, $r = 1, \ldots, n$. In the special case of a finite and known K,

$$N_0 = \sum_{k\geq 1} 1[Y_k = 0]$$

is also observable. The distribution of N_r, where $r = 0, 1, \ldots, n$, is discussed by many in more detail under the topic of occupancy problems. Both Johnson and Kotz (1977) and Kolchin, Sevastyanov, and Chistyakov (1978) offer an excellent coverage of the topic.

In an experiment of randomly allocating n balls into $K < \infty$ baskets according to a distribution $\mathbf{p} = \{p_k; k = 1, \ldots, K\}$, N_0 is the number of empty or unoccupied baskets among a total of K baskets. When N_0 is observable, it has a quite interesting statistical feature as stated in Theorem 1.16.

Theorem 1.16 *For a finite alphabet $\mathcal{X} = \{\ell_1, \ldots, \ell_K\}$ where $K \geq 2$, let $\mathcal{P}_K$ denote the collection of all possible probability distributions on $\mathcal{X}$. N_0 satisfies*

$$\min\{E_{\mathbf{p}}(N_0) : \mathbf{p} \in \mathcal{P}_K\} = K\left(1 - \frac{1}{K}\right)^n, \tag{1.39}$$

and $E_{\mathbf{p}}(N_0)$ attains its minimum value at $\mathbf{p}$ being uniform, that is, $p_k = 1/K$ for every $k = 1, \ldots, K$.

Proof: Since

$$E_{\mathbf{p}}(N_0) = \sum_{k=1}^{K} E(1[Y_k = 0]) = \sum_{k=1}^{K} (1 - p_k)^n, \tag{1.40}$$

it suffices to show that for any pair of probabilities, denoted by p_1 and p_2 (without loss of generality), if $p_1 \neq p_2$, then there would exist a $\mathbf{p}^* \in \mathcal{P}_K$ such that $E_{\mathbf{p}^*}(N_0) > E_{\mathbf{p}}(N_0)$. Toward that end, let

$$g(x_1, x_2) = (1 - x_1)^n + (1 - x_2)^n$$

and consider the behavior of

$$g(x_1, s-x_1) = (1-x_1)^n + [1-(s-x_1)]^n$$

for $x_1 \in [0, s]$, where $s = p_1 + p_2$ is held fixed:

$$\frac{dg(x_1, s-x_1)}{dx_1} = -n(1-x_1)^{n-1} + n(1-s+x_1)^{n-1}.$$

Since

$$\left.\frac{dg(x_1, s-x_1)}{dx_1}\right|_{x_1=0} = -n + n(1-s)^{n-1} \le 0,$$

$$\left.\frac{dg(x_1, s-x_1)}{dx_1}\right|_{x_1=s} = -n(1-s)^{n-1} + n \ge 0, \text{ and}$$

$$\frac{dg(x_1, s-x_1)}{dx_1} = 0 \text{ if and only if } x_1 = \frac{s}{2},$$

it is then clear that $g(x_1, x_2) = (1-x_1)^n + (1-x_2)^n$, subject to the constraint $x_1 + x_2 = p_1 + p_2$, reaches its minimum at $x_1 = x_2 = (p_1 + p_2)/2$. This implies $E_{\mathbf{p}^*}(N_0) < E_{\mathbf{p}}(N_0)$ where

$$\mathbf{p}^* = \left\{ \frac{p_1 + p_2}{2}, \frac{p_1 + p_2}{2}, p_3, \dots, p_K \right\}.$$

Therefore, $E_{\mathbf{p}}(N_0)$ is not minimized unless $p_k = 1/K$ for every k, $1 \le k \le K$, in which case, it takes on the value of $K(1-1/K)^n$. □

Theorem 1.16 suggests that, under the assumption of independent allocations, the uniform distribution would lead to the least expected number of empty baskets. Any other distribution would have a tendency for a larger number of empty baskets. If an observed N_0 differs significantly from $K(1-1/K)^n$, then either the uniformity assumption or the independence assumption, or both, would become questionable. This feature provides an interesting perspective to the statistical question of "goodness-of-fit."

For any $\mathbf{p} = \{p_k; k = 1, \dots, K\} \in \mathcal{P}_K$, in addition to $E_{\mathbf{p}}(N_0)$ given in (1.40),

$$\begin{aligned}
\text{Var}(N_0) &= E\left(N_0^2\right) - (E(N_0))^2 \\
&= E\left(\left(\sum_{1\le k\le K} 1[Y_k = 0]\right)^2\right) - (E(N_0))^2 \\
&= E\left(\sum_{1\le k\le K} 1[Y_k = 0]\right. \\
&\quad \left. + \sum_{i\ne j; i,j=1,\dots,K} 1[Y_i = 0]1[Y_j = 0]\right) - (E(N_0))^2
\end{aligned}$$

$$= \sum_{1\le k\le K} (1-p_k)^n + \sum_{i\ne j; i,j=1,\dots,K} (1-p_i-p_j)^n - (\mathrm{E}(N_0))^2. \quad (1.41)$$

Under the assumption of the uniform distribution, that is, $p_k = 1/K$ for every k, $1 \le k \le K$, (1.40) and (1.41) become

$$\mathrm{E}(N_0) = K\left(1-\frac{1}{K}\right)^n, \quad (1.42)$$

$$\mathrm{Var}(N_0) = K\left(1-\frac{1}{K}\right)^n + K(K-1)\left(1-\frac{2}{K}\right)^n - K^2\left(1-\frac{1}{K}\right)^{2n}. \quad (1.43)$$

Further suppose $n = \lambda K$ for some constant $\lambda > 0$, then (1.42) and (1.43) become

$$\mathrm{E}(N_0) = K\left(1-\frac{1}{K}\right)^{\lambda K}, \quad (1.44)$$

$$\mathrm{Var}(N_0) = K\left(1-\frac{1}{K}\right)^{\lambda K} + K(K-1)\left(1-\frac{2}{K}\right)^{\lambda K} - K^2\left(1-\frac{1}{K}\right)^{2\lambda K}. \quad (1.45)$$

More specifically, when $\lambda = 1$ or $K = n$, (1.42) and (1.45) become

$$\mathrm{E}(N_0) = K\left(1-\frac{1}{K}\right)^{K}, \quad (1.46)$$

$$\mathrm{Var}(N_0) = K\left(1-\frac{1}{K}\right)^{K} + K(K-1)\left(1-\frac{2}{K}\right)^{K} - K^2\left(1-\frac{1}{K}\right)^{2K}. \quad (1.47)$$

For any sufficiently large K, (1.42) and (1.47) become

$$\mathrm{E}(N_0) \approx K e^{-1}, \quad (1.48)$$

$$\mathrm{Var}(N_0) \approx K\left(e^{-1} - e^{-2}\right). \quad (1.49)$$

Let

$$P_0 = \frac{N_0}{K}, \quad (1.50)$$

that is, the proportion of baskets not occupied in an experiment of randomly allocating $n = K$ balls into K baskets, or more generally, the proportion of letters in $\mathscr{X}$ not observed in an *iid* sample under the uniform distribution, $p_k = 1/K$. By (1.46)–(1.49),

$$\mathrm{E}(P_0) = \left(1-\frac{1}{K}\right)^{K} \approx 1/e, \quad (1.51)$$

$$\begin{aligned} \mathrm{Var}(P_0) &= \frac{\mathrm{Var}(N_0)}{K^2} \\ &= \frac{1}{K}\left(1-\frac{1}{K}\right)^{K} + \frac{K-1}{K}\left(1-\frac{2}{K}\right)^{K} - \left(1-\frac{1}{K}\right)^{2K} \\ &\approx \frac{e-1}{Ke^2}. \end{aligned} \quad (1.52)$$

At this point, an interesting observation can be made. If $n = 100$ balls are randomly allocated into $K = 100$ baskets, one should expect to see approximately $100e^{-1} \approx 36.78\% \times 100 \approx 37$ empty baskets. The expected number of empty baskets in this experiment at 37 is counterintuitively high for most minds. The most common intuitive guess of $E(N_0)$ is 25, which is significantly below 37. It is perhaps more interesting to contemplate why an intuitive mind would arrive at a number so much lower than 37. An intuitive mind would try to balance the extreme event of a "perfectly uniform allocation" of one ball in each basket (with probability $P(N_0 = 0) = K!/K^K$) and another extreme event of a "completely skewed allocation" of all balls in one basket (with probability $P(N_0 = K - 1) = K/K^K$). The fact that the first event is so much more likely than the second event would pull an intuitive guess of $E(N_0)$ toward the lower end. Even without the notion of likelihoods of these events, the event of $N_0 = 0$ conforms much better to the notion of "uniformly distributed allocation" than that of $N_0 = 99$. Nevertheless, the fact remains that $E(N_0) \approx 37$.

In fact, in terms of proportion, that is, $P_0 = N_0/K$, the above-observed statistical feature becomes more pronounced since, as $K \to \infty$

$$E(P_0) \to e^{-1},$$

$$\sigma_{P_0} \simeq \frac{\sqrt{e-1}}{e\sqrt{K}} \to 0.$$

A goodness-of-fit statistical test may be devised based on the above-observed feature.

Given two positive integers, n and m, satisfying $n \geq m$, the Stirling number of the second kind is the number of ways of partitioning a set of n elements into m nonempty sets. For example, $n = 3$ elements in the set $\{\ell_1, \ell_2, \ell_3\}$ can be partitioned into $m = 3$ subsets in one way, that is, $\{\{\ell_1\}, \{\ell_2\}, \{\ell_3\}\}$; the same $n = 3$ elements can be partitioned into $m = 2$ subsets in three ways, that is,

$$\{\{\ell_1, \ell_2\}, \{\ell_3\}\}, \quad \{\{\ell_1, \ell_3\}, \{\ell_2\}\}, \quad \{\{\ell_2, \ell_3\}, \{\ell_1\}\};$$

and the same $n = 3$ elements can be partitioned into $m = 1$ subset in 1 way, that is, $\{\{\ell_2, \ell_2, \ell_3\}\}$. In general, the Stirling number of the second kind is often denoted and computed by

$$S(n, m) = \frac{1}{m!} \sum_{i=0}^{m} (-1)^i \binom{m}{i} (m - i)^n. \tag{1.53}$$

For a more detailed discussion, interested readers may refer to Stanley (1997).

In an experiment of randomly allocating n balls into K baskets, let x be the number of empty baskets. Subject to $1 \leq K - x \leq n$,

1) there are a total of K^n ways to place n balls in K baskets;
2) there are $\binom{K}{K-x}$ $(K - x)!$ ways to select and to arrange the filled baskets; and

3) there are $S(n, K-x)$ ways to completely occupy the selected $K-x$ filled baskets with n balls.

Therefore, under an *iid* allocation with the uniform distribution $p_k = 1/K$ for $k = 1, \ldots, K$, the probability of observing exactly x empty baskets is

$$P(N_0 = x) = \frac{\binom{K}{K-x}(K-x)!S(n, K-x)}{K^n}. \tag{1.54}$$

Simplifying (1.54) leads to

$$\begin{aligned} P(N_0 = x) &= \frac{K!}{x!(K-x)!K^n} \sum_{i=0}^{K-x} (-1)^i \binom{K-x}{i} (K-x-i)^n \\ &= \frac{K!}{x!(K-x)!K^n} \sum_{i=0}^{K-x} (-1)^i \frac{(K-x)!}{i!(K-x-i)!} (K-x-i)^n \\ &= \sum_{i=0}^{K-x} (-1)^i \frac{K!}{x!i!(K-x-i)!} \left(1 - \frac{x+i}{K}\right)^n. \end{aligned} \tag{1.55}$$

For simplicity in demonstration, (1.55) is reduced to the case of $n = K$, that is,

$$P(N_0 = x) = \sum_{i=0}^{K-x} (-1)^i \frac{K!}{x!i!(K-x-i)!} \left(1 - \frac{x+i}{K}\right)^K. \tag{1.56}$$

Table 1.3 is the exact probability distribution of N_0 represented by (1.56) when $K = 10$. Table 1.4 includes several effectively positive terms of the cumulative distribution function (*cdf*), whose probability mass function is represented by (1.56) for $K = 100$.

However, the numerical evaluation of (1.56) is computationally demanding for even moderately large values of K.

Theorem 1.16 and the explicit expression of (1.56) together enable a means of testing the hypothesis that a sample is *iid* under the uniform distribution. Example 1.11 below illustrates how a test based on N_0 may be used to quantify statistical evidence against such a hypothesis.

Table 1.3 Exact probabilities of (1.56) for $K = n = 10$

x	P ($N_0 = x$)	x	P ($N_0 = x$)
0	0.000362880	5	0.128595600
1	0.016329600	6	0.017188920
2	0.136080000	7	0.000671760
3	0.355622400	8	0.000004599
4	0.345144240	9	0.000000001

Table 1.4 Partial *cdf* of N_0 based on (1.56) for $K = n = 100$

x	P ($N_0 = x$)	x	P ($N_0 = x$)	x	P ($N_0 = x$)
24	0.000038799	35	0.362190439	42	0.970782939
25	0.000148654	36	0.487643439	43	0.986544139
26	0.000509259	37	0.614165439	44	0.994351189
27	0.001564749	38	0.729399439	45	0.997842829
28	0.004325109	39	0.824210739	46	0.999252139
29	0.010787039	40	0.894694539	47	0.999765198
30	0.024349639	41	0.942042739	48	0.999933542
31	0.049907939	33	0.159224739	49	0.999983287
32	0.093206939	34	0.249901439	50	0.999996511

Example 1.11 *The following data set contains $n = 100$ iid observations from $\mathscr{X} = \{\ell_1, \dots, \ell_{100}\}$ under the triangular distribution*

$$\mathbf{p} = \{p_k = k/5050; k = 1, \dots, 100\},$$

summarized by the observed letters and their corresponding frequencies, y_k.

letters	ℓ_3	ℓ_{14}	ℓ_{15}	ℓ_{16}	ℓ_{17}	ℓ_{18}	ℓ_{22}	ℓ_{25}	ℓ_{30}	ℓ_{33}	ℓ_{36}	ℓ_{37}
y_k	1	1	1	1	1	1	2	1	3	1	1	1
letters	ℓ_{38}	ℓ_{42}	ℓ_{44}	ℓ_{45}	ℓ_{46}	ℓ_{49}	ℓ_{50}	ℓ_{54}	ℓ_{56}	ℓ_{57}	ℓ_{58}	ℓ_{59}
y_k	2	1	1	1	1	1	2	1	1	1	2	2
letters	ℓ_{61}	ℓ_{64}	ℓ_{65}	ℓ_{66}	ℓ_{68}	ℓ_{69}	ℓ_{70}	ℓ_{71}	ℓ_{72}	ℓ_{73}	ℓ_{74}	ℓ_{75}
y_k	1	1	1	1	5	1	1	2	3	1	1	2
letters	ℓ_{76}	ℓ_{77}	ℓ_{78}	ℓ_{79}	ℓ_{81}	ℓ_{82}	ℓ_{83}	ℓ_{85}	ℓ_{86}	ℓ_{87}	ℓ_{88}	ℓ_{89}
y_k	3	2	1	2	2	2	3	2	3	4	1	3
letters	ℓ_{90}	ℓ_{91}	ℓ_{92}	ℓ_{93}	ℓ_{94}	ℓ_{95}	ℓ_{96}	ℓ_{97}	ℓ_{98}			
y_k	2	1	2	2	1	3	2	2	6			

Using the distribution of Table 1.4 to find the p-value in testing, the hypothesis H_0: $\mathbf{p}$ is uniform on $\mathscr{X}$.

Solution: Since it is known that $K = 100$, the number of missing letters in the sample is $n_0 = 100 - \sum_{k=1}^{100} 1[y_k \neq 0] = 100 - 57 = 43$. By Theorem 1.16, this is a one-sided test, and therefore

$$\text{the p-value} = \mathrm{P}(N_0 \geq 43|H_0) = 0.0135.$$

There is moderately strong statistical evidence against H_0 of the uniform distribution.

The deviation quantified in Example 1.11 is due to the difference between the true distribution **p** (the triangular distribution) and the hypothesized uniform distribution. The test could also be used to detect deviation from the assumption of independence between observations.

Consider a special Markov chain with memory m, that is,

$$X_0, X_1, X_2, \dots, X_{t-2}, X_{t-1}, X_t, \dots, \tag{1.57}$$

such that

1) X_0 is a random element on the state space $\mathscr{X} = \{\ell_1, \dots, \ell_K\}$, with the uniform distribution, that is, $P(X_0 = \ell_k) = 1/K$, for every k, $1 \le k \le K$;
2) X_t, $t \le m$, is a random element on the state space $\mathscr{X} = \{\ell_1, \dots, \ell_K\}$, such that, given $X_{t-1} = x_{t-1}, \dots, X_0 = x_0$, X_t is only allowed to visit the states, which have not been visited by any of X_0 to X_{t-1}, with equal probabilities;
3) X_t, $t > m$, is a random element on the state space $\mathscr{X} = \{\ell_1, \dots, \ell_K\}$, such that, given $X_{t-1} = x_{t-1}, \dots, X_{t-m} = x_{t-m}$, X_t is only allowed to visit the states, which have not been visited by any of X_{t-1} to X_{t-m}, with equal probabilities.

A simple argument based on symmetry would establish that, for every t, X_t is unconditionally a uniformly distributed random element on $\mathscr{X}$.

The following table includes a segment of $n = 100$ consecutive observations of a randomly generated sequence of X_t according to the above-described special Markov chain with memory $m = 10$ and $K = 100$.

ℓ_{90}	ℓ_{59}	ℓ_{63}	ℓ_{26}	ℓ_{40}	ℓ_{72}	ℓ_{36}	ℓ_{11}	ℓ_{68}	ℓ_{67}
ℓ_{29}	ℓ_{82}	ℓ_{30}	ℓ_{62}	ℓ_{23}	ℓ_{35}	ℓ_{02}	ℓ_{22}	ℓ_{58}	ℓ_{69}
ℓ_{67}	ℓ_{93}	ℓ_{56}	ℓ_{11}	ℓ_{42}	ℓ_{29}	ℓ_{73}	ℓ_{21}	ℓ_{19}	ℓ_{84}
ℓ_{37}	ℓ_{98}	ℓ_{24}	ℓ_{15}	ℓ_{70}	ℓ_{13}	ℓ_{26}	ℓ_{91}	ℓ_{80}	ℓ_{56}
ℓ_{73}	ℓ_{62}	ℓ_{96}	ℓ_{81}	ℓ_{05}	ℓ_{25}	ℓ_{84}	ℓ_{27}	ℓ_{36}	ℓ_{46}
ℓ_{29}	ℓ_{13}	ℓ_{57}	ℓ_{24}	ℓ_{95}	ℓ_{82}	ℓ_{45}	ℓ_{14}	ℓ_{67}	ℓ_{34}
ℓ_{64}	ℓ_{43}	ℓ_{50}	ℓ_{87}	ℓ_{08}	ℓ_{76}	ℓ_{78}	ℓ_{88}	ℓ_{84}	ℓ_{03}
ℓ_{51}	ℓ_{54}	ℓ_{99}	ℓ_{32}	ℓ_{60}	ℓ_{68}	ℓ_{39}	ℓ_{12}	ℓ_{26}	ℓ_{86}
ℓ_{94}	ℓ_{95}	ℓ_{70}	ℓ_{34}	ℓ_{78}	ℓ_{67}	ℓ_{01}	ℓ_{97}	ℓ_{02}	ℓ_{17}
ℓ_{92}	ℓ_{52}	ℓ_{56}	ℓ_{80}	ℓ_{86}	ℓ_{41}	ℓ_{65}	ℓ_{89}	ℓ_{44}	ℓ_{19}

Example 1.12 *In the above-described Markov chain with memory m, suppose it is known* a priori *that X_t for every t is unconditionally a uniformly distributed random element on $\mathscr{X}$ and it is of interest to test the hypothesis, $H_0 : m = 0$. Use the above-mentioned data to find the appropriate p-value for the test.*

Solution: Counting the number of missing letters in the data set gives $n_0 = 27$. *Since any positive value of m would tend to enlarge the coverage of* $\mathcal{X}$ *and hence to shrink* N_0, *a one-sided test is appropriate and correspondingly, by Table 1.4,*

$$\text{the p-value} = \mathrm{P}(N_0 \leq 27|H_0) = 0.0016.$$

There is strong statistical evidence in the data against H_0.

It is sometime conjectured that the digits of $\pi = 3.14159265358\ldots$ in nonoverlapping d-digit fragments, where d is any given positive integer, imitate well a sequence of *iid* uniform random variables on alphabet that contains all nonnegative integers from 0 to $\sum_{i=0}^{d-1} 9 \times 10^i$, that is, a number consisted of d nines. For example, if $d = 1$, then any single digit would be one of the 10 digits, ranging from 0 to 9. For another example, if $d = 3$, then any consecutive three digits would be an integer ranging from 0 to 999.

Example 1.13 *Use the sequence of the first one billion digits after the decimal point in* π *to evaluate the resemblance of this sequence to one with a billion iid uniform random digits from alphabet* $\mathcal{X} = \{0, 1, \ldots, 9\}$ *by Pearson's goodness-of-fit test statistic.*

Solution: There are at least two perspectives in which the resemblance to uniform random digits can be quantified. The first is a straightforward goodness-of-fit test for the distribution of $\mathrm{P}(0) = \mathrm{P}(1) = \cdots = \mathrm{P}(9) = 1/10$. *This test may be thought of as an "evenness" test, as it directly tests the equality in proportionality of the 10 digits. The second is a goodness-of-fit test for the distribution in Table 1.3. This test may be thought of as an "anti-evenness" test as it tests a particular pattern in Table 1.3, which deviates significantly from the "perfect" case of evenness at* $n_0 = 0$.

In the first case, the sequence of the first billion digits after the decimal point is hypothesized to be a realization of a sample of $n = 1$ *billion iid observations under the hypothesized uniform distribution, that is,* H_0: $\mathrm{P}(x) = 1/10$ *for* $x = 0, \ldots, 9$. *In the following, the observed and expected frequencies,* O_x *and* E_x, *are given.*

$n_0 = x$	Observed frequency (O_x)	Expected frequency (E_x)
0	99 993 942	100 000 000.0
1	99 997 334	100 000 000.0
2	100 002 410	100 000 000.0
3	99 986 911	100 000 000.0
4	100 011 958	100 000 000.0
5	99 998 885	100 000 000.0
6	100 010 387	100 000 000.0
7	99 996 061	100 000 000.0
8	100 001 839	100 000 000.0
9	100 000 273	100 000 000.0

Pearson's chi-squared test statistic is

$$\chi^2 = \sum_{x=0}^{9} \frac{(O_x - E_x)^2}{E_x} = 4.92$$

with degrees of freedom 9. The corresponding p-value is $\mathrm{P}(\chi^2 \geq 4.92) = 0.8412$ *where* χ^2 *is a chi-squared random variable with degrees of freedom 9. There is no statistical evidence in the sample against* H_0.

In the second case, the sequence of the first billion digits after the decimal point is viewed as a sequence of consecutive fragments or samples of size 10 digits, that is,

$$\{1, 4, 1, 5, 9, 2, 6, 5, 3, 5\}, \{8, 9, 7, 9, 3, 2, 3, 8, 4, 6\}, \ldots .$$

For each sample of size $n = 10$, *the number of unobserved digits* n_0 *is noted, resulting in the following observed/expected frequency table, where the expected frequencies are based on the distribution of Table 1.3.*

$n_0 = x$	Observed frequency (O_x)	Expected frequency (E_x)
0	36 270	36 288.0
1	1 633 243	1 632 960.0
2	13 609 335	13 608 000.0
3	35 556 620	35 562 240.0
4	34 518 949	34 514 424.0
5	12 859 799	12 859 560.0
6	1 718 714	1 718 892.0
7	66 617	67 176.0
8	453	459.9
9	0	0.1

Combining the frequencies of the last two categories, $n_0 = 8$ *and* $n_0 = 9$, *Pearson's chi-squared test statistic is*

$$\chi^2 = \sum_{x=0}^{8^*} \frac{(O_x - E_x)^2}{E_x} = 6.45$$

where $O_{8^*} = 453$ *and* $E_{8^*} = 460$, *with degrees of freedom 8. The corresponding p-value is* $\mathrm{P}(\chi^2 \geq 645) = 0.5971$ *where* χ^2 *is a chi-squared random variable with degrees of freedom 8. There is no statistical evidence in the sample against* H_0.

In summary, the first one billion decimal digits of π *seem to resemble well a randomly generated iid digits from 0 to 9 under the uniform distribution.*

Since by (1.51) and (1.52), $\mathrm{E}(P_0)$ and $\mathrm{Var}(P_0)$ are easily and precisely calculated, a large sample test may be devised for the hypothesis, H_0: the data are *iid* observations under the uniform distribution on $\mathscr{X}$. Consider the experiment in which a sample of n *iid* uniform random elements is drawn from an alphabet with K letters and the proportion of empty baskets, $P_0 = N_0/K$, is noted. Suppose such an experiment is repeated m times, resulting in, $P_{0,1}, \dots, P_{0,m}$, and denote the sample mean as $\overline{P}_0$. Then by the central limit theorem, provided that m is sufficiently large,

$$Z = \frac{\sqrt{m}\left[\overline{P}_0 - \left(1-\frac{1}{K}\right)^K\right]}{\sqrt{\frac{1}{K}\left(1-\frac{1}{K}\right)^K + \frac{K-1}{K}\left(1-\frac{2}{K}\right)^K - \left(1-\frac{1}{K}\right)^{2K}}} \tag{1.58}$$

converges weakly to a standard normal random variable under H_0.

Example 1.14 *Use (1.58) to test the hypothesis, H_0, that the sequence of the three-digit nonoverlapping segments starting from the first digit after the decimal point in π, that is,*

$$141{,}592, 653{,}589, 793{,}238, 462{,}643, 383{,}279, \dots$$

is a sequence of iid observations from alphabet $\mathscr{X} = \{0, 1, \dots, 999\}$ under the uniform distribution.

Solution: This is a case corresponding to $K = 1000$. The first sample of size $n = 1000$ consists of the first 3000 digits of π after the decimal point, breaking into 1000 nonoverlapping three-digit fragments. For this sample of size $n = 1000$, $P_0 = N_0/K$ is calculated. The next 3000 digits gave a second sample of size 1000, for which $P_0 = N_0/K$ is calculated. The step is repeated $m = 300000$ times, and the mean of these P_0s is calculated and denoted by $\overline{P}_0$. Finally, (1.58) takes a value $Z = 0.0036$, which indicates no evidence against H_0.

The idea of a goodness-of-fit test based on N_0 was perhaps first suggested by David (1950). There it is proposed that, for the hypothesis H_0 that an *iid* sample is taken from a population with known *cdf* $F(x)$, K points $z_0 = -\infty < z_1 < \cdots < z_{K-1} < z_K = \infty$ can be chosen such that $F(z_k) - F(z_{k-1}) = 1/K$, and that the test rejects H_0 if the p-value $\mathrm{P}(N_0 \geq n_0 | H_0)$, where n_0 is the observed number of intervals (z_{k-1}, z_k) containing no observation, is less than a prefixed $\alpha \in (0, 1)$.

1.6 Remarks

Turing's formula is often perceived to project a counterintuitive impression. This impression is possibly linked to the implied utility of Turing's formula in making *nonparametric* inferences regarding certain distributional characteristics *beyond the data range*. To better discern the spirit of Turing's formula, it is perhaps helpful to, at least pedagogically, consider the formula in several different perspectives.

To begin, one may be benefited by a brief revisit to a dilemma in the practice of statistical estimation. Primarily consider a simple example in which the population is assumed to be $N(\mu, \sigma^2)$ with an unknown μ and a known $\sigma^2 > 0$. An *iid* sample of size n, $\mathbf{X} = \{X_1, \dots, X_n\}$, is to be taken from the population, and the objective is to estimate the unknown population mean μ based on the sample. In the realm of Statistics, this process is carried out in two stages corresponding to two distinct chronological time points, t_1 and t_2, that is, $t_1 < t_2$.

At time t_1, before the sampling experiment is conducted, the observations of $\mathbf{X} = \{X_1, \dots, X_n\}$ are random, and so is the estimator $\hat{\mu} = \overline{X}$. One contemplates the statistical behavior of the random quantity $\overline{X}$, establishes all of its good mathematical properties, derives the distributional characteristics of $\sqrt{n}(\overline{X} - \mu)/\sigma$, and finally claims that the true value of μ is likely, say with probability $1 - \alpha = 0.99$, contained in the random interval

$$\left(\overline{X} - z_{\alpha/2}\frac{\sigma}{\sqrt{n}}, \overline{X} + z_{\alpha/2}\frac{\sigma}{\sqrt{n}} \right).$$

This is a scientifically meaningful and mathematically correct statement. Unfortunately an issue is left unresolved: the occurrence of the event that the random interval contains the true value of μ is not verifiable, not at time t_1 before the sampling experiment is conducted nor at time t_2 after the sampling experiment is conducted.

At time t_2, a realization of the sample $\mathbf{x} = \{x_1, \dots, x_n\}$ is obtained, but all the scientific statements made about $\overline{X}$ at time t_1 become much less relevant. First, one does not know whether the true value of μ is contained in $\overline{x} \pm z_{\alpha/2}\sigma/\sqrt{n}$, and therefore one does not know how close μ is to the observed $\overline{x}$. Second, one cannot even claim that μ is "probably" in the interval $\overline{x} \pm z_{\alpha/2}\sigma/\sqrt{n}$ since the meaning of probability ends as the sampling experiment ends. At this point, one has no other choices but to believe that the true value of μ is "close" to $\overline{x}$ in a very vague sense. However, it is important to remember that such a chosen belief is essentially faith-based.

A parallel argument can be put forth for a more general statistical model, where a parameter with an unknown value, say θ, is estimated by an estimator, say $\hat{\theta}(\mathbf{X})$. A statement at time t_1, for example,

$$P\left(\theta \in (a(\mathbf{X}), b(\mathbf{X}))\right) = 1 - \alpha$$

and the question whether $\theta \in (a(\mathbf{x}), b(\mathbf{x}))$ at time t_2 are somewhat disconnected.

However, whether this is a valid issue or, if so, how awkward a role this dilemma plays in the logic framework of statistical estimation is largely philosophical and is, therefore, in addition to the points of the current discussion. The relevant points here are as follows.

1) The interpretation of a statistic could change significantly from t_1 to t_2.

2) In the usual statistical situation, the estimand θ is unambiguously defined and invariantly meaningful at both time points t_1 and t_2.

This is however not the case for the estimand of Turing's formula since

$$\pi_0 = \sum_{k\geq 1} p_k 1[Y_k = 0]$$

is a random variable, which makes the statistical implication somewhat unusual.

To make the subsequent discussion clearer, consider the following three different sampling experiments with three different sample spaces and hence three different perspectives, involving four different time points, t_1, t_2, t_3, and t_4, in a chronological order.

Experiment 1: Draw an *iid* sample of size n from $\mathscr{X}$ at time t_2.
Experiment 2: Draw an *iid* sample of size m from $\mathscr{X}$ at time t_4.
Experiment 3: A two-step experiment:
Step 1: Draw an *iid* sample of size n from $\mathscr{X}$ at time t_2, and then
Step 2: Draw an additional *iid* sample of size m from $\mathscr{X}$ at time t_4.

At time t_1, one anticipates Experiment 1, and both Turing's formula T_1 and its estimand π_0 are random variables. However, π_0 is not a probability of any event in the σ-algebra entailed by Experiment 1 and its sample space $\mathscr{X}^n$. As a random variable, π_0 has some unusual statistical properties, for example, even at time t_3 when the sample has an observed realization $\{x_1, \dots, x_n\}$,

$$\mathring{\pi}_0 = \sum_{k\geq 1} p_k 1[y_k = 0] \tag{1.59}$$

is unobservable under the assumption that $\{p_k\}$ is unknown (however, see Exercise 15). Since the target of the estimation is random at t_1, it is not clear what a good practical interpretation of π_0 could be.

At time t_2, Experiment 1 is conducted and a realization of a sample $\{x_1, \dots, x_n\}$ is observed.

At time t_3, the estimand π_0 takes a fixed value in the form of $\mathring{\pi}_0$. $\mathring{\pi}_0$ becomes a probability, not a probability associated with Experiment 1 and its σ-algebra, but a probability of an event in the σ-algebra associated with Experiment 2 (if Experiment 2 is to be conducted with $m = 1$) and its sample space $\mathscr{X}^1$. The fact that the estimand π_0 with a vague practical interpretation at time t_1 transforms itself into the probability of a very interesting and meaningful event associated with Experiment 2 may, in part, be the reason for the oracle-like impression about Turing's formula.

In addition to π_0, there are several interesting variants of π_0 worth noting. The first is the expected value of π_0, denoted by $\zeta_{1,n}$,

$$\zeta_{1,n} = \mathrm{E}(\pi_0) = \mathrm{E}\left(\sum_{k\geq 1} p_k 1[Y_k = 0]\right) = \sum_{k\geq 1} p_k(1-p_k)^n.$$

Incidentally $\zeta_{1,n}$ has a probabilistic interpretation in Experiment 3 with $m = 1$. Let X_i, $i = 1, \dots, n$, be the *iid* observations of the sample in the first step of Experiment 3, and X_{n+1} be the random letter taken by the single observation in the second step of Experiment 3. Let E_1 be the event that X_{n+1} is not one of the letters taken into the sample of size n in the first step. Then

$$\begin{aligned}
\mathrm{P}(E_1) &= \sum_{k\geq 1} \mathrm{P}(X_1 \neq \ell_k, \dots, X_n \neq \ell_k, X_{n+1} = \ell_k) \\
&= \sum_{k\geq 1} \mathrm{P}(X_1 \neq \ell_k, \dots, X_n \neq \ell_k | X_{n+1} = \ell_k)\mathrm{P}(X_{n+1} = \ell_k) \\
&= \sum_{k\geq 1} p_k \prod_{i=1}^{n} \mathrm{P}(X_i \neq \ell_k) = \sum_{k\geq 1} p_k(1-p_k)^n = \zeta_{1,n}.
\end{aligned} \tag{1.60}$$

Note that $\zeta_{1,n} = \mathrm{P}(E_1)$ is a probability associated with the σ-algebra of Experiment 3 and its sample space $\mathscr{X}^{n+1}$, not to be confused with $\mathring{\pi}_0$.

Furthermore, consider Experiment 3 yet again with $m = n$. Let the two subsamples in the first and the second steps be represented by the two rows as follows:

$$\begin{aligned}
&X_1, \quad X_2, \quad \dots, \quad X_n, \\
&X_{n+1}, X_{n+2}, \dots, X_{n+n}.
\end{aligned}$$

Let E_j be the event that X_{n+j} is not one of the letters taken into the sample of size n in the first step. Clearly $\mathrm{P}(E_j) = \sum_{k\geq 1} p_k(1-p_k)^n = \zeta_{1,n}$ for every j, $j = 1, 2, \dots, n$. The expected number of observations in the second subsample, whose observed letters are not included in the first subsample, is then

$$\tau_n := \mathrm{E}\left(\sum_{j=1}^{n} 1[E_j]\right) = \sum_{j=1}^{n} \mathrm{P}(E_j) = n\sum_{k\geq 1} p_k(1-p_k)^n = n\zeta_{1,n}.$$

Both $\zeta_{1,n}$ and $\tau_n = n\zeta_{1,n}$ will play fundamental roles in the subsequent chapters.

1.7 Exercises

1 Consider the following three events when tossing different number of balanced dice:
  a) A: observing at least one six when tossing six dice.
  b) B: observing at least two sixes when tossing 12 dice.
  c) C: observing at least three sixes when tossing 18 dice.

Which event has the highest probability?

2 A random variable X is said to have a Poisson distribution if $p_k = \mathrm{P}(X = k) = e^{-\lambda}\lambda^k/k!$, where $\lambda > 0$ is a parameter, for $k = 0, 1, \ldots$.
   a) Show that when λ is not an integer, p_k has a unique maximum. Find that maximum.
   b) When λ is a positive integer, show that p_k attains its maximum at two distinct values of k. Find these two values of k.

3 One hundred families live in a remote village. The following table summarizes the proportions of families with various numbers of school-aged children.

Number of children	0	1	2	3	4
Proportion of families	0.3	0.3	0.2	0.1	0.1

All school-aged children from the village go to the same village school and all students of the school are from the village.
   a) Let X be the number of children in a randomly selected family in the village. Find $\mathrm{E}(X)$.
   b) Let Y be the number of children in the family of a randomly selected student in the village school. Find $\mathrm{E}(Y)$.

4 In a sequence of identically distributed and independent Bernoulli trials, each of which is with probability of "success" $p > 0$, a geometric random variable X is such that $X = k$, $k = 1, 2, \ldots$, if and only if the first "success" comes on the kth trial, that is,

$$p_k = \mathrm{P}(X = k) = (1-p)^{k-1}p,$$

for $k = 1, \ldots$. Show that $\mathrm{E}(X) = 1/p$ and that $\mathrm{Var}(X) = (1-p)/p^2$.

5 Show
   a) $\mathrm{E}\,(T_1) = \sum_k p_k(1-p_k)^{n-1}$ and
   b) $\mathrm{E}\,(\pi_0) = \sum_k p_k(1-p_k)^n$.

6 Show $T_1 \xrightarrow{p} 0$ and $\pi_0 \xrightarrow{p} 0$.

7 Show $\mathrm{E}\,(\pi_0) \sim c(1-p_\wedge)^n$ for some $c > 0$, where $\sim$ represents equality in rate of divergence or convergence and $p_\wedge = \min\{p_k; k \geq 1\}$ for any probability distribution $\{p_k; k \geq 1\}$ on a finite alphabet.

8 Prove Lemma 1.3.

9 Let X_n be a sequence of random variables satisfying
   a) $\mathrm{P}(X_n \geq 0) = 1$ and
   b) $\mathrm{E}(X_n) \to 0$ as $n \to \infty$.

Then $X_n \xrightarrow{p} 0$.

10 Show $\lim_{n\to\infty} \sigma_n^2 = 0$ where $\sigma_n^2 = \mathrm{Var}(T_1 - \pi_0)$ under any probability distribution.

11 Let c_1 and c_2 be as in Condition 1.1, and let $\hat{c}_1$ and $\hat{c}_2$ be as in (1.14). Show that, if $\{p_k; k \geq 1\}$ satisfies Condition 1.1,

$$\hat{c}_1 \xrightarrow{p} c_1 \text{and } \hat{c}_2 \xrightarrow{p} c_1.$$

(Hint: $\mathrm{E}(\hat{c}_1) \to c_1$, $\mathrm{Var}(\hat{c}_1) \to 0$, $\mathrm{E}(\hat{c}_2) \to c_2$, and $\mathrm{Var}(\hat{c}_2) \to 0$.)

12 Show that the statement of Example 1.5 is true.

13 Show that, for any real number $x \in (0, 1/2)$, $\frac{1}{1-x} < 1 + 2x$.

14 If $\{p_k\}$ is such that $p_k = c_0\, e^{-k}$ where $c_0 = e - 1$ for all $k \geq 1$, then there exists an $M > 0$ such that $\mathrm{E}\,(N_1) + \mathrm{E}\,(N_2) < M$ for all $n \geq 1$.

15 Describe a sampling scheme with which $\mathring{\pi}_0$ in (1.59) is observed even under the assumption that $\{p_k\}$ is unknown. (Hint: Consider the case of a known finite K.)

16 Rewrite Theorem 1.8 into Theorem 1.9.

17 Prove Theorem 1.10.

18 Let $\{p_k\}$ be such that $p_k = Ck^{-\lambda}$ for $k > k_0$, where C, λ, and a positive integer k_0 are unknown parameters. Show that

$$\mathrm{P}(\min\{k; Y_k = r - 1\} \geq k_0) \to 1$$

as n increases indefinitely.

19 Show that under Condition 1.1,

$$\frac{n\,(T^{\sharp\sharp} - \pi_0)}{\sqrt{E(N_1) + 2E(N_2)}} \xrightarrow{L} N(0, 1).$$

20 Verify $S(3, 1) = 1$, $S(3, 2) = 3$, and $S(3, 3) = 1$ by (1.53).

21 Let $\mathscr{X} = \{\ell_1, \ell_2, \ell_3, \ell_4\}$. Verify $S(4, 2) = 7$ by (1.53) and list all possible partitions.

22 Show that each individual random element X_t, $t = 0, \ldots,$ in the special Markov chain with memory m, defined in (1.57), follows (unconditionally) the uniform distribution on the alphabet $\mathscr{X}$.

23 Provide a proof for each part of Corollary 1.3. (Hint: For Part 1, use the delta method, but consider first an augmented Turing's formula $T = T_1 + n^{-n}$ so that $\ln T$ is well defined.)

24 Use Part 2 of Corollary 1.3 to verify (1.23).

2

Estimation of Simpson's Indices

Consider a probability distribution $\{p_k; k \geq 1\}$ on a countable alphabet $\mathscr{X} = \{\ell_k; k \geq 1\}$ where each letter ℓ_k may be viewed as a species in a biological population and each p_k as the proportion of kth species in the population. Simpson (1949) defined a biodiversity index

$$\lambda = \sum_{k=1}^{K} p_k^2$$

for a population with a finite number of species K, which has an equivalent form

$$\zeta_{1,1} = 1 - \lambda = \sum_{k=1}^{K} p_k(1 - p_k). \tag{2.1}$$

$\zeta_{1,1}$ assumes a value in $[0, 1)$ with a higher level of $\zeta_{1,1}$ indicating a more diverse population and is widely used across many fields of study. $\zeta_{1,1}$ is also known as the Gini–Simpson index, an attribution to Corrado Gini for his pioneer work in Gini (1912).

2.1 Generalized Simpson's Indices

Simpson's biodiversity index can be naturally and beneficially generalized in two directions. First, the size of the alphabet may be extended from a finite K to infinity. Second, $\zeta_{1,1}$ may be considered as a special member of the following family:

$$\zeta_{u,v} = \sum_{k\geq 1} p_k^u (1 - p_k)^v \tag{2.2}$$

where $u \geq 1$ and $v \geq 0$ are two integers. Since

$$\zeta_{u,v} = \sum_{k\geq 1} \left[p_k^{u-1}(1 - p_k)^{v-1}\right] p_k(1 - p_k),$$

Statistical Implications of Turing's Formula, First Edition. Zhiyi Zhang.

it may be viewed as a weighted version of (2.1) with the weight for each k being $w_k = p_k^{u-1}(1-p_k)^{v-1}$. For examples, $\zeta_{1,2}$ loads higher weight on minor species (those with smaller p_ks); and $\zeta_{2,1}$ loads higher weight on major species, etc. Equation (2.2) is proposed by Zhang and Zhou (2010) and is referred to as the *generalized Simpson's indices.*

As u assumes each positive integer value and v assumes each nonnegative integer value, (2.2) may be expressed as a panel in (2.3). This panel of indices is the centerpiece of many of the statistical procedures based on Turing's perspective:

$$\{\zeta_{u,v}\} = \{\zeta_{u,v}; u \geq 1, v \geq 0\} = \begin{pmatrix} \zeta_{1,0} & \zeta_{1,1} & \zeta_{1,2} & \cdots & \zeta_{1,v} & \cdots \\ \zeta_{2,0} & \zeta_{2,1} & \zeta_{2,2} & \cdots & \zeta_{2,v} & \cdots \\ \vdots & \vdots & \vdots & \vdots\vdots\vdots & \vdots & \vdots\vdots\vdots \\ \zeta_{u,0} & \zeta_{u,1} & \zeta_{u,2} & \cdots & \zeta_{u,v} & \cdots \\ \vdots & \vdots & \vdots & \vdots\vdots\vdots & \vdots & \vdots\vdots\vdots \end{pmatrix} \tag{2.3}$$

For any probability sequence $\mathbf{p} = \{p_k; k \geq 1\}$ on $\mathcal{X}$, let $\mathbf{p}_\downarrow$ be the non-increasingly rearranged $\mathbf{p}$ and let $\{\zeta_{u,0}; u \geq 1\}$ be the first column of the index panel (2.3).

Lemma 2.1 *For any given probability distribution* $\mathbf{p} = \{p_k; k \geq 1\}$ *on* $\mathcal{X}$, $\mathbf{p}_\downarrow$ *and* $\{\zeta_{u,0}; u \geq 1\}$ *uniquely determine each other.*

Proof: Without loss of generality, assume $\mathbf{p} = \{p_k; k \geq 1\}$ is non-increasingly ordered. It is obvious that $\{p_k; k \geq 1\}$ uniquely determines the sequence $\{\zeta_{u,0}; u \geq 1\}$. It suffices to show that $\{\zeta_{u,0;}\}$ uniquely determines $\{p_k; k \geq 1\}$.

Toward that end, suppose that there exists another nonincreasingly ordered probability sequence, $\{q_k\}$, on $\mathcal{X}$ satisfying

$$\sum_{k\geq 1} p_k^u = \sum_{k\geq 1} q_k^u$$

for all $u \geq 1$.

Let $k_0 = \min\{k : p_k \neq q_k\}$. If k_0 does not exist, then $\{p_k\} = \{q_k\}$. If k_0 exists, then

$$\sum_{k\geq k_0} p_k^u = \sum_{k\geq k_0} q_k^u \tag{2.4}$$

for each and every integer $u \geq 1$. It can be easily shown that

$$\lim_{u\to\infty} \frac{\sum_{k\geq k_0} p_k^u}{p_{k_0}^u} = r_p \geq 1 \quad \text{and} \quad \lim_{u\to\infty} \frac{\sum_{k\geq k_0} q_k^u}{q_{k_0}^u} = r_q \geq 1 \tag{2.5}$$

where r_p and r_q are, respectively, multiplicities of the value p_{k_0} in $\{p_k\}$ and of the value q_{k_0} in $\{q_k\}$. But by (2.4),

$$\frac{\sum_{k\geq k_0} p_k^u}{p_{k_0}^u} = \frac{\sum_{k\geq k_0} q_k^u}{q_{k_0}^u}\left(\frac{q_{k_0}}{p_{k_0}}\right)^u. \tag{2.6}$$

The right-hand side of (2.6) approaches 0 or ∞ if $p_{k_0} \neq q_{k_0}$, which contradicts (2.5). Therefore, k_0 does not exist and $\{p_k\} = \{q_k\}$. □

Since $\{\zeta_{u,0}; u \geq 1\}$ is a subset of $\{\zeta_{u,v}; u \geq 1, v \geq 0\}$, the following theorem is an immediate consequence of Lemma 2.1.

Theorem 2.1 *For any given probability distribution* $\mathbf{p} = \{p_k; k \geq 1\}$ *on* $\mathscr{X}$, $\mathbf{p}_\downarrow$ *and* $\{\zeta_{u,v}; u \geq 1, v \geq 0\}$ *uniquely determine each other.*

Theorem 2.1 has an intriguing implication: the complete knowledge of $\{p_k\}$ up to a permutation and the complete knowledge of $\{\zeta_{u,v}\}$ are equivalent. In other words, all the generalized Simpson's indices collectively and uniquely determine the underlying distribution $\{p_k\}$ upto a permutation on the index set $\{k; k \geq 1\}$. This implication is another motivation for generalizing Simpson's diversity index beyond $\zeta_{1,1}$.

Theorem 2.2 *For any given probability distribution* $\mathbf{p} = \{p_k; k \geq 1\}$ *on* $\mathscr{X}$, $\mathbf{p}_\downarrow$ *and* $\{\zeta_{1,v}; v \geq 0\}$ *uniquely determine each other.*

Proof: By Theorem 2.1, it suffices to show that $\{\zeta_{1,v}; v \geq 0\}$ uniquely determines $\{\zeta_{u,v}; u \geq 1, v \geq 0\}$. For every pair of (u, v),

$$\begin{aligned}
\zeta_{u,v} &= \sum_{k\geq 1} p_k^u (1-p_k)^v \\
&= \sum_{k\geq 1} p_k^{u-1}[1-(1-p_k)](1-p_k)^v \\
&= \sum_{k\geq 1} p_k^{u-1}(1-p_k)^v - \sum_{k\geq 1} p_k^{u-1}(1-p_k)^{v+1} \\
&= \zeta_{u-1,v} - \zeta_{u-1,v+1}.
\end{aligned}$$

That is to say that every $\zeta_{u,v}$ in (2.3) may be expressed as a linear combination of two indices in the row above, namely $\zeta_{u-1,v}$ and $\zeta_{u-1,v+1}$, which in turn may be expressed as linear combinations of indices in the row further above. Therefore, each $\zeta_{u,v}$ may be expressed as a linear combination of, and hence is uniquely determined by, the indices in the top row, which is denoted as

$$\{\zeta_v; v \geq 0\} := \{\zeta_{1,v}; v \geq 0\} = \{\zeta_{1,0}, \zeta_{1,1}, \zeta_{1,2}, \dots\}.$$

□

Theorem 2.2 suggests that the complete knowledge of $\{p_k\}$ is essentially captured by the complete knowledge of $\{\zeta_v\}$. In this sense, $\{\zeta_v\}$ may be considered as a reparameterization of $\{p_k\}$. This is in fact the basis for the statistical perspective offered by Turing's formula.

2.2 Estimation of Simpson's Indices

Let $X_1, \dots, X_i, \dots, X_n$ be an *iid* sample under $\{p_k\}$ from $\mathscr{X}$. X_i may be written as $X_i = (X_{i,k}; k \geq 1)$ where for every i, $X_{i,k}$ takes 1 only for one k and 0 for all other k values. Let $Y_k = \sum_{i=1}^{n} X_{i,k}$ and $\hat{p}_k = Y_k/n$. Y_k is the number of observations of the kth species found in the sample.

For a fixed pair (u, v) and an *iid* sample of size $m = u + v$, let $Y_k^{(m)}$ be the frequency of ℓ_k in the sample for every $k \geq 1$ and

$$N_u^{(m)} = \sum_{k\geq 1} 1\left[Y_k^{(m)} = u\right] \tag{2.7}$$

the number of species each of which is represented exactly u times in the sample of size $m = u + v$. It may be easily verified that

$$\hat{\zeta}_{u,v}^{(m)} = \binom{u+v}{u}^{-1} N_u^{(m)} \tag{2.8}$$

is an unbiased estimator of $\zeta_{u,v}$ (see Exercise 8).

Remark 2.1 *It is important to note that counting the number of different letters in a sample with exactly r observed frequency, N_r, and using that to estimate various other quantities is the very essence of the perspective brought forward by Turing's formula. To honor Alan Turing's contribution in this regard, any statistical estimator based on N_r for $r \geq 1$ in the subsequent text will be referred to as an estimator in* Turing's perspective. *The fact that (2.8) is an unbiased estimator of $\zeta_{u,v}$, though straightforward, is one of the most important consequences of Turing's perspective.*

Next follow a U-statistics construction by considering (2.8) for each of the $\binom{n}{u+v}$ possible subsamples of size $m = u + v$ from the original sample of size n. Each subsample gives an unbiased estimator of $\zeta_{u,v}$, and therefore the average of these $\binom{n}{u+v}$ unbiased estimators

$$Z_{u,v} = \binom{n}{u+v}^{-1} \sum_{*} \hat{\zeta}_{u,v}^{(m)} = \binom{n}{u+v}^{-1} \binom{u+v}{u}^{-1} \sum_{*} N_u^{(m)}, \tag{2.9}$$

where the summation $\sum_*$ is over all possible size-m subsamples of the original size-n sample, is in fact an unbiased estimator.

On the other hand, noting (2.7) and the fact

$$\sum_{*} N_u^{(m)} = \sum_{k\geq 1}\left(\sum_{*} 1\left[Y_k^{(m)} = u\right]\right),$$

$\sum_* N_u^{(m)}$ is simply the total number of times exactly u observations are found in a same species among all possible subsamples of size $m = u + v$ taken from the sample of size n.

In counting the total number of such events, it is to be noted that, for any fixed u, only for species that are represented in the sample u times or more can such an event occur. Therefore,

$$\sum_* N_u^{(m)} = \sum_{k\geq 1} 1\,[Y_k \geq u] \binom{Y_k}{u}\binom{n-Y_k}{v}.$$

Furthermore, assuming $v \geq 1$,

$$\begin{aligned}
&1\,[Y_k \geq u]\binom{Y_k}{u}\binom{n-Y_k}{v} \\
&= 1\,[Y_k \geq u]\left[\frac{Y_k(Y_k-1)\cdots(Y_k-u+1)}{u!}\right] \\
&\quad \times \left[\frac{(n-Y_k)(n-Y_k-1)\cdots(n-Y_k-v+1)}{v!}\right] \\
&= \frac{n^{u+v}}{u!v!} 1\left[\hat{p}_k \geq \frac{u}{n}\right]\left[\hat{p}_k\left(\hat{p}_k - \frac{1}{n}\right)\cdots\left(\hat{p}_k - \frac{u}{n} + \frac{1}{n}\right)\right] \\
&\quad \times \left[(1-\hat{p}_k)\left(1-\hat{p}_k-\frac{1}{n}\right)\cdots\left(1-\hat{p}_k-\frac{v}{n}+\frac{1}{n}\right)\right] \\
&= \frac{n^{u+v}}{u!v!} 1\left[\hat{p}_k \geq \frac{u}{n}\right]\left[\prod_{i=0}^{u-1}\left(\hat{p}_k - \frac{i}{n}\right)\right]\left[\prod_{j=0}^{v-1}\left(1-\hat{p}_k-\frac{j}{n}\right)\right]
\end{aligned}$$

and

$$\begin{aligned}
\binom{n}{u+v}^{-1}\binom{u+v}{u}^{-1} &= \frac{(n-u-v)!(u+v)!}{n!}\frac{u!v!}{(u+v)!} \\
&= \frac{(n-u-v)!u!v!}{n!}.
\end{aligned}$$

Therefore, for every pair of (u, v) where $u \geq 1$ and $v \geq 1$,

$$\begin{aligned}
Z_{u,v} &= \frac{[n-(u+v)]!n^{u+v}}{n!} \\
&\quad \times \sum_{k\geq 1}\left\{1\left[\hat{p}_k \geq \frac{u}{n}\right]\left[\prod_{i=0}^{u-1}\left(\hat{p}_k - \frac{i}{n}\right)\right]\left[\prod_{j=0}^{v-1}\left(1-\hat{p}_k-\frac{j}{n}\right)\right]\right\}.
\end{aligned} \tag{2.10}$$

In case of $u \geq 1$ and $v = 0$, a similar argument leads to

$$Z_{u,v} = \left[\prod_{j=1}^{u-1}\left(\frac{n}{n-j}\right)\right]\sum_{k\geq 1}\left\{1\left[\hat{p}_k \geq \frac{u}{n}\right]\left[\prod_{i=0}^{u-1}\left(\hat{p}_k - \frac{i}{n}\right)\right]\right\} \tag{2.11}$$

(see Exercise 9).

For a few special pairs of u and v, $Z_{u,v}$ reduces to

$$Z_{1,1} = \frac{n}{n-1}\sum_{k\geq 1}\hat{p}_k(1-\hat{p}_k)$$

$$Z_{2,0} = \frac{n}{n-1}\sum_{k\geq 1} 1[\hat{p}_k \geq 2/n]\hat{p}_k(\hat{p}_k - 1/n)$$

$$Z_{3,0} = \frac{n^2}{(n-1)(n-2)}\sum_{k\geq 1} 1[\hat{p}_k \geq 3/n]\hat{p}_k(\hat{p}_k - 1/n)(\hat{p}_k - 2/n)$$

$$Z_{2,1} = \frac{n^2}{(n-1)(n-2)}\sum_{k\geq 1} 1[\hat{p}_k \geq 2/n]\hat{p}_k(\hat{p}_k - 1/n)(1-\hat{p}_k)$$

$$Z_{1,2} = \frac{n^2}{(n-1)(n-2)}\sum_{k\geq 1} 1[\hat{p}_k \geq 1/n]\hat{p}_k(1-\hat{p}_k)(1-1/n-\hat{p}_k). \qquad (2.12)$$

$Z_{u,v}$ is an unbiased estimator of $\zeta_{u,v}$. This fact is established by the U-statistic construction of the above-mentioned estimator. In fact, $Z_{u,v}$ is a *uniformly minimum variance unbiased estimator* (umvue) of $\zeta_{u,v}$ when the cardinality of $\mathcal{X}$, K, is finite. Since $Z_{u,v}$ is unbiased, by the Lehmann–Scheffé theorem, it suffices to note that $\{\hat{p}_k\}$ is a set of complete and sufficient statistics under $\{p_k\}$ (see Exercise 12).

2.3 Normal Laws

The same U-statistic construction paves the path for establishing the asymptotic normality of $Z_{u,v} - \zeta_{u,v}$. For a review of the theory of U-statistics, readers may wish to refer to Serfling (1980) or Lee (1990).

Let

1) $X_1, \dots, X_n$ be an *iid* sample under a distribution F;
2) $\theta = \theta(F)$ be a parameter of interest;
3) $h(X_1, \dots, X_m)$ where $m < n$ be a symmetric kernel satisfying

$$E_F\{h(X_1, \dots, X_m)\} = \theta(F);$$

4) $U_n = U(X_1, \dots, X_n) = \binom{n}{m}^{-1} \sum_* h(X_{i_1}, \dots, X_{i_m})$ where the summation $\sum_*$ is over all possible subsamples of size m from the original sample of size n;
5) $h_1(x_1) = E_F(h(x_1, X_2, \dots, X_m))$ be the conditional expectation of h given $X_1 = x_1$; and
6) $\sigma_1^2 = \mathrm{Var}_F(h_1(X_1))$.

The following lemma is due to Hoeffding (1948).

Lemma 2.2 *If* $\mathrm{E}_F(h^2) < \infty$ *and* $\sigma_1^2 > 0$, *then*

$$\sqrt{n}(U_n - \theta) \xrightarrow{L} N(0, m^2\sigma_1^2).$$

To apply Lemma 2.2 to the U-statistic $Z_{u,v}$, the two conditions $\mathrm{E}_F(h^2) < \infty$ and $\sigma_1^2 > 0$ must be checked.

Let

$$\begin{aligned} m &= u + v, \\ h &= h(X_1, \dots, X_m) = \binom{m}{u}^{-1} N_u^{(m)}, \\ F &= \{p_k\}. \end{aligned}$$

Clearly in this case, $\mathrm{E}_F(h^2) < \infty$ since $N_u^{(m)}$ is bounded above by every m. Therefore, only the second condition of Lemma 2.2, namely $\sigma_1^2 > 0$, needs to be checked.

Toward that end, suppose $u \geq 1$ and $v \geq 1$. Given

$$\begin{aligned} X_1 = x_1 &= (x_{1,1}, x_{1,2}, \dots, x_{1,k}, \cdots), \\ \binom{m}{u} h_1(x_1) &= \binom{m}{u} \mathrm{E}_F(h(x_1, X_2, \dots, X_m)) \\ &= \mathrm{E}_F\,(N_u^{(m)} | X_1 = x_1) \\ &= \sum_{k\geq 1} 1\,[x_{1,k} = 1] \binom{m-1}{u-1} p_k^{u-1}(1-p_k)^v \\ &\quad + \sum_{k\geq 1} 1\,[x_{1,k} = 0] \binom{m-1}{u} p_k^u(1-p_k)^{v-1} \\ &= \sum_{k\geq 1} \binom{m-1}{u} p_k^u(1-p_k)^{v-1} \\ &\quad + \sum_{k\geq 1} 1[x_{1,k} = 1] \binom{m-1}{u} p_k^{u-1}(1-p_k)^{v-1} \left[(1-p_k)\frac{u}{v} - p_k\right] \\ &= \binom{m-1}{u} \sum_{k\geq 1} p_k^u(1-p_k)^{v-1} \\ &\quad + \binom{m-1}{u} \sum_{k\geq 1} 1[x_{1,k} = 1] p_k^{u-1}(1-p_k)^{v-1} \left[(1-p_k)\frac{u}{v} - p_k\right], \end{aligned}$$

or

$$\begin{aligned} h_1(x_1) &= \frac{v}{u+v} \sum_{k\geq 1} p_k^u(1-p_k)^{v-1} \\ &\quad + \frac{v}{u+v} \sum_{k\geq 1} 1[x_{1,k} = 1] p_k^{u-1}(1-p_k)^{v-1} \left[(1-p_k)\frac{u}{v} - p_k\right], \end{aligned}$$

and therefore

$$
\begin{aligned}
\sigma_1^2(u,v) &= \mathrm{Var}_F(h_1(X_1))\\
&= \left(\frac{v}{u+v}\right)^2 \mathrm{Var}_F\left\{\sum_{k\geq 1} 1[X_{1,k}=1]p_k^{u-1}(1-p_k)^{v-1}\left[(1-p_k)\frac{u}{v}-p_k\right]\right\}\\
&= \left(\frac{v}{u+v}\right)^2 \left\{\mathrm{E}_F\left\{\sum_{k\geq 1} 1[X_{1,k}=1]p_k^{u-1}(1-p_k)^{v-1}\left[(1-p_k)\frac{u}{v}-p_k\right]\right\}^2\right.\\
&\quad \left. -\left\{\sum_{k\geq 1} p_k^{u}(1-p_k)^{v-1}\left[(1-p_k)\frac{u}{v}-p_k\right]\right\}^2\right\}\\
&= \left(\frac{v}{u+v}\right)^2 \left\{\sum_{k\geq 1} p_k^{2u-1}(1-p_k)^{2v-2}\left[(1-p_k)\frac{u}{v}-p_k\right]^2\right.\\
&\quad \left. -\left\{\sum_{k\geq 1} p_k^{u}(1-p_k)^{v-1}\left[(1-p_k)\frac{u}{v}-p_k\right]\right\}^2\right\}\\
&= \frac{u^2}{v^2}\left(\frac{v}{u+v}\right)^2 \sum_{k\geq 1} p_k^{2u-1}q_k^{2v} - \frac{2u}{v}\left(\frac{v}{u+v}\right)^2 \sum_{k\geq 1} p_k^{2u}q_k^{2v-1}\\
&\quad + \left(\frac{v}{u+v}\right)^2 \sum_{k\geq 1} p_k^{2u+1}(1-p_k)^{2v-2}\\
&\quad - \left(\frac{v}{u+v}\right)^2 \left[\frac{u}{v}\sum_{k\geq 1} p_k^{u}(1-p_k)^{v} - \sum_{k\geq 1} p_k^{u+1}(1-p_k)^{v-1}\right]^2\\
&= \frac{u^2}{v^2}\left(\frac{v}{u+v}\right)^2 \zeta_{2u-1,2v} - \frac{2u}{v}\left(\frac{v}{u+v}\right)^2 \zeta_{2u,2v-1}\\
&\quad + \left(\frac{v}{u+v}\right)^2 \zeta_{2u+1,2v-2}\\
&\quad - \left(\frac{v}{u+v}\right)^2 \left(\frac{u}{v}\zeta_{u,v} - \zeta_{u+1,v-1}\right)^2\\
&= \frac{u^2}{(u+v)^2}\zeta_{2u-1,2v} - \frac{2uv}{(u+v)^2}\zeta_{2u,2v-1}\\
&\quad + \frac{v^2}{(u+v)^2}\zeta_{2u+1,2v-2} - \frac{v^2}{(u+v)^2}\left(\frac{u}{v}\zeta_{u,v} - \zeta_{u+1,v-1}\right)^2 \geq 0. \qquad (2.13)
\end{aligned}
$$

The last inequality in (2.13) becomes an equality if and only if $h(X_1)$ is a constant that occurs when all the positive probabilities of $\{p_k\}$ are equal.

Suppose $u \geq 1$ and $v = 0$ and therefore $\binom{m}{u} = 1$. It is easy to see that

$$h_1(x_1) = \sum_{k \geq 1} 1[x_{1,k} = 1] p_k^{u-1}$$

and

$$\sigma_1^2(u, 0) = \text{Var}_F(h_1(X_1)) = \zeta_{2u-1,0} - \zeta_{u,0}^2 \geq 0 \tag{2.14}$$

(see Exercise 10). The strict inequality holds for all cases except when $\{p_k\}$ is uniform.

Thus, the following theorem is established.

Theorem 2.3 *For any given pair of u and v,*

1) if $\{p_k\}$ is such that $\sigma_1^2(u, v) > 0$, then

$$\sqrt{n}(Z_{u,v} - \zeta_{u,v}) \xrightarrow{L} N(0, (u+v)^2 \sigma_1^2(u, v));$$

2) if $\{p_k\}$ is such that $\sigma_1^2(u, 0) > 0$, then

$$\sqrt{n}(Z_{u,0} - \zeta_{u,0}) \xrightarrow{L} N(0, u^2 \sigma_1^2(u, 0)).$$

Theorem 2.3 immediately implies consistency of $Z_{u,v}$ to $\zeta_{u,v}$ and the consistency of $Z_{u,0}$ to $\zeta_{u,0}$ for any $u \geq 1$ and $v \geq 1$ under the stated condition.

By the last expressions of (2.13) and (2.14), and Theorem 2.3, it is easily seen that when $u \geq 1$ and $v \geq 1$,

$$\begin{aligned} \hat{\sigma}_1^2(u, v) = {} & \frac{u^2}{(u+v)^2} Z_{2u-1,2v} \\ & - \frac{2uv}{(u+v)^2} Z_{2u,2v-1} \\ & + \frac{v^2}{(u+v)^2} Z_{2u+1,2v-2} \\ & - \frac{v^2}{(u+v)^2} \left(\frac{u}{v} Z_{u,v} - Z_{u+1,v-1} \right)^2 \end{aligned} \tag{2.15}$$

and

$$\hat{\sigma}_1^2(u, 0) = Z_{2u-1,0} - Z_{u,0}^2 \tag{2.16}$$

are consistent estimators of $\sigma_1^2(u, v)$ and of $\sigma_1^2(u, 0)$, respectively, and hence the following corollary is established.

Corollary 2.1 *Suppose, for any given pair of u and v, the condition of Theorem 2.3 for each of the two parts is respectively satisfied. Then the following* *cases occur:*

1) *if* $\{p_k\}$ *is such that* $\sigma_1^2(u,v) > 0$, *then*

$$\frac{\sqrt{n}(Z_{u,v} - \zeta_{u,v})}{(u+v)\hat{\sigma}_1(u,v)} \xrightarrow{L} N(0,1) \tag{2.17}$$

2) *if* $\{p_k\}$ *is such that* $\sigma_1^2(u,0) > 0$, *then*

$$\frac{\sqrt{n}(Z_{u,0} - \zeta_{u,0})}{u\hat{\sigma}_1(u,0)} \xrightarrow{L} N(0,1), \tag{2.18}$$

where $\hat{\sigma}_1^2(u,v)$ *and* $\hat{\sigma}_1^2(u,0)$ *are as in (2.15) and (2.16) respectively.*

Remark 2.2 *The condition* $\sigma_1^2(u,v) > 0$ *of Part 1 for both Theorem 2.3 and Corollary 2.1 is difficult to verify in practice. However under the following two special cases, the condition has an equivalent form.*

1) *When* $u = v = 1$, $\sigma_1^2(u,v) > 0$ *if and only if* $\{p_k\}$ *is not a uniform distribution on* $\mathscr{X}$.
2) *When* $u \geq 1$ *and* $v = 0$, $\sigma_1^2(u,0) > 0$ *if and only if* $\{p_k\}$ *is not a uniform distribution on* $\mathscr{X}$.

As a case of special interest when $u = v = 1$, the computational formula of $Z_{1,1}$ is given in (2.12) and

$$\frac{\sqrt{n}(Z_{1,1} - \zeta_{1,1})}{2\hat{\sigma}_1(1,1)} \xrightarrow{L} N(0,1) \tag{2.19}$$

where $\hat{\sigma}_1(1,1)$ is such that

$$4\hat{\sigma}_1^2(1,1) = Z_{1,2} - 2Z_{2,1} + Z_{3,0} - (Z_{1,1} - Z_{2,0})^2$$

and $Z_{1,2}$, $Z_{2,1}$, $Z_{3,0}$, and $Z_{2,0}$ are all given in (2.12). Equation (2.19) may be used for large sample inferences with respect to Simpson's index, $\zeta_{1,1}$, whenever the nonuniformity of the underlying distribution is deemed reasonable.

$Z_{u,v}$ is *asymptotically efficient* when K is finite. This fact is established by recognizing first that $\{\hat{p}_k\}$ is the maximum likelihood estimator (mle) of $\{p_k\}$, second that $\hat{\zeta}_{u,v}^* = \sum \hat{p}_k^u (1-\hat{p}_k)^v$ is the mle of $\zeta_{u,v}$, and third that $\sqrt{n}(Z_{u,v} - \hat{\zeta}_{u,v}^*) \xrightarrow{p} 0$.

In simplifying the expression of $Z_{u,v}$ in (2.10), let

$$Z_{u,v} = a_n(u+v)Z_{u,v}^*$$

where

$$a_n(u+v) = \frac{[n-(u+v)]!\,n^{u+v}}{n!}$$

and

$$Z^*_{u,v} = \sum_{k=1}^{K} 1\left[\hat{p}_k \geq \frac{u}{n}\right] \left[\prod_{i=0}^{u-1}\left(\hat{p}_k - \frac{i}{n}\right)\prod_{j=0}^{v-1}\left(1 - \hat{p}_k - \frac{j}{n}\right)\right].$$

Since

$$\sqrt{n}(Z_{u,v} - \hat{\zeta}^*_{u,v}) = \sqrt{n}(a_n(u+v) - 1)Z^*_{u,v} + \sqrt{n}(Z^*_{u.v} - \hat{\zeta}^*_{u,v}), \tag{2.20}$$

the third fact may be established by showing that each of the above two terms converges to zero in probability, respectively. Toward that end, consider the following steps:

Step 1. $Z^*_{u,v} \leq 1$. This is so because every summand in $Z^*_{u,v}$ is nonnegative and bounded above by $\hat{p}_k$ and $\sum_{k=1}^{K}\hat{p}_k = 1$.

Step 2. $\sqrt{n}(a_n(u+v) - 1) \to 0$ as $n \to \infty$. Let $m = u + v$ be any fixed positive integer. This statement is proven by induction. If $m = 1$, then

$$\sqrt{n}(a_n(1) - 1) = \sqrt{n}\left[\frac{(n-1)!n}{n(n-1)!} - 1\right] = 0.$$

Assume that for any positive integer $m - 1$, $\sqrt{n}(a_n(m-1) - 1) \to 0$ as $n \to \infty$. Then

$$\begin{aligned}\sqrt{n}(a_n(m) - 1) &= \sqrt{n}\left[\frac{(n-m)!n^m}{n!} - 1\right] \\ &= \sqrt{n}\left\{\frac{[n-(m-1)]!n^{m-1}n}{n![n-(m-1)]} - 1\right\} \\ &= \sqrt{n}\left[\frac{a_n(m-1)n}{(n-m+1)} - a_n(m-1) + a_n(m-1) - 1\right] \\ &= \sqrt{n}a_n(m-1)\frac{m-1}{n-m+1} + \sqrt{n}(a_n(m-1) - 1).\end{aligned}$$

Noting $a_n(m-1) \to 1$ and, by the inductive assumption $\sqrt{n}(a_n(m - 1) - 1) \to 0$, it follows that $\sqrt{n}(a_n(m) - 1) \to 0$. At this point, it has been established that the first term in (2.20) converges to zero in probability, that is, $\sqrt{n}(a_n(u+v) - 1)Z^*_{u,v} \xrightarrow{p} 0$. It only remains to show $\sqrt{n}(Z^*_{u.v} - \hat{\zeta}^*_{u,v})$.

Step 3. Re-expressing the mle of $\zeta_{u,v}$ in

$$\begin{aligned}\hat{\zeta}^*_{u,v} &= \sum_{k=1}^{K} 1\left[\hat{p}_k \geq \frac{u}{n}\right]\hat{p}_k^u(1-\hat{p}_k)^v + \sum_{k=1}^{K} 1\left[\hat{p}_k < \frac{u}{n}\right]\hat{p}_k^u(1-\hat{p}_k)^v \\ &=: \hat{\zeta}^{(1)}_{u,v} + \hat{\zeta}^{(2)}_{u,v}.\end{aligned}$$

It suffices to show that

$$\sqrt{n}\left\{\sum_{k=1}^{K} 1\left[\hat{p}_k \geq \frac{u}{n}\right]\left[\prod_{i=0}^{u-1}\left(\hat{p}_k - \frac{i}{n}\right)\prod_{j=0}^{v-1}\left(1-\hat{p}_k-\frac{j}{n}\right)\right] - \hat{\zeta}_{u,v}^{(1)}\right\}$$
$$- \sqrt{n}\hat{\zeta}_{u,v}^{(2)}$$
$$= \sqrt{n}\left\{\sum_{k=1}^{K} 1\left[\hat{p}_k \geq \frac{u}{n}\right]\left[\prod_{i=0}^{u-1}\left(\hat{p}_k - \frac{i}{n}\right)\prod_{j=0}^{v-1}\left(1-\hat{p}_k-\frac{j}{n}\right)\right.\right.$$
$$\left.\left. - \hat{p}_k^u(1-\hat{p}_k)^v\right]\right\}$$
$$- \sqrt{n}\hat{\zeta}_{u,v}^{(2)} \xrightarrow{p} 0, \tag{2.21}$$

or to show that each of the above-mentioned two terms converges to zero in probability.

Step 4. Consider first $v = 0$, in which case

$$\prod_{j=0}^{v-1}\left(1-\hat{p}_k-\frac{j}{n}\right) = 1.$$

$\prod_{i=0}^{u-1}\left(\hat{p}_k - \frac{i}{n}\right)$ may be written as a sum of $\hat{p}_k^u$ and finitely many other terms each of which has the following form:

$$\frac{s_1}{n^{s_2}}\hat{p}_k^{s_3}$$

where s_1, $s_2 \geq 1$, and $s_3 \geq 1$ are finite fixed integers. Since

$$0 \leq \sqrt{n}\sum_{k=1}^{K} 1\left[\hat{p}_k \geq \frac{u}{n}\right]\frac{|s_1|}{n^{s_2}}\hat{p}_k^{s_3} \leq \sqrt{n}\sum_{k=1}^{K}\frac{|s_1|}{n^{s_2}}\hat{p}_k = \sqrt{n}\frac{|s_1|}{n^{s_2}} \to 0$$

as $n \to \infty$, the first term of (2.21) converges to zero in probability. The second term of (2.21) converges to zero when $u = 1$ is an obvious case since $\hat{\zeta}_{u,v}^{(2)} = 0$. It also converges to zero in probability when $u \geq 2$ since

$$0 \leq \sqrt{n}\sum_{k=1}^{K} 1\left[\hat{p}_k < \frac{u}{n}\right]\hat{p}_k^u \leq \sqrt{n}\sum_{k=1}^{K}\left(\frac{u}{n}\right)\hat{p}_k \leq \frac{u}{\sqrt{n}} \to 0.$$

Step 5. Next consider the case of $v \geq 1$. $\prod_{i=0}^{u-1}\left(\hat{p}_k - \frac{i}{n}\right)\prod_{j=0}^{v-1}\left(1-\hat{p}_k-\frac{j}{n}\right)$ may be written as a sum of $\hat{p}_k^u(1-\hat{p}_k)^v$ and finitely many other terms each of which has the following form:

$$\frac{s_1}{n^{s_2}}\hat{p}_k^{s_3}(1-\hat{p}_k)^{s_4}$$

where s_1, $s_2 \geq 1$, $s_3 \geq 1$, and $s_4 \geq 1$ are finite fixed integers. Since

$$\begin{aligned} 0 &\leq \sqrt{n} \sum_{k=1}^{K} 1\left[\hat{p}_k \geq \frac{u}{n}\right] \frac{|s_1|}{n^{s_2}} \hat{p}_k^{s_3} (1 - \hat{p}_k)^{s_4} \\ &\leq \sqrt{n} \sum_{k=1}^{K} \frac{|s_1|}{n^{s_2}} \hat{p}_k < \sqrt{n} \frac{|s_1|}{n^{s_2}} \to 0 \end{aligned}$$

as $n \to \infty$, the first term of (2.21) converges to zero in probability. The second term of (2.21) remains the same as in Step 4 and therefore converges to zero in probability.

Thus, the asymptotic efficiency of $Z_{u,v}$ is established.

2.4 Illustrative Examples

In practice, it is often of interest to have a statistical tool to test the equality of diversity between two populations against a difference across time or space. Let the said diversity be measured by $\zeta_{u,v}$. The appropriate hypothesis to be tested is often of the form:

$$H_0 : \zeta_{1,1}^{(1)} - \zeta_{1,1}^{(2)} = D_0 \text{ versus } H_a : \zeta_{1,1}^{(1)} - \zeta_{1,1}^{(2)} > D_0 \text{ (or } < D_0 \text{ or } \neq D_0)$$

where D_0 is a prefixed known constant, and $\zeta_{1,1}^{(1)}$ and $\zeta_{1,1}^{(2)}$ are, respectively, the true $\zeta_{u,v}$ values for the two underlying populations. Similarly, the superscripts (1) and (2) are to index all mathematical objects associated with the two populations whenever appropriate.

Suppose there are available two large independent *iid* samples of sizes n_1 and n_2. Based on Corollary 2.1, under H_0,

$$Z = \frac{\left(Z_{u,v}^{(1)} - Z_{u,v}^{(2)}\right) - D_0}{(u+v)\sqrt{\frac{\left(\hat{\sigma}_1^{(1)}(u,v)\right)^2}{n_1} + \frac{\left(\hat{\sigma}_1^{(2)}(u,v)\right)^2}{n_2}}} \xrightarrow{L} N(0,1) \tag{2.22}$$

(see Exercise 15) where $(\hat{\sigma}_1^{(i)}(u,v))^2$ is as given in (2.15) with the ith sample, $i = 1, 2$. In regard to the hypothesis, one would reject H_0 in favor of H_a if the test statistic Z assumes a value deemed statistically unusual.

Wang and Dodson (2006) offered an analysis on several aspects of dinosaur diversity based on an interesting fossil data set, which includes two subsamples, one early Cretaceous and one late Cretaceous. The early Cretaceous set contains $n_1 = 773$ pieces of fossil belonging to 139 genera; the late Cretaceous set contains $n_2 = 1348$ pieces of fossil belonging to 237 genera. Let r be the number of observations in each observed genus and N_r be the number of observed genera containing exactly r observations, as in (1.1). The two subsamples are presented in Tables 2.1 and 2.2 in terms of r and N_r.

Table 2.1 Early Cretaceous

r	1	2	3	4	5	6	7	8
n_r	90	19	12	3	2	1	1	1
r	9	13	14	15	30	38	124	300
n_r	2	1	2	1	1	1	1	1

Table 2.2 Late Cretaceous

r	1	2	3	4	5	6	7	8	9	10
n_r	136	19	17	7	13	6	4	2	2	1
r	11	12	13	16	17	18	20	24	25	26
n_r	1	4	3	2	1	2	3	1	2	1
r	29	30	33	40	41	43	51	81	201	
n_r	1	2	1	1	1	1	1	1	1	

Table 2.3 $Z_{u,v}$ of early Cretaceous

$z_{1,1}^{(e)}$	$z_{2,0}^{(e)}$	$z_{3,0}^{(e)}$	$z_{2,1}^{(e)}$	$z_{1,2}^{(e)}$
0.8183	0.1817	0.0623	0.1193	0.6990

Suppose it is of interest to detect a decrease in the Gini–Simpson diversity index $\zeta_{1,1}$ from the early Cretaceous period to the late Cretaceous period. Denote the true Simpson's diversity indices at the early and the late Cretaceous periods by $\zeta_{1,1}^{(e)}$ and $\zeta_{1,1}^{(l)}$, respectively. The appropriate hypothesis to be tested is

$$H_0 : \zeta_{1,1}^{(e)} - \zeta_{1,1}^{(l)} = 0 \text{ versus } H_a : \zeta_{1,1}^{(e)} - \zeta_{1,1}^{(l)} > 0. \tag{2.23}$$

Assuming the two fossil samples are *iid*, the value of the test statistic Z in (2.22) is calculated to complete the test. Toward that end, the respective values of $Z_{1,1}$, $Z_{2,0}$, $Z_{3,0}$, $Z_{2,1}$, and $Z_{1,2}$ in (2.12) for the two samples are summarized in Tables 2.3 and 2.4.

By (2.15),

$$\begin{aligned}(\hat{\sigma}_1^{(e)}(1,1))^2 &= \frac{1}{4} \times 0.6990 - \frac{1}{2} \times 0.1193 + \frac{1}{4} \times 0.0623 \\ &\quad - \frac{1}{4} \times (0.8183 - 0.1817)^2 \\ &= 0.0293\end{aligned}$$

Table 2.4 $Z_{u,v}$ of late Cretaceous

$Z^{(l)}_{1,1}$	$Z^{(l)}_{2,0}$	$Z^{(l)}_{3,0}$	$Z^{(l)}_{2,1}$	$Z^{(l)}_{1,2}$
0.9642	0.0358	0.0037	0.0321	0.9321

and

$$\begin{aligned}(\hat{\sigma}_1^{(l)}(1,1))^2 &= \frac{1}{4}\times 0.9321 - \frac{1}{2}\times 0.0321 + \frac{1}{4}\times 0.0037 \\ &\quad - \frac{1}{4}\times (0.9642 - 0.0358)^2 \\ &= 0.0024\end{aligned}$$

which lead to, by (2.22),

$$Z = \frac{0.8183 - 0.9642}{2\sqrt{\frac{0.0293}{773} + \frac{0.0024}{1348}}} = -11.5706,$$

which provides no statistical evidence of a decrease in diversity, as measured by the Gini–Simpson index, from the early Cretaceous period to the late Cretaceous period. In fact, the highly negative value of the test statistic may even suggest an increase in diversity over that time period.

Simpson's index was originally used to measure the diversity of subspecies of the dinosaur population hundreds of million years ago. The idea was to infer from fossil data whether the diversity of dinosaur population had a gradual diminishing period prior to its sudden extinction about 65 million years ago. If such a diminishing period existed, then the extinction of dinosaurs may be a consequence of natural evolution. If not, then the hypothesis of the Cretaceous extinction, due to a sudden extraterrestrial impact, such as a huge asteroid or a massive round of volcanoes, may be deemed more creditable.

In practice, an appropriate choice of specific pair of integer values (u, v) often varies according to situational need. In general, a higher value of v assigns more weight to species with smaller proportions, that is, with smaller p_ks, and a higher value of u assigns more weight to species with larger proportions, that is, with larger p_ks. However, one may choose to estimate $\zeta_{u,v}$ for a set of (u, v) values, also known as a profile. For example,

$$\{\zeta_{1,v}; v = 1, 2, \ldots\} \tag{2.24}$$

is a profile and it may be considered as a function of v on $\mathbb{N}$. Using profiles to compare the diversity of two ecological communities was first discussed by Taillie (1979) and Patil and Taillie (1982), among others, and has since become an important tool for ecologists.

Denote the two profiles of the two underlying populations and the difference between them as, respectively,

$$\left\{\zeta_{1,v}^{(1)}\right\}, \quad \left\{\zeta_{1,v}^{(2)}\right\}, \quad \{\delta_v\} = \left\{\zeta_{1,v}^{(1)} - \zeta_{1,v}^{(2)}\right\},$$

where $\{\delta_v\} = \{\delta_v; v \geq 1\}$ is sometimes called a difference profile.

$$\hat{\delta}_v = Z_{1,v}^{(1)} - Z_{1,v}^{(2)} \tag{2.25}$$

gives an unbiased estimator of δ_v for every v, $v = 1, \dots, \min\{n_1 - 1, n_2 - 1\}$. Furthermore, (2.22) with $u = 1$ gives an approximate $(1 - \alpha) \times 100\%$ point-wise confidence band

$$\hat{\delta}_v \pm z_{\alpha/2}(v+1)\sqrt{\frac{\left(\hat{\sigma}_1^{(1)}(1,v)\right)^2}{n_1} + \frac{\left(\hat{\sigma}_1^{(2)}(1,v)\right)^2}{n_2}}. \tag{2.26}$$

To illustrate, two cases of authorship attribution are considered next. The concept of species diversity in an ecological population is often applied to quantitative linguistics as that of vocabulary diversity or lexical richness. It is often thought that different authors share a common alphabet, say the English vocabulary, but that each has a propensity for using certain word types over others and hence a unique distribution of word types. If this notion is deemed reasonable, then the difference between the two authors' distributions of word types may be captured by certain diversity measures or a profile of them.

Consider the following three text samples:

1) The collection of 154 sonnets, which are commonly believed to have been penned by William Shakespeare, in a book titled "SHAKESPEARES SONNETS" published by Thomas Thorpe in 1609. This collection contains $n = 17539$ word types and is used as the corpus in the two cases of authorship attribution.
2) The collection of three sonnets in the script of Romeo and Juliet,
   a) the prologue to Act I;
   b) Act I, Scene 5, "If I profane with my unworthiest hand … Thus from my lips, by thine, my sin is purg'd."; and
   c) the prologue to Act II.

   This collection contains $n = 339$ word types.
3) The collection of the first four sonnets from Philip Sidney's Astrophel and Stella sequence,
   a) "Loving in truth, … My Muse said to me 'Fool, look in your heart and write.'";
   b) "Not at first sight, … While with sensitive art I depict my self in hell.";
   c) "Let dainty wits cry on the Sisters nine, … Is to copy what Nature has written in her."; and
   d) "Virtue, … Virtue, will be in love with her.".

   This collection contains $n = 1048$ word types.

Case 1 consists of a comparison between the corpus of 154 Shakespearean sonnets, indexed by (1), and the sample of three Shakespearean sonnets in the script of Romeo and Juliet, indexed by (2). Case 2 consists of a comparison between the same corpus of Shakespearean sonnets, indexed by (1), and the sample of the first four sonnets in Philip Sidney's and Stella sequence, index by (2). Let the profile $\{\zeta_{1,v}; v \geq 1\}$ of each underlying population of interest be restricted to a subset, that is, $\{\zeta_{1,v}; 1 \leq v \leq r\}$, and consequently consider only the estimation of the correspondingly restricted difference profile $\{\delta_v; 1 \leq v \leq r\}$, where $r = 50$.

Using (2.25) and (2.26) with $\alpha = 0.05$, the estimated profile differences and their corresponding 95% point-wise confidence bands are represented graphically in Figures 2.1 and 2.2, where the curve in the middle represents the profile differences while the upper and lower curves are, respectively, the upper bound and lower bound of the confidence band. It may be interesting to note that, in Figure 2.1, the 95% confidence band well contains the zero line (the dashed line) for all v, $1 \leq v \leq 50$, indicating no statistical evidence of a difference, as one would expect. However, in Figure 2.2, the 95% confidence band mostly bounds away from the zero line, indicating a difference in pattern of word usage, particularly the less commonly used words as they are assigned with higher weights by higher values of v.

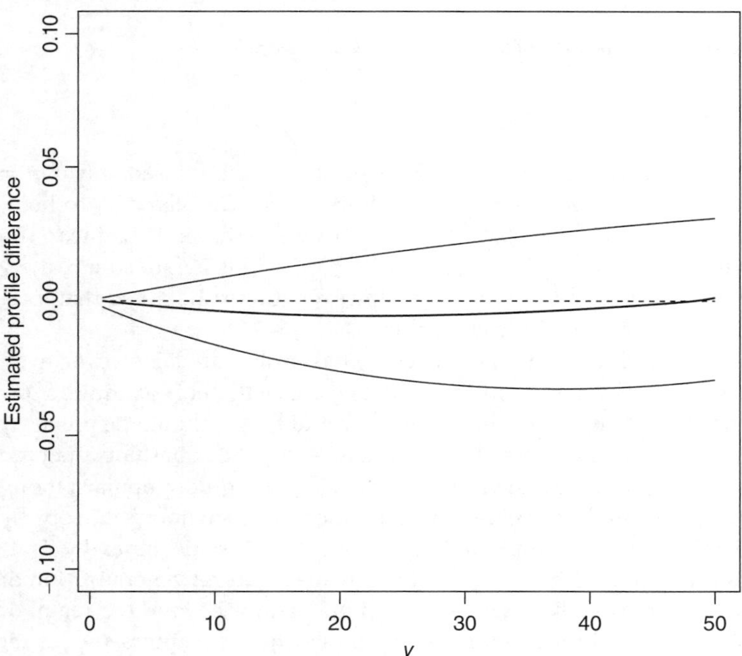

Figure 2.1 Estimated difference profile: Shakespeare versus Shakespeare.

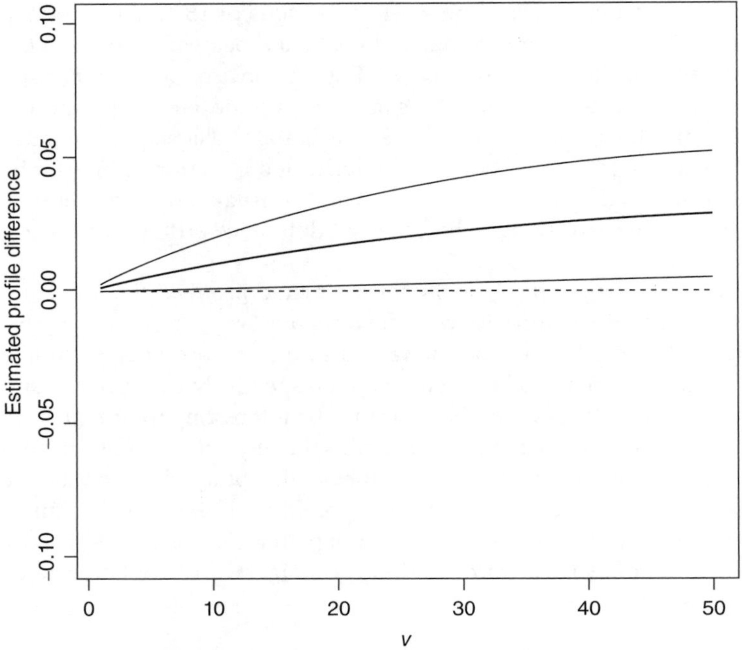

Figure 2.2 Estimated difference profile: Shakespeare versus Sidney.

2.5 Remarks

By the U-statistic construction, $Z_{u,v}$ of (2.10) exists and is unbiased only if $m = u + v$ is less or equal to the sample size n. However, for Corollary 2.1 to hold, $m = u + v$ must further satisfy $2u + 2v - 1 \leq n$ or $u + v \leq (n + 1)/2$ due to the required existence of $Z_{2u-1,2v}$, $Z_{2u,2v-1}$, and $Z_{2u+1,2v-2}$. This is indeed a restriction on the choices of u and v in practice. However, it must be noted that for a sufficiently large n, any one $\zeta_{u,v}$ is estimable.

Since Simpson (1949), research on diversity has proliferated. However, even today diversity or biodiversity is still an unsettled concept. There are quite a few different schools of thought on what diversity is and how it should be precisely defined. Consequently, a universally accepted diversity index has not yet come to be although several well-known diversity indices are quite popular among practitioners, for example, the Gini–Simpson index $\zeta_{1,1}$, Shannon's entropy by Shannon (1948), Rényi's entropy by Rényi (1961), and Tsallis' index by Tsallis (1988), among others. The indices in (2.3), while a natural generalization of Simpson's diversity index, have far-reaching implications beyond the realm of diversity. For reasons to be seen more clearly in subsequent chapters, the subset of the panel, namely $\{\zeta_v; v \geq 1\}$, captures all the information about the underlying distribution $\{p_k\}$, particularly the tail information. Although this fact is made clear in Theorem 2.2, the importance of estimability of ζ_v, up to $v = n - 1$, cannot be overstated.

The perspective on $\{p_k; k \geq 1\}$ by way of $\{\zeta_v; v \geq 1\}$ offers several advantages in certain situations. However, when it comes to statistical estimation based on a sample, it is not without a cost. For example, for every k, there exists an unbiased estimator of p_k, namely $\hat{p}_k = Y_k/n$. (In fact, it is an *unvue*.) In other words, the entire distribution $\{p_1, \ldots\}$ is estimable. This is however not the case for $\{\zeta_v; v \geq 1\}$. Given an *iid* sample of size n, only the first $n - 1$ components in $\{\zeta_v; v \geq 1\}$ are estimable. For example, ζ_n is not estimable. The following is an argument to support the claim.

For any n,

$$\begin{aligned}
\zeta_n &= \sum_{k\geq 1} p_k(1-p_k)^n \\
&= \sum_{k\geq 1} p_k(1-p_k)^{n-1} - \sum_{k\geq 1} p^2(1-p_k)^{n-1} \\
&= \sum_{k\geq 1} p_k(1-p_k)^{n-1} - \sum_{k\geq 1} p_k^2(1-p_k)^{n-2} + \sum_{k\geq 1} p_k^3(1-p_k)^{n-2} \\
&= \sum_{v=1}^{n} \left[(-1)^{v-1} \sum_{k\geq 1} p_k^v (1-p_k)^{n-v} \right] + (-1)^{n+1} \sum_{k\geq 1} p_k^{n+1}. \qquad (2.27)
\end{aligned}$$

By the existence of $Z_{n,0}, Z_{n-1,1}, \ldots, Z_{1,n-1}$, the sum in the first term of the last expression of (2.27) is estimable. Therefore, if ζ_n were estimable, then the last term of (2.27), $\zeta_{n+1,0} = \sum_{k\geq 1} p_k^{n+1}$, would be estimable. In the following text, it is shown that $\zeta_{n+1,0}$ is not estimable based on a sample of size n. In fact, it is to be shown, more generally, that $\zeta_{n+1,0}^{(c)} = \sum_{k\geq 1} c_k p_k^{n+1}$, where $\{c_k; k \geq 1\}$ are known coefficients with at least one nonzero element, is not estimable based on a sample of size n.

First consider a finite alphabet containing $K \geq 2$ letters. Suppose $\zeta_{n+1,0}^{(c)}$ were estimable, that is, there existed an unbiased estimator $\hat{\zeta}_{n+1,0}^{(c)}(\mathbf{X})$ where $\mathbf{X} = (X_1, \ldots, X_n)'$ and each X_i takes one of the K letters in the alphabet according to $\{p_k\}$, such that

$$\mathrm{E}\left(\hat{\zeta}_{n+1,0}^{(c)}(\mathbf{X})\right) = \sum_{k\geq 1} c_k p_k^{n+1}. \qquad (2.28)$$

The left-hand side of (2.28) is of the form

$$\sum_{*} \hat{\zeta}_{n+1,0}^{(c)}(\mathbf{x}) \mathrm{P}(\mathbf{X} = \mathbf{x}), \qquad (2.29)$$

where the summation $\sum_*$ is over all K^n possible configurations of a sample of size n and $\mathbf{x}$ is a realization of $\mathbf{X}$. Equation (2.29) is a K-variable $(p_1, \ldots, p_K)$ polynomial of degree n. For (2.28) to hold for all possible $\{p_k; k \geq 1\}$ on the finite alphabet, the coefficients of the corresponding terms must be equal. In particular, $(c_1, \ldots, c_K)' = (0, \ldots, 0)'$, which contradicts the assumption that at least one of them is nonzero. This establishes the claim that $\zeta_n = \sum_{k\geq 1} p_k(1-p_k)^n$ is not estimable based on an *iid* sample of size n when the alphabet is finite. This result can be immediately extended from the case of a finite alphabet to the case of a countable alphabet since the former is a special case of the latter.

With a cost of the perspective by way of $\{\zeta_v; v \geq 1\}$ established earlier, why would one then prefer $\{\zeta_v; v \geq 1\}$ over $\{p_k; k \geq 1\}$? The answer to the question lies in the fact that, in Statistics, often the object of estimation is not $\mathbf{p} = \{p_k; k \geq 1\}$ itself but a functional of it, say $F(\mathbf{p})$. In estimating certain types of $F(\mathbf{p})$, the estimability of the entire $\mathbf{p}$ becomes much less important than that of the first $n-1$ component of the $\{\zeta_v; v \geq 1\}$.

For an example, suppose the object of estimation is the Gini–Simpson index, that is,

$$F(\mathbf{p}) = \zeta_1 = \sum_{k\geq 1} p_k(1-p_k).$$

The estimability of $\{p_k; k \geq 1\}$ does not directly translate to that of $F(\mathbf{p})$, by means of, say, the plug-in

$$\hat{F}(\mathbf{p}) = F(\hat{\mathbf{p}}) = \sum_{k\geq 1} \hat{p}_k(1-\hat{p}_k),$$

which is an estimator with a sizable negative bias. It is not surprising that the plug-in estimator has a sizable bias because when the cardinality of the alphabet, K, is relatively large compared to the sample size n, many (possibly infinite) letters in the alphabet would not be represented in the sample. Consequently, the corresponding $\hat{p}_k$s would be zero and therefore causing $\hat{p}_k(1-\hat{p}_k) = 0$, all the while the sum of $p_k(1-p_k)$s of the corresponding indices ks could accumulate to be a significant value. Therefore, the plug-in $F(\hat{\mathbf{p}})$ underestimates $F(\mathbf{p})$.

On the other hand, the partial estimability of $\{\zeta_v; v \geq 1\}$ directly translates to that of $F(\mathbf{p})$, that is, $\mathrm{E}(Z_{1,1}) = F(\mathbf{p}) = \zeta_1$.

In summary, Turing's perspective by way of $\{\zeta_v; v \geq 1\}$ is not necessarily a better perspective than that by way of $\{p_k; k \geq 1\}$ in general. However, in situations where tail characteristics play a greater role, Turing's perspective leads to a more efficient way to amplify and to extract tail-relevant statistical information.

2.6 Exercises

1 Let X be a Bernoulli random variable with parameter $p \in (0, 1)$, that is, $\mathrm{P}(X = 1) = p$ and $\mathrm{P}(X = 0) = 1 - p$. Find the first to the fourth central moments of X.

2 Let X be a Poisson random variable with parameter $\lambda > 0$, that is, $p_k = \mathrm{P}(X = k) = e^{-\lambda}\lambda^k/k!$ for $k = 0, 1, \cdots$. Find

$$\mathrm{E}[X(X-1)(X-2)\cdots(X-k+1)]$$

for any positive integer k.

3 An experiment involves repeatedly tossing together a fair coin and balanced die until a "3" appears on the die. Let Y be the number of tosses needed to see the first "3" on the die, and let X be the total number of heads observed on the coin when the experiment ends.
a) Find the probability distribution of Y and calculate $E(Y)$.
b) Find the probability distribution of X and calculate $E(X)$.

4 Let X_1 be a binomial random variable with parameters $n_1 = 5$ and $p_1 = 1/2$; and let X_2 be an independent binomial random variable with parameters $n_2 = 5$ and $p_2 = 1/4$. Find the conditional probability $P(X_1 = 2|X_1 + X_2 = 4)$.

5 If the effective cardinality of $\mathscr{X}$ is finite, that is, $K < \infty$, show that the Gini–Simpson index $\zeta_{1,1}$ of (2.1) is maximized at the uniform distribution, that is, $p_k = 1/K$ for every k, $k = 1, \dots, K$.

6 Derive (2.5).

7 For any probability sequence $\{p_k; \geq 1\}$ on $\mathscr{X}$, show that $\{p_k; \geq 1\}$ is equivalent in permutation to $\{\zeta_{u,v}; u \geq 1\}$ where v is any fixed nonnegative integer.

8 Show that $\hat{\zeta}^{(m)}_{u,v}$ given in (2.8) is an unbiased estimator of $\zeta_{u,v}$.

9 Show that, in case of $u \geq 1$ and $v = 0$, $Z_{u,v}$ of (2.10) reduces to $Z_{u,v}$ of (2.11).

10 Verify (2.14).

11 Show that for any pair of $u \geq 1$ and $v \geq 0$, as $n \to \infty$,

$$\frac{[n-(u+v)]!n^{u+v}}{n!} \to 1.$$

12 Suppose the cardinality of $\mathscr{X}$ is a known and finite integer $K \geq 2$. Show that $\{\hat{p}_k; k \geq 1\}$ is a set of jointly complete and sufficient statistics.

13 For every pair of (u, v), where $u \geq 1$ and $v \geq 0$, show that

$$\zeta_{u,v} = \sum_{i=1}^{u} \binom{u-1}{i-1} (-1)^{u-i} \zeta_{1,u+v-i}.$$

14 The estimator $\hat{\zeta}^*_{u,v}$ of $\zeta_{u,v}$ is a mle when the cardinality K is finite and known. When K is finite but unknown, the estimator $\hat{\zeta}^*_{u,v}$ is still well defined, but is no longer an mle of $\zeta_{u,v}$ (why?). In this case, it is called the plug-in estimator of $\zeta_{u,v}$. Show in case of an unknown but finite K, the estimator retains its asymptotic efficiency.

15 Verify (2.22).

16 Verify the entries of Table 2.3.

17 Verify the entries of Table 2.4.

18 Show that the plug-in estimator

$$\hat{\zeta}^*_{1,1} = \sum_{k\geq 1} \hat{p}_k(1 - \hat{p}_k)$$

of $\zeta_{1,1}$ is a biased estimator.

19 Show that
a) when $u = v = 1$, $\sigma_1^2(u, v) > 0$ if and only if $\{p_k\}$ is not a uniform distribution on $\mathscr{X}$;
b) when $u \geq 1$ and $v = 0$, $\sigma_1^2(u, 0) > 0$ if and only if $\{p_k\}$ is not a uniform distribution on $\mathscr{X}$.
(Hint: Consider the case when $h_1(x_1)$ is a constant.)

20 Starting at 0, a random walk on the x-axis takes successive one-unit steps, going to the right with probability p and to the left with probability $1 - p$. The steps are independent. Let X denote the position after n steps. Find the distribution of X.

21 Let X_1 and X_2 be, respectively, binomial random variables with parameters (n, p_1) and (n, p_2). Show that, if $p_1 < p_2$,

$$\mathrm{P}(X_1 \leq k) \geq \mathrm{P}(X_2 \leq k)$$

for every k, $k = 0, 1, \ldots, n$.

22 Let $X_1, \ldots, X_m, \ldots, X_n$ be a sample of *iid* random variable X with $\mathrm{E}(X) = \mu < \infty$. Let $S_m = \sum_{i=1}^{m} X_i$ and $S_n = \sum_{i=1}^{n} X_i$, where $n \geq m$.
a) Find $\mathrm{E}(S_n | S_m = s_m)$.
b) Find $\mathrm{E}(S_m | S_n = s_n)$.

23 Suppose that X is a Poisson random variable with intensity parameter $\lambda_X > 0$ and that, given $X = k$, Y is a binomial random variable with parameters $n = k$ and $p > 0$. Show that Y is a Poisson random variable with intensity parameter $\lambda_Y = \lambda_X p$.

3

Estimation of Shannon's Entropy

Let $\mathscr{X} = \{\ell_k; k \geq 1\}$ be a countable alphabet and $\mathbf{p} = \{p_k; k \geq 1\}$ be an associated probability distribution of the alphabet. Shannon (1948) defined entropy of $\mathbf{p}$ in the form of $H_S = -\sum_{k\geq 1} p_k \log_2 p_k$. Often Shannon's entropy is also expressed with the natural logarithm,

$$H = -\sum_{k\geq 1} p_k \ln p_k, \tag{3.1}$$

which differs from H_S by a constant factor, $H = H_S \ln 2$. Throughout this book, H of (3.1) is referred to as Shannon's entropy. H is of profound importance in many branches of modern data sciences. In recent decades as entropy finds its way into a wider range of applications, the estimation of entropy also becomes increasingly important. The volume of literature on the general topic of entropy estimation is quite large and covers a wide range of models with different stochastic structures. The two most commonly cited review articles are by Beirlant, Dudewicz, Györfi, and Meulen (2001) and Paninski (2003). While Beirlant *et al.* (2001) include results under a broader range of models, Paninski (2003) mainly focuses on the issues of entropy estimation on a countable alphabet, the most basic and therefore the most fundamental model for entropy estimation.

Entropy estimation is generally considered a difficult problem regardless of the domain type for the underlying distribution. The cause of difficulty is twofold. First, Shannon's entropy is a measure sensitive to small perturbations on the probabilities in the tail, as evidenced by

$$\frac{d(-p\ln p)}{dp} = -\ln p - 1,$$

which assumes a large positive value for small p. This sensitivity could amplify ordinary sampling errors at lower probability region of the sampling space into significant variability in estimation. Second, a sample of size n, however large it may be, could often have a hopelessly poor coverage of the alphabet. If the cardinality of $\mathscr{X}$ is infinite, then that of the subset not covered by a sample is always infinite. Even for a finite alphabet, the cardinality of the subset not

Statistical Implications of Turing's Formula, First Edition. Zhiyi Zhang.

covered by a sample is often much larger than that of the subset covered by the sample. In light of this fact, it is clear that a good entropy estimator must find a way to utilize certain characteristics of the probability distribution over the not-covered subset of the alphabet.

Two fundamental questions are immediately raised. The first question is whether such information exists in a finite sample; and the second question is, if it does, whether it can be extracted nonparametrically. After an overview on the mainlines of research in entropy estimation during the past decades, a nonparametric entropy estimator in Turing's perspective is constructed; some of its statistical properties are established; and, in turn, the nonparametric entropy estimator in Turing's perspective will give an affirmative answer to each of the above-mentioned two questions.

3.1 A Brief Overview

Let $\{Y_k\} = \{Y_k; k \geq 1\}$ be the sequence of observed counts of letters of the alphabet in an *iid* sample of size n and let $\{\hat{p}_k = Y_k/n\}$. The general nonparametric estimator that plays a central role in the literature is commonly known as the plug-in estimator and is given by

$$\hat{H} = -\sum_{k\geq 1} \hat{p}_k \ln \hat{p}_k. \tag{3.2}$$

The plug-in estimator is appealing to practitioners since it is simple, intuitive, easily calculated, and asymptotically efficient in some cases (when it is a maximum likelihood estimator). However, the plug-in estimator has a well-established issue in large and complex bias. For example, according to Harris (1975), when $\mathscr{X}$ is of finite cardinality $K < \infty$, the bias of $\hat{H}$ up to the second order is of the form

$$\mathrm{E}\left(\hat{H} - H\right) = -\frac{K-1}{2n} + \frac{1}{12n^2}\left(1 - \sum_{k=1}^{K} \frac{1}{p_k}\right) + \mathcal{O}\left(n^{-3}\right). \tag{3.3}$$

The two immediate features of the first-order term in (3.3) are as follows:

1) $\hat{H}$ tends to underestimate H since the term is negative; and
2) in case of a large K relative to the sample size n, the magnitude of the bias could be substantial.

There are plenty of situations in application where a relatively large K meets a relatively small n. For example, there are more than 50000 ($K > 50000$) common English word types in current usage and a typical sample text, for example, a web page, often contains less than 300 words ($n < 300$). For another example, there are approximately 35000 gene types ($K \approx 35000$) in the human genome and genetic studies often have sample sizes in hundreds ($100 < n < 500$).

In fact, the complexity of the bias in (3.3) is beyond the first-order term. Just for a moment, suppose the value of K is finite and known *a priori*. Then one could adjust $\hat{H}$ by letting

$$\hat{H}^{+} = \hat{H} + \frac{K-1}{2n}$$

to obtain an estimator with a bias decaying at the rate of $\mathcal{O}\,(n^{-2})$. Unfortunately such an adjustment does not solve the general problem because the next leading term is

$$\frac{1}{12n^2}\left(1 - \sum_{k=1}^{K} \frac{1}{p_k}\right)$$

which, given n, could assume a hugely negative value if just for one index k, $0 < p_k \ll 1$. Similar observations can be made into higher order terms of (3.3) for $K < \infty$. When K is countably infinite, much less is known about the bias of the plug-in estimator in (3.2).

Historically much of the research on entropy estimation has been inspired by the practical need to reduce the bias of the plug-in estimator. A large number of estimators have been proposed. Many of these proposed estimators may be roughly classified into three categories: (i) correcting the bias of $\hat{H}$ by estimating K, (ii) jackknife, and (iii) Bayesian.

By estimating K and assuming a satisfactory estimator $\hat{K}$ can be found, the plug-in estimator may be naturally adjusted by means of an additive term according to (3.3), that is,

$$\hat{H}_{*} = \hat{H} + \frac{\hat{K}-1}{2n}.$$

However estimating K is a difficult statistical problem on its own. There are no unbiased nonparametric estimators of K under general models; therefore, the first-order term in (3.3) cannot be entirely eliminated by such a simple adjustment. In addition, the very action of estimating K implies the assumption that the cardinality K of the alphabet $\mathscr{X}$ is finite. This assumption puts a conceptual limitation on the models for which such estimator can be justifiably used. Nevertheless, the general concept of such approaches is practically sound. For an introduction to the statistical problem of estimating K, interested readers may wish to refer to Chao (1987) and Bunge, Willis, and Walsh (2014).

A popular and frequently cited entropy estimator in this category is the one proposed by Miller (1955), also known as the Miller–Madow estimator, which uses

$$\hat{m} = \sum_{k\geq 1} 1[Y_k > 0],$$

the number of covered letters in the sample, in place of $\hat{K}$, that is,

$$\hat{H}_{MM} = \hat{H} + \frac{\hat{m}-1}{2n}. \tag{3.4}$$

The jackknife methodology is another commonly used technique to reduce bias of the plug-in estimator. The basic elements of the jackknife estimator are

1) to construct n plug-in estimators based on a subsample of size $n-1$ by leaving one (the ith) observation out, $\hat{H}^{(i)}$, which has, based on (3.3), a bias with a leading first-order term

$$-\frac{K-1}{2(n-1)};$$

2) to obtain $\hat{H}_{(i)} = n\hat{H} - (n-1)\hat{H}^{(i)}$ for $i = 1, \dots, n$; and then
3) to define the jackknife estimator by

$$\hat{H}_{JK} = \frac{\sum_{i=1}^{n} \hat{H}_{(i)}}{n} = \hat{H} + \frac{n-1}{n} \sum_{i=1}^{n} (\hat{H} - \hat{H}^{(i)}) . \tag{3.5}$$

The jackknife methodology was first applied to the plug-in estimator by Zahl (1977). It may be verified that the bias of $\hat{H}_{JK}$ is

$$\mathrm{E}\,(\hat{H}_{JK} - H) = -\frac{1}{12n(n-1)} \left(1 - \sum_{k=1}^{K} \frac{1}{p_k}\right) + \mathcal{O}\,(n^{-3}), \tag{3.6}$$

specifically noting that the leading term is positive and decays at the rate of $\mathcal{O}\,(n^{-2})$ (see Exercise 4). Another noteworthy methodology in this area is that by Strong, Koberle, de Ruyter van Steveninck, and Bialek (1998), where an interesting extension of the jackknife procedure can be found to correct bias simultaneously for several terms of different orders in (3.3).

Another important group of entropy estimators is based on Bayesian perspectives. Consider the underlying probability distribution $\{p_1, \dots, p_K\}$ as a point in the K-dimensional unit cube satisfying the following constraints:

1) $p_k \in (0, 1)$ for $i = 1, \dots, K$, and
2) $\sum_{k=1}^{K} p_k = 1$.

A prior distribution is imposed on $\{p_1, \dots, p_K\}$ in the form of a Dirichlet density function

$$f(p_1, \dots, p_K; \alpha_1, \dots, \alpha_K) = \frac{1}{B(\alpha)} \prod_{k=1}^{K} p_k^{\alpha_k - 1} \tag{3.7}$$

where $\alpha = (\alpha_1, \dots, \alpha_K)^\tau$ is such that $\alpha_k > 0$ for each k, $k = 1, \dots, K$;

$$B(\alpha) = \frac{\prod_{k=1}^{K} \Gamma(\alpha_k)}{\Gamma\left(\sum_{k=1}^{K} \alpha_k\right)};$$

and $\Gamma(\cdot)$ is the gamma function. Under the prior distribution in (3.7), the posterior is also a Dirichlet distribution in the form of

$$f(p_1, \dots, p_K | y_1, \dots, y_K) = \frac{1}{B(\mathbf{y} + \alpha)} \prod_{k=1}^{K} p_k^{(y_k+\alpha_k)-1} \tag{3.8}$$

where $\mathbf{y} = (y_1, \dots, y_K)^\tau$ is a realization of the observed letter frequencies in the sample. Under (3.8), the posterior mean for each p_k is

$$\hat{p}_k^* = \frac{y_k + \alpha_k}{n + \sum_{k=1}^{K} \alpha_k},$$

and, in turn, the Bayes estimator of entropy is given by the plug-in form

$$\hat{H}_{Bayes} = -\sum_{k=1}^{K} \hat{p}_k^* \ln \hat{p}_k^*. \tag{3.9}$$

For various sets of prefixed $\{\alpha_1, \dots, \alpha_K\}$, this group of entropy estimators includes those proposed by Krichevsky and Trofimov (1981), Holste, Große, and Herzel (1998), and Schürmann and Grassberger (1996), among many others.

The Dirichlet prior, while a natural conjugate to $\{p_k; k = 1, \dots, K\}$, is by no means the only reasonable family of priors to be used. It may be modified or extended in many ways. For example, Nemenman, Shafee, and Bialek (2002) adopted a hierarchical Bayesian perspective by

1) reducing (3.7) to a one-parameter Dirichlet model by setting $\beta = \alpha_k$ for all k, and
2) imposing a hyper-prior on the single hyper-parameter β, designed to reduce the impact of observed data $\mathbf{y}$ when n is small.

The resulting estimator is known as the NSB estimator.

There is a long list of other notable entropy estimators that do not necessarily fall into these categories. Examples are Hausser and Strimmer (2009) for James-Stein type of shrinkage estimators; Chao and Shen (2003) and Vu, Yu, and Kass (2007) for coverage-adjusted estimators; and Valiant and Valiant (2011) and Zhang and Grabchak (2013) for general bias reduction. There is also an interesting unpublished manuscript by Montgomery-Smith and Schürmann (2007) that established an unbiased entropy estimator under a sequential sampling scheme. In summary, there is no shortage of proposed entropy estimators in the literature and many more are certainly to come.

There are however several fundamental issues that are not addressed sufficiently and satisfactorily by the existing research on entropy estimation.

1) Convergence rates of most of the proposed estimators are largely unknown. The majority of the research thus far focuses on bias reduction, and the primary methodology of evaluating the performance is by simulations. While simulations are important tools, they cannot replace theoretical

assessment of statistical procedures. The lack of understanding of various convergence rates for most of the proposed estimators, for example, in distribution, in bias, in variance, in mean squared error, etc., leaves users vulnerable to potentially poor performances, with n of all sizes. As certain as the fact that most of the proposed entropy estimators have their respective merits, the best of them could be easily proven inadequate under some underlying distributions $\{p_k\}$.

2) Distributional characteristics of most of the proposed estimators are largely unknown. It may serve as a reminder to practitioners everywhere that only when the sampling distribution of an estimator is known (or can be approximated in some sense) can one reasonably believe that the behavior of that estimator is statistically understood.
3) What happens if the cardinality of the alphabet $\mathscr{X}$ is countably infinite, that is, $K = \sum_{k\geq 1} 1[p_k > 0] = \infty$? The inspiration of bias reduction comes mainly from the bias expression of (3.3). Consequently the probabilistic platforms for developing bias-reduced estimators implicatively assume a finite K. In fact, all of the estimators in the Bayesian category with a Dirichlet prior require a known K for the argument to make sense. If $K = \infty$, it is not clear how the bias can be expressed and how it behaves under various distributional classes.

These statistical issues are at the very center of entropy estimation and hence of information theory. Future research efforts must move toward resolving them.

3.2 The Plug-In Entropy Estimator

The plug-in entropy estimator of (3.2) is a natural reference estimator for all others. Despite its notoriety associated with a large bias when n is not sufficiently large, the plug-in estimator has quite a lot to offer. Firstly, when K is finite and known, it is a maximum likelihood estimator. As such, it is asymptotically efficient, that is essentially saying that when the sample is sufficiently large, no other estimator could perform much better. Secondly, it offers a relatively less convoluted mathematical form and therefore makes the establishing of its statistical properties relatively easy. Thirdly, its various statistical properties would shed much light on those of other estimators. It is fair to say that understanding the statistical properties of the plug-in is a first step toward understanding the statistical properties of many other estimators.

3.2.1 When K Is Finite

Several statistical properties of the plug-in estimator of (3.2) are summarized in this section.

Theorem 3.1 *Let H be as in (3.1) and let $\hat{H}$ be as in (3.2) based on an iid sample of size n. Suppose $K < \infty$. Then*

1) $\mathrm{E}(\hat{H} - H) = -\frac{K-1}{2n} + \frac{1}{12n^2}\left(1 - \sum_{k=1}^{K} \frac{1}{p_k}\right) + \mathcal{O}(n^{-3})$;

2) $\mathrm{Var}(\hat{H}) = \frac{1}{n}\left(\sum_{k=1}^{K} p_k \ln^2 p_k - H^2\right) + \frac{K-1}{2n^2} + \mathcal{O}(n^{-3})$; *and*

3) $\mathrm{E}(\hat{H} - H)^2 = \frac{1}{n}\left(\sum_{k=1}^{K} p_k \ln^2 p_k - H^2\right) + \frac{K^2-1}{4n^2} + \mathcal{O}(n^{-3})$.

The proof of Theorem 3.1 is found in Harris (1975). Theorems 3.2 and 3.3 are due to Miller and Madow (1954).

Theorem 3.2 *Let H be as in (3.1) and let $\hat{H}$ be as in (3.2) based on an iid sample of size n. Suppose $K < \infty$ and $\{p_k\}$ is a nonuniform distribution. Then*

$$\sqrt{n}(\hat{H} - H) \xrightarrow{L} N(0, \sigma^2) \tag{3.10}$$

where

$$\sigma^2 = \sum_{k=1}^{K} p_k \left(\ln p_k + H\right)^2. \tag{3.11}$$

Proof: Let $\hat{\mathbf{p}}^- = (\hat{p}_1, \dots, \hat{p}_{K-1})'$ and $\mathbf{p}^- = (p_1, \dots, p_{K-1})'$. Noting

$$\sqrt{n}(\hat{\mathbf{p}}^- - \mathbf{p}^-) \xrightarrow{L} MVN(\emptyset, \Sigma)$$

where $\emptyset$ is a $(K-1)$-dimensional vector of zeros, $\Sigma = (\sigma_{i,j})$ is a $(K-1) \times (K-1)$ covariance matrix with diagonal and off-diagonal elements being, respectively,

$$\sigma_{i,i} = p_i(1 - p_i),$$
$$\sigma_{i,j} = -p_i p_j$$

for $i = 1, \dots, K-1$, $j = 1, \dots, K-1$, and $i \neq j$. The result of the theorem follows an application of the delta method with function $H = g(\mathbf{p}^-)$. (The details of the proof are left as an exercise; see Exercise 7.) □

The condition of $\{p_k\}$ being a nonuniform distribution is required by Theorem 3.2 because otherwise σ^2 in (3.11) would be zero (see Exercise 6) and consequently the asymptotic distribution degenerates.

Theorem 3.3 *Let H be as in (3.1) and let $\hat{H}$ be as in (3.2) based on an iid sample of size n. Suppose $K < \infty$ and $p_k = 1/K$ for each k, $k = 1, \dots, K$. Then*

$$2n(H - \hat{H}) \xrightarrow{L} \chi^2_{K-1} \tag{3.12}$$

where χ^2_{K-1} is a chi-squared random variable with degrees of freedom $K - 1$.

Proof: Noting the proof of Theorem 3.2, the result of the theorem follows an application of the second-order delta method with function $H = g(\mathbf{p}^-)$. (The details of the proof are left as an exercise; see Exercise 8.) □

Theorem 3.2 is important not only in its own right but also in passing asymptotic normality to other estimators beyond the plug-in. Many bias-adjusted estimators in the literature have a general form of

$$\hat{H}_{adj} = \hat{H} + \hat{B}. \tag{3.13}$$

Corollary 3.1 *Let $\hat{H}_{adj}$ be as in (3.13). If $\sqrt{n}\hat{B} \xrightarrow{p} 0$, then*

$$\frac{\sqrt{n}\left(\hat{H}_{adj} - H\right)}{\sqrt{\sum_{k=1}^{K} p_k \ln^2 p_k - H^2}} \xrightarrow{L} N(0,1). \tag{3.14}$$

Proof: Noting first

$$\sqrt{n}\left(\hat{H}_{adj} - H\right) = \sqrt{n}(\hat{H} - H) + \sqrt{n}(\hat{B}) \xrightarrow{L} N(0, \sigma^2)$$

and second

$$\begin{aligned}
\sigma^2 &= \sum_{k=1}^{K} p_k (\ln p_k + H)^2 \\
&= \sum_{k=1}^{K} p_k (\ln^2 p_k + 2 \ln p_k H + H^2) \\
&= \sum_{k=1}^{K} p_k \ln^2 p_k - H^2,
\end{aligned}$$

the desired result follows Slutsky's lemma. □

Example 3.1 *Consider the Miller–Madow estimator*

$$\hat{H}_{MM} = \hat{H} + \frac{\hat{m} - 1}{2n}$$

where the adjusting term of (3.13) is $\hat{B} = (\hat{m} - 1)/(2n) \geq 0$. Noting, as $n \to \infty$,

$$0 \leq \sqrt{n}\hat{B} = \sqrt{n}\left(\frac{\hat{m} - 1}{2n}\right) \leq \frac{K - 1}{2\sqrt{n}} \to 0,$$

$\hat{B} \xrightarrow{a.s.} 0$, $\hat{B} \xrightarrow{p} 0$, and therefore the Miller–Madow estimator admits the asymptotic normality of Corollary 3.1.

In order to use (3.14) of Corollary 3.1 for inference about entropy, one must have not only a consistent estimator of entropy but also a consistent estimator of

$$H_2 = \sum_{k=1}^{K} p_k \ln^2 p_k. \tag{3.15}$$

As will be seen in subsequent text, H_2 appears in the asymptotic variance of many entropy estimators and therefore is a key element of statistical inference about H. It is sufficiently important to deserve a defined name.

Definition 3.1 *For a probability distribution $\{p_k; k \geq 1\}$ on alphabet $\mathcal{X}$ and a positive integer r, the rth entropic moment is defined to be*

$$H_r = (-1)^r \sum_{k \geq 1} p_k \ln^r p_k. \tag{3.16}$$

To give an intuitive justification for the names of entropic moments of Definition 3.1, one may consider a random variable **H** defined by a given $\{p_k\}$, with no multiplicities among the probabilities, in the following tabulated form:

H	$-\ln p_1$	$-\ln p_2$	$\cdots$	$-\ln p_k$	$\cdots$
P(**H**)	p_1	p_2	$\cdots$	p_k	$\cdots$

The first moment of **H** gives entropy $H = \mathrm{E}(\mathbf{H}) = -\sum_{k\geq 1} p_k \ln p_k$ and the second moment of **H** gives $H_2 = \mathrm{E}(\mathbf{H}^2) = \sum_{k\geq 1} p_k \ln^2 p_k$, etc.

Remark 3.1 *A cautionary note must be taken with the random variable* **H**. *If there is no multiplicity among the probabilities in $\{p_k\}$, then*

$$\mathrm{Var}(\mathbf{H}) = \sum_{k\geq 1} p_k \ln^2 p_k - H^2.$$

However, if there exists at least one pair of (i, j), $i \neq j$, such that $p_i = p_j$, then **H** *is not a well-defined random variable, and neither therefore are the notions of* E(**H**) *and* Var(**H**). *This point is often overlooked in the existing literature.*

To estimate H_2 is however not easy. Its plug-in form of

$$\hat{H}_2 = \sum_{k\geq 1} \hat{p}_k \ln^2 \hat{p}_k$$

suffers all the same drawbacks of $\hat{H}$ in estimating H. In fact, the drawbacks are much worse, exacerbated by the heavier weights of $w_k = \ln^2 p_k$ in comparison to $w_k = -\ln p_k$ in the weighted average of the probabilities, that is, $\sum_{k\geq 1} w_k p_k$. In a relatively small sample, the smaller sample coverage causes tendency in $\hat{H}_2$

to underestimate H_2, to underestimate the asymptotic variance, and in turn to inflate the rate of Type I error (false positive) in hypothesis testing.

In summary, sound inferential procedures about entropy must be supported by good estimators of not only entropy H but also the second entropic moment H_2.

The effective cardinality $K = \sum_{k\geq 1} 1[p_k > 0]$ of $\{p_k; k = 1, \dots, K\}$ may be extended to $K(n)$ of $\{p_{n,k}; k = 1, \dots, K(n)\}$, where $K(n)$ changes dynamically as n increases. In this case, the corresponding entropy is more appropriately denoted by H_n as a quantity varying with n. Some of the statistical properties established above for fixed K may also be extended accordingly. The following theorem is due to Zubkov (1973).

Theorem 3.4 *Let $\{p_{n,k}; k = 1, \dots, K(n)\}$ be a probability distribution on $\mathscr{X}$ and let H_n be its entropy. Suppose there exist two fixed constants ε and δ, satisfying $\varepsilon > \delta > 0$, such that, as $n \to \infty$,*

1)

$$\frac{n^{1-\varepsilon}}{K(n)} \left(\sum_{k=1}^{K(n)} p_{n,k} \ln^2 p_{n,k} - H_n^2 \right) \to \infty,$$

2)

$$\max_{1\leq k\leq K(n)} \left\{ \frac{1}{np_{n,k}} \right\} = \mathcal{O}\left(\frac{K(n)}{n^{1-\delta}} \right), \quad \textit{and}$$

3)

$$\frac{K(n)}{\sqrt{n \left(\sum_{k=1}^{K(n)} p_{n,k} \ln^2 p_{n,k} - H_n^2 \right)}} \to 0.$$

Then

$$\frac{\sqrt{n}\left(\hat{H} - H_n\right)}{\sqrt{\sum_{k=1}^{K(n)} p_{n,k} \ln^2 p_{n,k} - H_n^2}} \xrightarrow{L} N(0, 1). \tag{3.17}$$

Paninski (2003) established the same asymptotic normality under a slightly weaker sufficient condition as stated in the theorem below.

Theorem 3.5 *Let $\{p_{n,k}; k = 1, \dots, K(n)\}$ be a probability distribution on $\mathscr{X}$ and let H_n be its entropy. Suppose there exists a fixed constant $\varepsilon > 0$, such that, as $n \to \infty$,*

1)

$$\frac{K(n)}{\sqrt{n}} \to 0, \quad and$$

2)

$$\liminf_{n\to\infty} n^{1-\varepsilon}\left(\sum_{k=1}^{K(n)} p_{n,k}\ln^2 p_{n,k} - H_n^2\right) > 0.$$

Then

$$\frac{\sqrt{n}(\hat{H} - H_n)}{\sqrt{\sum_{k=1}^{K(n)} p_{n,k}\ln^2 p_{n,k} - H_n^2}} \xrightarrow{L} N(0,1). \tag{3.18}$$

The proofs of both Theorems 3.4 and 3.5 are not presented here. Interested readers may refer to Zubkov (1973) and Paninski (2003), respectively, for details.

3.2.2 When K Is Countably Infinite

When there are infinitely many positive probabilities in $\{p_k; k \geq 1\}$, the first issue encountered is whether entropy is finite. The convergence of

$$-\sum_{k=1}^{m} p_k \ln p_k$$

depends on the rate of decay of p_k as k increases. For H to be finite, $-p_k \ln p_k$ must decay sufficiently fast or equivalently p_k must decay sufficiently fast.

Example 3.2 *Let, for* $k = 1, 2, \ldots,$

$$p_k = \frac{C_\lambda}{k^\lambda},$$

where $\lambda > 1$ *and* $C_\lambda > 0$ *are two fixed real numbers. In this case,* $H < \infty$.

The details of the argument for Example 3.2 are left as an exercise (see Exercise 9).

Example 3.3 *Let, for* $k = 1, 2, \ldots,$

$$p_k = \frac{1}{C(k+1)\ln^2(k+1)}$$

where the constant $C > 0$ is such that $\sum_{k\geq 1} p_k = 1$:

$$\begin{aligned} H &= -\sum_{k\geq 1} p_k \ln p_k \\ &= \sum_{k\geq 1} (-\ln p_k) p_k \\ &= \sum_{k\geq 1} \frac{\ln(C(k+1)\ln^2(k+1))}{C(k+1)\ln^2(k+1)} \\ &= \sum_{k\geq 1} \frac{\ln C + \ln(k+1) + 2\ln\ln(k+1)}{C(k+1)\ln^2(k+1)} \\ &= \ln C + \sum_{k\geq 1} \frac{1}{C(k+1)\ln(k+1)} \\ &\quad + \sum_{k\geq 1} \frac{2\ln\ln(k+1)}{C(k+1)\ln^2(k+1)} \\ &= \left(\ln C + \frac{\ln\ln 2}{C\ln^2 2}\right) + \sum_{k\geq 1} \frac{1}{C(k+1)\ln(k+1)} \\ &\quad + \sum_{k\geq 2} \frac{2\ln\ln(k+1)}{C(k+1)\ln^2(k+1)}. \end{aligned}$$

Noting that

$$\ln C + \frac{\ln\ln 2}{C\ln^2 2}$$

is a fixed number,

$$\sum_{k\geq 2} \frac{2\ln\ln(k+1)}{C(k+1)\ln^2(k+1)}$$

is a sum of positive terms, and by the integral test for series convergence,

$$\lim_{m\to\infty} \sum_{k=1}^{m} \frac{1}{(k+1)\ln(k+1)} = \infty.$$

The result $H = \infty$ follows.

However, for the purpose of statistical estimation of entropy, only the distributions with finite entropy are considered. In subsequent text, unless otherwise specified, the general discussion is within the class of all $\{p_k; k \geq 1\}$ with finite entropy, that is,

$$\mathscr{P} = \left\{ \{p_k; k \geq 1\} : H = -\sum_{k\geq 1} p_k \ln p_k < \infty \right\}.$$

Several important statistical properties of $\hat{H}$ are established in Antos and Kontoyiannis (2001) and are summarized in the following theorem without proofs.

Theorem 3.6 *For any* $\{p_k\} \in \mathscr{P}$, *as* $n \to \infty$,

1) $\mathrm{E}(\hat{H} - H)^2 \to 0$;
2) *for all* n, $\mathrm{E}(\hat{H}) \le H$; *and*
3) *for all* n,

$$\mathrm{Var}(\hat{H}) \le \frac{\ln^2 n}{n \ln^2 2}. \tag{3.19}$$

Part 1 of Theorem 3.6 implies comforting assurance that $\hat{H}$ converges to H in mean square error (MSE), which in turn implies $|\mathrm{E}(\hat{H} - H)| \to 0$, but does not say anything about its convergence rate. Part 3 gives an upper bound of $\mathrm{Var}(\hat{H})$, which decays at the rate of $\mathcal{O}(\ln^2 n/n)$. Under the general class $\mathscr{P}$, there does not seem to be an established upper bound for $\mathrm{E}(\hat{H} - H)^2$ in the existing literature. Given the upper bound for $\mathrm{Var}(\hat{H})$ in Part 3, it is easy to see that the lack of upper bound for $\mathrm{E}(\hat{H} - H)^2$ is due to that for the bias, $|\mathrm{E}(\hat{H} - H)|$.

Antos and Kontoyiannis (2001) also provided another interesting result as stated in Theorem 3.7.

Theorem 3.7 *Let* $\hat{H}^*$ *be any entropy estimator based on an iid sample of size* n. *Let* $a_n > 0$ *be any sequence of positive numbers converging to zero. Then there exists a distribution* $\{p_k\} \in \mathscr{P}$ *such that*

$$\limsup_{n\to\infty} \frac{\mathrm{E}(|\hat{H}^* - H|)}{a_n} = \infty.$$

Theorem 3.7 may be stated in a more intuitive way: given a positive sequence $0 < a_n \to 0$ with a decaying rate however slow, there exists a distribution $\{p_k\} \in \mathscr{P}$ such that $\mathrm{E}(|\hat{H}^* - H|)$ decays with a slower rate. This theorem is sometimes known as the *slow convergence theorem of entropy estimation*. It may be beneficial to call attention to two particular points of the theorem. First, the estimator $\hat{H}^*$ is referring to not only the plug-in estimator but any arbitrary estimator. In this sense, the so-called slow convergence is not a consequence of the estimator $\hat{H}^*$ but of the estimand H. Second, the criterion of convergence used is the *mean absolute error* (MAE) $\mathrm{E}(|\hat{H}^* - H|)$, and not the bias $|\mathrm{E}(\hat{H}^* - H)|$. Therefore, the nonexistence of a uniformly decaying upper bound for $\mathrm{E}(|\hat{H}^* - H|)$ on $\mathscr{P}$ does not necessarily imply the same for the bias $|\mathrm{E}(\hat{H}^* - H)|$. Of course this by no means suggests that such an upper bound would exist for some $\hat{H}^*$.

Naturally, convergence rates could be better described in subclasses of $\mathscr{P}$. Antos and Kontoyiannis (2001) studied distributions of the types $p_k = Ck^{-\lambda}$

where $\lambda > 1$ and $p_k = C(k\ln^\lambda k)^{-1}$ where $\lambda > 2$ and gave their respective rates of convergence.

Zhang and Zhang (2012) established a normal law for the plug-in estimator under a distribution $\{p_k\}$ on a countably infinite alphabet. The result is stated in Theorem 3.8.

Theorem 3.8 *Let $\{p_k; k \geq 1\} \in \mathscr{P}$ be a probability distribution satisfying $\sigma^2 = \sum_{k\geq 1} p_k \ln^2 p_k - H^2 < \infty$. If there exists an integer-valued function $K(n)$ such that as, $n \to \infty$,*

1) *$K(n) \to \infty$,*
2) *$K(n) = o(\sqrt{n})$, and*
3) *$\sqrt{n}\sum_{k\geq K(n)} p_k \ln p_k \to 0$,*

then

$$\frac{\sqrt{n}(\hat{H} - H)}{\sigma} \xrightarrow{L} N(0, 1).$$

Let

$$\hat{\sigma}^2 = \sum_{k\geq 1} \hat{p}_k \ln^2 \hat{p}_k - \hat{H}^2 < \infty \tag{3.20}$$

or any other consistent estimator of $\sigma^2 = \sum_{k\geq 1} p_k \ln^2 p_k - H^2 < \infty$. The following corollary, also due to Zhang and Zhang (2012), provides a means of large sample inference on H.

Corollary 3.2 *Under the same condition of Theorem 3.8,*

$$\frac{\sqrt{n}(\hat{H} - H)}{\hat{\sigma}} \xrightarrow{L} N(0, 1).$$

Example 3.4 *Let $p_k = C_\lambda k^{-\lambda}$. The sufficient condition of Theorem 3.8 holds for $\lambda > 2$ but not for $\lambda \in (1, 2)$.*

To support the claim of Example 3.4, consider

$$\begin{aligned}
&\sqrt{n} \sum_{k\geq K(n)} (-p_k \ln p_k) \\
&\quad \sim -\sqrt{n} \int_{K(n)}^{\infty} \frac{C_\lambda}{x^\lambda} \ln\left(\frac{C_\lambda}{x^\lambda}\right) dx \\
&\quad = \frac{C_\lambda \lambda}{(\lambda - 1)} \frac{\sqrt{n} \ln K(n)}{(K(n))^{\lambda-1}} + \left[\frac{C_\lambda \lambda}{(\lambda - 1)^2} - \frac{C_\lambda \ln C_\lambda}{\lambda - 1}\right] \frac{\sqrt{n}}{(K(n))^{\lambda-1}} \\
&\quad \text{(see Exercise 10)} \\
&\quad \sim \frac{C_\lambda \lambda}{(\lambda - 1)} \frac{\sqrt{n} \ln K(n)}{(K(n))^{\lambda-1}},
\end{aligned}$$

where "$\sim$" indicates equality in convergence or divergence rate of two sequences.

If $\lambda > 2$, letting $K(n) \sim n^{1/\lambda}$,

$$\frac{C_\lambda \lambda}{(\lambda-1)} \frac{\sqrt{n}\ln K(n)}{(K(n))^{\lambda-1}} \sim \frac{C_\lambda}{(\lambda-1)} \frac{\ln n}{n^{1/2-1/\lambda}} \to 0.$$

If $1 < \lambda \le 2$, for any $K(n) \sim o(\sqrt{n})$ and a sufficiently large n,

$$\frac{C_\lambda \lambda}{(\lambda-1)} \frac{\sqrt{n}\ln K(n)}{(K(n))^{\lambda-1}} > \frac{C_\lambda \lambda}{(\lambda-1)} \frac{\sqrt{n}}{n^{1/2-1/\lambda}} = \frac{C_\lambda \lambda}{(\lambda-1)} n^{1/\lambda} \to \infty.$$

Example 3.5 *Let $p_k = C_\lambda e^{-\lambda k}$. The sufficient condition of Theorem 3.8 holds.*

This is so because, letting $K(n) \sim \lambda^{-1}\ln n$, for a sufficiently large n,

$$\begin{aligned}
&\sqrt{n}\sum_{k\ge K(n)}(-p_k \ln p_k)\\
&\quad\sim -\sqrt{n}\int_{\ln n^{1/\lambda}}^{\infty} C_\lambda e^{-\lambda x}\ln\left(C_\lambda e^{-\lambda x}\right)dx\\
&\quad\sim \frac{C_\lambda}{\lambda}\frac{\ln n}{\sqrt{n}} \to 0.
\end{aligned}$$

Example 3.6 *Let $p_k = C(k^2\ln^2 k)^{-1}$. The sufficient condition of Theorem 3.8 holds.*

Let $K(n) \sim \sqrt{n}/\ln\ln n$. For a sufficiently large n, consider

$$\begin{aligned}
&\sqrt{n}\sum_{k\ge K(n)}(-p_k \ln p_k)\\
&\quad\sim \sqrt{n}C\int_{K(n)}^{\infty}\frac{2\ln x + 2\ln\ln x - \ln C}{x^2\ln^2 x}dx\\
&\quad\sim 2\sqrt{n}C\int_{K(n)}^{\infty}\frac{1}{x^2\ln x}dx\\
&\quad\le \frac{2C\sqrt{n}}{K(n)\ln K(n)} \to 0.
\end{aligned}$$

Example 3.7 *Let $p_k = C(k^2\ln k)^{-1}$. The sufficient condition of Theorem 3.8 does not hold.*

For any $K(n) \sim o(\sqrt{n})$ and a sufficiently large n,

$$\begin{aligned}
&\sqrt{n} \sum_{k \geq K(n)} (-p_k \ln p_k) \\
&\sim \sqrt{n} C \int_{K(n)}^{\infty} \frac{2 \ln x + 2 \ln \ln x - \ln C}{x^2 \ln^2 x} dx \\
&> \sqrt{n} C \int_{K(n)}^{\infty} \frac{2}{x^2} dx \\
&= \frac{2C\sqrt{n}}{K(n)} \to \infty.
\end{aligned}$$

3.3 Entropy Estimator in Turing's Perspective

An entropy estimator in Turing's perspective is introduced in this section. The centerpiece of that estimator is an alternative representation of entropy, which is different from that of Shannon's. Consider any distribution $\{p_k; k \geq 1\} \in \mathcal{P}$ and its entropy:

$$H = -\sum_{k \geq 1} p_k \ln p_k, \tag{3.21}$$

$$H = \sum_{k \geq 1} p_k \sum_{v=1}^{\infty} \frac{1}{v} (1 - p_k)^v, \tag{3.22}$$

$$H = \sum_{v=1}^{\infty} \frac{1}{v} \sum_{k \geq 1} p_k (1 - p_k)^v. \tag{3.23}$$

The first equation in (3.21) is Shannon's definition of entropy. The second equation in (3.22) is due to an application of Taylor's expansion of $-\ln p_k$ for each and every k at $p_0 = 1$. The third equation in (3.23) is due to Fubini's theorem , which allows the interchange of the summations. The two conditions required by Fubini's theorem are satisfied here because

a) the summation H, as in Shannon's entropy, is finite by assumption, and
b) the summands are all positive.

Entropy in the form of (3.23) is a representation in Turing's perspective since it is a linear combination of elements in $\{\zeta_v; v \geq 1\}$ where

$$\zeta_v = \sum_{k \geq 1} p_k (1 - p_k)^v$$

for every nonnegative integer v as described in Chapter 2. Therefore, entropy in Turing's perspective is

$$H = \sum_{v=1}^{\infty} \frac{1}{v} \zeta_v, \tag{3.24}$$

a weighted linear form of $\{\zeta_v\}$.

An interesting heuristic comparison can be made between (3.21) and (3.24). The representation of Shannon's entropy series may be viewed as a stack of positive pieces $-p_k \ln p_k$ over k; and the representation of (3.24) may also be viewed as a stack of positive pieces ζ_v/v over v.

Consider the following two side-by-side comparisons.

1) *Parts versus Whole.*
   a) For each k, the summand $-p_k \ln p_k$ pertains to only the probability of one letter ℓ_k of $\mathcal{X}$, and says nothing about other probabilities or the whole distribution.
   b) For each v, the summand $\zeta_v/v = \sum_{k\geq 1} p_k(1-p_k)^v/v$ is a characteristic of the whole underlying distribution $\{p_k\}$.
2) *Estimability.*
   a) For each k, there is no unbiased estimator of $-p_k \ln p_k$.
   b) For each v, the estimator $Z_{1,v}$ of (2.4) is an uniformly minimum variance unbiased estimator (*umvue*) of ζ_v, provided $v \leq n-1$, and hence $Z_{1,v}/v$ is *umvue* of ζ_v/v.

The estimability of ζ_v up to $v = n-1$ implies that the first part of (3.25) is estimable.

$$H = \sum_{v=1}^{n-1} \frac{1}{v} \zeta_v + \sum_{v=n}^{\infty} \frac{1}{v} \zeta_v. \tag{3.25}$$

That is, letting

$$Z_{1,v} = \frac{n^{1+v}[n-(1+v)]!}{n!} \sum_{k\geq 1} \left[\hat{p}_k \prod_{j=0}^{v-1} \left(1 - \hat{p}_k - \frac{j}{n}\right)\right], \tag{3.26}$$

$$\hat{H}_z = \sum_{v=1}^{n-1} \frac{1}{v} Z_{1,v} \tag{3.27}$$

is an unbiased estimator of $\sum_{v=1}^{n-1} \frac{1}{v} \zeta_{1,v}$. As an estimator of H, the second additive part in (3.25) is the bias of (3.27),

$$|B_n| = H - \mathrm{E}(\hat{H}_z) = \sum_{v=n}^{\infty} \frac{1}{v} \zeta_v. \tag{3.28}$$

The explicit expression of the bias of $\hat{H}_z$ under general distributions enables easier computation and evaluation and is one of many advantages brought about by Turing's perspective.

Remark 3.2 *To express entropy as a series via Taylor's expansion for bias reduction of entropy estimation was perhaps first discussed in the literature by Blyth (1959). There an application of Taylor's expansion was used for $p_k \ln p_k$ at $p_0 = 0.5$ under the assumption that $K < \infty$, an estimator was proposed, and its* *reduced bias was demonstrated. However, the center of the expansion $p_0 = 0.5$ was somehow arbitrary. The argument provided by Blyth (1959) would hold for any reasonable value of $p_0 \in (0, 1)$. For some reason, that line of research was not rigorously pursued in the history. On the contrary, the expansion used above is for $-\ln p_k$ and is at $p_0 = 1$. There are two important consequences of such an expansion. First, the polynomials in the expansion has a particular form, namely $p_k(1 - p_k)^v$, which transforms the entire problem to a local binary platform. The binary platform allows statistical inferences to be made based on whether the letter ℓ_k is either observed or not observed in an iid sample. In fact, this is the very reason the perspective is named in honor of Alan Turing although it is not known whether Alan Turing actually thought about the problem of entropy estimation in his living years. Second, the coefficient of the expansion in Turing's perspective is of a simple form, namely $1/v$, which makes the derivation of mathematical properties of the corresponding estimator $\hat{H}_z$ much easier. The said ease is perhaps one of the main reasons why many good properties of the perspective are established, as will be seen throughout the subsequent text.*

3.3.1 When K Is Finite

When the effective cardinality of $\mathscr{X}$ is finite, that is, $K < \infty$, an immediate benefit of Turing's perspective is stated in the following theorem.

Theorem 3.9 *If $K = \sum_{k=1}^{K} 1[p_k > 0] < \infty$, then*

$$|B_n| = H - \mathrm{E}(\hat{H}_z) \le \frac{(1 - p_\wedge)^n}{np_\wedge}$$

where $p_\wedge = \min\{p_k > 0; k = 1, \ldots, K\}$.

Proof: By (3.28),

$$\begin{aligned} |B_n| &= \sum_{v=n}^{\infty} \frac{1}{v} \sum_{k=1}^{K} p_k(1 - p_k)^v \\ &\le \frac{1}{n} \sum_{v=n}^{\infty} \sum_{k=1}^{K} p_k(1 - p_\wedge)^v \\ &= \frac{1}{n} \sum_{v=n}^{\infty} (1 - p_\wedge)^v \\ &= \frac{(1 - p_\wedge)^n}{np_\wedge}. \end{aligned}$$

□

Three important points are to be noted here. First, the exponential decay of the bias of $\hat{H}_z$ is a qualitative improvement from the plug-in estimator, which has a $\mathcal{O}(n^{-1})$ decaying rate in n. Second, the upper bound depends on the distribution $\{p_k\}$ via $p_\wedge$, and therefore it is not a universal upper bound even within the class of distributions on a finite alphabet since $p_\wedge > 0$ could be arbitrarily small. Third, when n is small and $p_\wedge > 0$ is very small, the bias could still be sizable.

Several asymptotic properties of $\hat{H}_z$, when $K < \infty$, were established in Zhang (2013b). Let H_2 be as in Definition 3.1.

Theorem 3.10 *Let $\{p_k; 1 \le k \le K\}$ be a nonuniform probability distribution on a finite alphabet $\mathscr{X}$. Then*

$$\frac{\sqrt{n}\,(\hat{H}_z - H)}{\sigma} \xrightarrow{L} N(0,1)$$

where $\sigma^2 = H_2 - H^2$.

Toward proving Theorem 3.10, let

$$w(v+1,n) = \frac{n^{v+1}[n-(v+1)]!}{n!} \tag{3.29}$$

and consider the following re-expression of $\hat{H}_z$,

$$\begin{aligned} \hat{H}_z &= \sum_{k=1}^{K} \left\{ \hat{p}_k \left\{ \sum_{v=1}^{n-1} \frac{1}{v} w(v+1,n) \left[\prod_{j=0}^{v-1} \left(1 - \hat{p}_k - \frac{j}{n}\right) \right] \right\} \right\} \\ &=: \sum_{k=1}^{K} \{\hat{p}_k\, \{\hat{g}_{k,n}\}\,\}\,. \end{aligned} \tag{3.30}$$

Of first interest is an asymptotic normal law of $\hat{p}_k \hat{g}_{k,n}$ for each k. For simplicity, consider first a binomial distribution with parameters n and $p \in (0,1)$ and functions

$$g_n(p) = \sum_{v=1}^{n-1} \left\{ \frac{1}{v} w(v+1,n) \left[\prod_{j=0}^{v-1} \left(1 - p - \frac{j}{n}\right) \right] 1[v \le n(1-p)+1] \right\}$$

and $h_n(p) = p g_n(p)$. Let $h(p) = -p\ln(p)$.

The following two lemmas are needed to prove Theorem 3.10. Lemma 3.1 is easily proved by induction (see Exercise 11). A proof of Lemma 3.2 is given in Appendix at the end of this chapter.

Lemma 3.1 *Let a_j, $j = 1, \ldots, n$, be real numbers satisfying $|a_j| \le 1$ for every j. Then $\left|\prod_{j=1}^{n} a_j - 1\right| \le \sum_{j=1}^{n} |a_j - 1|$.*

Lemma 3.2 *Let $\hat{p} = X/n$ where X is a binomial random variable with parameters n and p.*

1) $\sqrt{n}[h_n(p) - h(p)] \to 0$ uniformly in $p \in (c,1)$ for any c, $0 < c < 1$.

2) $\sqrt{n}|h_n(p) - h(p)| < A(n) = \mathcal{O}(n^{3/2})$ *uniformly in* $p \in [1/n, c]$ *for any* c, $0 < c < p$.
3) $\mathrm{P}(\hat{p} \le c) < B(n) = \mathcal{O}\,(n^{-1/2}e^{-nC})$ *where* $C = (p-c)^2/[p(1-p)]$ *for any* $c \in (0, p)$.

Proof of Theorem 3.10: Without loss of generality, consider the sample proportions of the first two letters of the alphabet $\hat{p}_1$ and $\hat{p}_2$ in an *iid* sample of size n. $\sqrt{n}\,(\hat{p}_1 - p_1, \hat{p}_2 - p_2)' \xrightarrow{L} N(0, \Sigma)$, where $\Sigma = (\sigma_{ij})$, $i, j = 1, 2$, $\sigma_{ii} = p_i(1-p_i)$, and $\sigma_{ij} = -p_ip_j$ when $i \neq j$. Write

$$\begin{aligned}
&\sqrt{n}\,[(h_n(\hat{p}_1) + h_n(\hat{p}_2)) - (-p_1 \ln p_1 - p_2 \ln p_2)] \\
&\quad = \sqrt{n}\,[(h_n(\hat{p}_1) + h_n(\hat{p}_2)) - (h(\hat{p}_1) + h(\hat{p}_2))] \\
&\qquad + \sqrt{n}\,[(h(\hat{p}_1) + h(\hat{p}_2)) - (-p_1 \ln p_1 - p_2 \ln p_2)] \\
&\quad = \sqrt{n}\,(h_n(\hat{p}_1) - h(\hat{p}_1)) + \sqrt{n}\,(h_n(\hat{p}_2) - h(\hat{p}_2)) \\
&\qquad + \sqrt{n}\,[(h(\hat{p}_1) + h(\hat{p}_2)) - (-p_1 \ln p_1 - p_2 \ln p_2)] \\
&\quad = \sqrt{n}\,[h_n(\hat{p}_1) - h(\hat{p}_1)]\ 1[\hat{p}_1 \le p_1/2] \\
&\qquad + \sqrt{n}\,(h_n(\hat{p}_2) - h(\hat{p}_2))\ 1[\hat{p}_2 \le p_2/2] \\
&\qquad + \sqrt{n}\,(h_n(\hat{p}_1) - h(\hat{p}_1))\ 1[\hat{p}_1 > p_1/2] \\
&\qquad + \sqrt{n}\,(h_n(\hat{p}_2) - h(\hat{p}_2))\ 1[\hat{p}_2 > p_2/2] \\
&\qquad + \sqrt{n}\,[(h(\hat{p}_1) + h(\hat{p}_2)) - (-p_1 \ln p_1 - p_2 \ln p_2)]\ .
\end{aligned}$$

The third and fourth terms in the last expression above converge to zero almost surely by Part 1 of Lemma 3.2. The last term, by the delta method, converges in law to $N(0, \tau^2)$ where after a few algebraic steps

$$\begin{aligned}
\tau^2 &= (\ln p_1 + 1)^2 p_1(1-p_1) + (\ln p_2 + 1)^2 p_2(1-p_2) \\
&\quad -2(\ln p_1 + 1)(\ln p_2 + 1)p_1p_2 \\
&= (\ln p_1 + 1)^2 p_1 + (\ln p_2 + 1)^2 p_2 \\
&\quad -[(\ln p_1 + 1)p_1 + (\ln p_1 + 1)p_1]^2.
\end{aligned}$$

It remains to show that the first term (the second term will admit the same argument) converges to zero in probability. However, this fact can be established by the following argument. By Part 2 and then Part 3 of Lemma 3.2,

$$\begin{aligned}
&\mathrm{E}\,(\sqrt{n}\ |h_n(\hat{p}_1) - h(\hat{p}_1)|\ 1[\hat{p}_1 \le p_1/2])\,) \\
&\quad \le A(n)P(\hat{p}_1 \le p_1/2) \\
&\quad \le A(n)B(n) \\
&\quad = \mathcal{O}(n^{3/2})\mathcal{O}(n^{-1/2}e^{-nC}) \to 0
\end{aligned}$$

for some positive constant C. This fact, noting that $\sqrt{n}\ |h_n(\hat{p}_1) - h(\hat{p}_1)| \geq 0$, gives immediately the desired convergence in probability, that is,

$$\sqrt{n}\ |h_n(\hat{p}_1) - h(\hat{p}_1)|\ 1[\hat{p}_1 \leq p_1/2] \xrightarrow{p} 0.$$

In turn, it gives the desired weak convergence for

$$\sqrt{n}\ [(h_n(\hat{p}_1) + h_n(\hat{p}_2)) - (-p_1 \ln p_1 - p_2 \ln p_2)]\ .$$

The desired result follows a generalization to K terms. □

To use Theorem 3.10 for inferences, one must find a consistent estimator for $\sigma^2 = H_2 - H^2$. Here the plug-in estimators for both H and H_2, as well as any other consistent estimators, in principle could serve the purpose, but could perform poorly due to their respective heavy biases.

Let

$$\hat{H}_{2z} = \sum_{v=1}^{n-1} \left\{ \left[\sum_{i=1}^{v-1} \frac{1}{i(v-i)} \right] w(v+1, n) \sum_{k=1}^{K} \left[\hat{p}_k \prod_{j=0}^{v-1} \left(1 - \hat{p}_k - \frac{j}{n}\right) \right] \right\}. \tag{3.31}$$

It may be verified that (Exercise 12), for any given distribution with finite K, $|\mathrm{E}(\hat{H}_{2z} - H_2)|$ decays exponentially in n.

Corollary 3.3 *Let $\{p_k; 1 \leq k \leq K\}$ be a nonuniform probability distribution on a finite alphabet and $\hat{H}_{2z}$ be as in (3.31). Then*

$$\sqrt{n} \left(\frac{\hat{H}_z - H}{\sqrt{\hat{H}_{2z} - \hat{H}_z^2}} \right) \xrightarrow{L} N(0, 1).$$

Toward proving Corollary 3.3, let

$$C_v = \sum_{i=1}^{v-1} \frac{1}{i(v-i)} \tag{3.32}$$

for $v \geq 2$ (and define $C_1 = 0$) and write

$$\hat{H}_{2z} = \sum_{v=1}^{n-1} C_v Z_{1,v}.$$

For better clarity in proving Corollary 3.3, a few more notations and two well-known lemmas in U-statistics are first given.

For each i, $1 \leq i \leq n$, let X_i be a random element such that $X_i = \ell_k$ indicates the event that the kth letter of the alphabet is observed and $\mathrm{P}(X_i = \ell_k) = p_k$. Let $X_1, \ldots, X_n$ be an *iid* sample, and denote $x_1, \ldots, x_n$ as the corresponding sample realization. A U-statistic is an n-variable function obtained by averaging the values of an m-variable function (kernel of degree m, often denoted by ψ) over all $n!/[m!(n-m)!]$ possible subsets of m observations from the set of n observations. Interested readers may refer to Lee (1990) for an introduction. In Zhang and Zhou (2010), it is shown that $Z_{1,v}$ is a U-statistic with kernel ψ being Turing's formula with degree $m = v + 1$. Let

$$\psi_c(x_1, \dots, x_c) = \mathrm{E}(\psi(x_1, \dots, x_c, X_{c+1}, \dots, X_m))$$

and

$$\sigma_c^2 = \mathrm{Var}(\psi_c(X_1, \dots, X_c)).$$

Lemmas 3.3 and 3.4 are due to Hoeffding (1948).

Lemma 3.3 *Let U_n be a U-statistic with kernel ψ of degree m.*

$$\mathrm{Var}(U_n) = \binom{n}{m}^{-1} \sum_{c=1}^{m} \binom{m}{c} \binom{n-m}{m-c} \sigma_c^2.$$

Lemma 3.4 *Let U_n be a U-statistic with kernel ψ of degree m.*
For any two nonnegative integers, c and d, satisfying $0 \le c \le d \le m$,

$$\frac{\sigma_c^2}{c} \le \frac{\sigma_d^2}{d}.$$

Lemma 3.5

$$\mathrm{Var}(Z_{1,v}) \le \frac{1}{n}\zeta_{1,v} + \frac{v+1}{n}\zeta_{1,v-1}^2.$$

A proof of Lemma 3.5 is given in the Appendix at the end of this chapter.

Proof of Corollary 3.3: Since $\mathrm{E}(Z_{1,v}) = \zeta_v$,

$$\begin{aligned}
\mathrm{E}(\hat{H}_{2z}) &= \sum_{v=1}^{n-1} C_v \sum_{k=1}^{K} p_k(1-p_k)^v \\
&= \sum_{k=1}^{K} p_k \sum_{v=1}^{n-1} C_v(1-p_k)^v \\
&\to \sum_{k=1}^{K} p_k(-\ln p_k)^2 = \sum_{k=1}^{K} p_k \ln^2 p_k.
\end{aligned}$$

It only remains to show $\text{Var}(\hat{H}_{2z}) \to 0$.

$$\begin{aligned}\text{Var}(\hat{H}_{2z}) &= \sum_{v=1}^{n-1}\sum_{w=1}^{n-1} C_v C_w \text{Cov}(Z_{1,v}, Z_{1,w}) \\ &\le \sum_{v=1}^{n-1}\sum_{w=1}^{n-1} C_v C_w \sqrt{\text{Var}(Z_{1,v})\text{Var}(Z_{1,w})} \\ &= \left(\sum_{v=1}^{n-1} C_v \sqrt{\text{Var}(Z_{1,v})}\right)^2.\end{aligned}$$

Noting

$$\begin{aligned}C_v &= \sum_{i=1}^{v-1} \frac{1}{i(v-i)} \le \sum_{i=1}^{v-1} \frac{1}{v-1} = 1, \\ \zeta_v &\le \zeta_{v-1}, \\ \zeta_{v-1}^2 &\le \zeta_{v-1}, \\ \zeta_{v-1} &= \sum_{k=1}^{K} p_k(1-p_k)^{v-1} \le \sum_{k=1}^{K} p_k(1-p_\wedge)^{v-1} = (1-p_\wedge)^{v-1},\end{aligned}$$

where $p_\wedge = \min\{p_k > 0; k = 1, \ldots, K\}$, and by Lemma 3.5 for $v \ge 2$,

$$\sqrt{\text{Var}(Z_{1,v})} \le \frac{1}{\sqrt{n}}\sqrt{(v+2)\zeta_{1,v-1}} \le \frac{\sqrt{2}v^{1/2}}{\sqrt{n}}(1-p_\wedge)^{(v-1)/2}.$$

As $n \to \infty$,

$$\begin{aligned}\sum_{v=1}^{n-1} C_v \sqrt{\text{Var}(Z_{1,v})} &\le \frac{\sqrt{2}}{\sqrt{n}} \sum_{v=1}^{n} v^{1/2} \left(\sqrt{1-p_\wedge}\right)^{v-1} \\ &= \frac{\sqrt{2}}{\sqrt{n}} \sum_{v=1}^{\lfloor n^{1/4} \rfloor} v^{1/2} \left(\sqrt{1-p_\wedge}\right)^{v-1} \\ &\quad + \frac{\sqrt{2}}{\sqrt{n}} \sum_{v=\lfloor n^{1/4} \rfloor+1}^{n} v^{1/2} \left(\sqrt{1-p_\wedge}\right)^{v-1} \\ &\le \frac{\sqrt{2}}{\sqrt{n}} n^{1/4} \left(n^{1/4}\right)^{1/2} \\ &\quad + \frac{\sqrt{2}}{\sqrt{n}} n^{1/2} \left(\sqrt{1-p_\wedge}\right)^{\lfloor n^{1/4} \rfloor} \frac{1}{1-\sqrt{1-p_\wedge}} \\ &= \sqrt{2}n^{-1/8} + \sqrt{2}\left(\sqrt{1-p_\wedge}\right)^{\lfloor n^{1/4} \rfloor} \frac{1}{1-\sqrt{1-p_\wedge}} \to 0,\end{aligned}$$

and $\text{Var}(\hat{H}_{2z}) \to 0$ follows. Hence $\hat{H}_{2z} \xrightarrow{p} H_2$. The fact of $\hat{H}_z \xrightarrow{p} H$ is implied by Theorem 3.10. Finally, the corollary follows Slutsky's lemma. □

Theorem 3.11 *Let $\{p_k; 1 \le k \le K\}$ be a nonuniform probability distribution on a finite alphabet $\mathscr{X}$. Then $\hat{H}_z$ is asymptotically efficient.*

Proof: First consider the plug-in estimator $\hat{H}$. It can be verified that $\sqrt{n}(\hat{H} - H) \to N(0, \sigma^2)$ where $\sigma^2 = \sigma^2(\{p_k\})$ is as in Theorem 3.10. It is of interest to show first that $\hat{H}$ is asymptotically efficient in two separate cases: (i) when K is known and (ii) when K is unknown. If K is known, then the underlying model $\{p_k; 1 \le k \le K\}$ is a $(K-1)$-parameter multinomial distribution, and therefore $\hat{H}$ is the maximum likelihood estimator of H, which implies that it is asymptotically efficient. Since the estimator $\hat{H}$ takes the same value, given a sample, regardless whether K is known or not, its asymptotic variance is the same. Therefore, $\hat{H}$ must be asymptotically efficient when K is finite but unknown, or else, it would contradict the fact that $\hat{H}$ is asymptotically efficient when K is known. The asymptotic efficiency of $\hat{H}_z$ follows from the fact that $\sqrt{n}(\hat{H}_z - H)$ and $\sqrt{n}(\hat{H} - H)$ have identical limiting distribution. □

3.3.2 When *K* Is Countably Infinite

The following lemma gives a necessary and sufficient condition for entropy H to be finite.

Lemma 3.6 *For a given $\{p_k\}$, $H = -\sum_k p_k \ln(p_k) < \infty$ if and only if there exists a strictly increasing divergent sequence of positive real numbers $\{a_n\}$ such that $\sum_{n=1}^{\infty} (na_n)^{-1} < \infty$ and, as $n \to \infty$,*

$$a_n \sum_{k=1}^{\infty} p_k(1-p_k)^n \to 1.$$

Proof: If an a_n satisfying the conditions exists, then there also exists a positive integer n_0 such that, for all $n > n_0$, $a_n \sum_k p_k(1-p_k)^n < 2$. Therefore,

$$\begin{aligned}
&\sum_{v=1}^{\infty} \frac{1}{v} \sum_k p_k(1-p_k)^{v-1} \\
&\quad = \sum_{v=1}^{n_0} \frac{1}{v} \sum_k p_k(1-p_k)^{v-1} + \sum_{v=n_0+1}^{\infty} \frac{1}{v} \sum_k p_k(1-p_k)^{v-1} \\
&\quad \le \sum_{v=1}^{n_0} \frac{1}{v} \sum_k p_k(1-p_k)^{v-1} + 2 \sum_{v=n_0+1}^{\infty} \frac{1}{va_v} < \infty.
\end{aligned}$$

On the other hand, if $\sum_{v=1}^{\infty}(1/v)\zeta_v < \infty$, then letting $a_n = 1/\zeta_n$ the following arguments show that all conditions of Lemma 3.6 are satisfied. First, ζ_n is strictly decreasing and therefore a_n is strictly increasing. Second, $\zeta_n = \sum_k p_k(1-p_k)^n \leq \sum_k p_k = 1$; and by the dominated convergence theorem, $\zeta_n \to 0$ and hence $a_n \to \infty$. Third, $\sum_{v=1}^{\infty} 1/(va_v) = \sum_{v=1}^{\infty}(1/v)\zeta_v < \infty$ and $a_n\zeta_n = 1$ by assumption and definition. □

Lemma 3.6 serves two primary purposes. First, it encircles all the discrete distributions that may be of interest with regard to entropy estimation, that is, distributions with finite entropy. Second, it provides a characterization of the tail of a distribution in terms of a sequence a_n and its conjugative relationship to ζ_n. The rate of divergence of a_n characterizes the rate of tail decay of the underlying distribution $\{p_k\}$ as k increases. A faster (slower) rate of divergence of a_n signals a thinner (thicker) probability tail.

Let

$$M_n = \frac{5}{n}\left(\sum_{v=1}^{n-1}\frac{1}{v}\zeta_{v-1}^{1/2}\right)^2. \tag{3.33}$$

The next lemma provides an upper bound for $\mathrm{Var}(\hat{H}_z)$ under the general conditions and plays a central role in establishing many of the subsequent results.

Lemma 3.7 *For any probability distribution* $\{p_k\}$, $\mathrm{Var}(\hat{H}_z) \leq M_n$.

The proof of Lemma 3.7 is based on a U-statistic argument. The details of the proof are found in Zhang (2012).

Corollary 3.4 *For any probability distribution* $\{p_k\}$,

$$\mathrm{Var}(\hat{H}_z) < 5\left\{\frac{[1+\ln(n-1)]^2}{n}\right\} = \mathcal{O}(\ln^2 n/n).$$

Proof: Referring to (3.33) and noting $\zeta_v \leq 1$ and that $\sum_{k=1}^{n-1} 1/v$ is the harmonic series hence with upper bound $1+\ln(n-1)$,

$$M_n < 5\left\{\frac{[1+\ln(n-1)]^2}{n}\right\} = \mathcal{O}(\ln^2 n/n).$$

□

Corollary 3.5 *For any probability distribution* $\{p_k\}$ *with finite entropy* H,

$$\mathrm{Var}(\hat{H}_z) < 5(1+H)\left[\frac{1+\ln(n-1)}{n}\right] = \mathcal{O}(\ln n/n).$$

A proof of Corollary 3.5 is given in the Appendix of Chapter 3.

The implications of Corollary 3.5 are quite significant. The upper bound in Corollary 3.5 for every distribution in the entire class of $\mathcal{P}$ decays faster than the upper bound for $\text{Var}(\hat{H})$, that is, $\mathcal{O}(\ln^2 n/n)$, established in Antos and Kontoyiannis (2001) for the same class of distributions by a factor of $\ln n$. The improvement in variance suggests that entropy estimation in view of $\hat{H}_z$ is fundamentally more efficient than that of the plug-in, $\hat{H}$, in some sense. This may also be viewed as an advantage brought about by Turing's perspective. Nevertheless, the fact that the upper bound is proportional to $1 + H$ suggests that, for any fixed sample size n, a distribution with finite entropy can be found to have an upper bound arbitrarily large since H can be arbitrarily large.

Lemma 3.7 implies the following general statements about $\hat{H}_z$ under the condition of finite entropy. Theorem 3.12 is stated for a convenient comparison to Theorem 3.6 regarding the plug-in estimator $\hat{H}$.

Theorem 3.12 *For any $\{p_k\} \in \mathcal{P}$, as $n \to \infty$,*

1) $\text{E}(\hat{H}_z - H)^2 \to 0$;
2) for all n, $\text{E}(\hat{H}_z) \le H$; and
3) for all n,

$$\text{Var}(\hat{H}_z) \le 5(1 + H)\left(\frac{1 + \ln(n-1)}{n}\right). \tag{3.34}$$

Proof: For Part 1, since $\text{E}(\hat{H}_z - H)^2 = \text{Var}(\hat{H}_z) + |B_n|^2$, it suffices to show $\text{Var}(\hat{H}_z) \to 0$, which is implied by Corollary 3.5, and $|B_n| \to 0$ which is implied by H being finite. Part 2 is true simply because the bias $\sum_{v=n}^{\infty} \zeta_v/v > 0$. Part 3 is a restatement of Corollary 3.5. □

In different subclasses of $\mathcal{P}$, the convergence rates of $\hat{H}_z$ could be better described. Naturally $\mathcal{P}$ may be partitioned into layers characterized by the rate of divergence of ζ_n or equivalently by $t_n = n\ \zeta_n$. Consider the following conditions.

Condition 3.1 *For a probability sequence $\{p_k\} \in \mathcal{P}$, there exists a constant $C > 0$ such that $n\ \zeta_n \le C$ for all n.*

Condition 3.2 *For a probability sequence $\{p_k\} \in \mathcal{P}$,*

1) $n\ \zeta_n \to \infty$ as $n \to \infty$, and
2) $\sqrt{n}\ \zeta_n \le C$ for some constant $C > 0$ and all n.

Condition 3.3 *For a probability sequence $\{p_k\} \in \mathcal{P}$,*

1) $\sqrt{n}\ \zeta_n \to \infty$ as $n \to \infty$, and
2) $n^{\delta}\ \zeta_n \le C$ for a constant $\delta \in (0, 1/2)$, some constant $C > 0$ and all n.

Conditions 3.1–3.3 are mutually exclusive conditions. Let $\mathscr{P}_i$, $i = 1, 2, 3$, be the subclasses of $\mathscr{P}$ under Conditions 3.1–3.3, respectively. Let $\mathscr{P}_4 = (\mathscr{P}_1 \cup \mathscr{P}_2 \cup \mathscr{P}_3)^c$ where the complement is with respect to $\mathscr{P}$.

It may be instructive to make several observations at this point. First, $\mathscr{P}_i$, $1 \le i \le 4$, are defined in an order of the tail decaying rate of the underlying distribution with $\mathscr{P}_1$ having the fastest decaying rate. Second, it can be verified that $p_k = c_\lambda e^{-\lambda k}$ where $\lambda > 0$ satisfies $n\zeta_n < C$ for some $C > 0$ and hence Condition 3.1. Condition 3.1 is therefore satisfied by all distributions with faster decaying tails than that of $p_k = c_\lambda e^{-\lambda k}$, including the distributions on any finite alphabet. Third, it can also be verified that, for $p_k = c_\lambda k^{-\lambda}$ where $\lambda > 1$, $n^\delta \zeta_n \to C > 0$ where $\delta = 1 - 1/\lambda$. It is seen here that $p_k = c_\lambda k^{-\lambda}$ where $\lambda \ge 2$ belongs to $\mathscr{P}_2$ and that $p_k = c_\lambda k^{-\lambda}$ where $1 < \lambda < 2$ belongs to $\mathscr{P}_3$. $\mathscr{P}_4$ holds all other very thick-tailed distributions whose ζ_n's converge so slowly that they cannot be bounded above by a sequence $b_n \to 0$ at a rate of $\mathcal{O}(n^{-\varepsilon})$ for any small $\varepsilon > 0$.

Theorem 3.13 *For any probability distribution* $\{p_k\}$,

1) *if* $\{p_k\} \in \mathscr{P}_1 \cup \mathscr{P}_2$, *then there exists* $M_1(n) = \mathcal{O}(n^{-1})$ *such that* $E(\hat{H}_z - H)^2 \le M_1(n)$; *and*
2) *if* $\{p_k\} \in \mathscr{P}_3$, *then there exists* $M_2(n) = \mathcal{O}(n^{-2\delta})$ *such that* $E(\hat{H}_z - H)^2 \le M_2(n)$.

Proof: By Lemma 3.7, $E(\hat{H}_z - H)^2 \le M_n + |B_n|^2$ in general where M_n is as in (3.33) and $B_n = \sum_{v=n}^{\infty} \zeta_v / v$. For Part 1, if Condition 3.1 holds, then

$$
\begin{aligned}
M_n &= \frac{5}{n} \sum_{v=1}^{n-1} \frac{1}{v}\, \zeta_{v-1}^{1/2} \\
&= \frac{5}{n}\, \zeta_0^{1/2} + \frac{5}{n} \sum_{v=2}^{n-1} \frac{1}{v}\, \zeta_{v-1}^{1/2} \\
&\le \frac{5}{n} + \frac{5\sqrt{C}}{n} \sum_{v=2}^{n-1} \frac{1}{v(v-1)^{1/2}} =: V_1(n) = \mathcal{O}(n^{-1}).
\end{aligned}
$$

On the other hand,

$$
B_n = \sum_{v=n}^{\infty} \frac{1}{v}\, \zeta_v \le C \sum_{v=n}^{\infty} \frac{1}{v^2} =: B_1(n) = \mathcal{O}(n^{-1}).
$$

Letting $M_1(n) = V_1(n) + B_1^2(n) = \mathcal{O}(n^{-1})$ establishes the desired result.

Similarly if Condition 3.2 holds, then

$$
\begin{aligned}
M_n &= \frac{5}{n}\, \zeta_0^{1/2} + \frac{5}{n} \sum_{v=2}^{n-1} \frac{1}{v}\, \zeta_{v-1}^{1/2} \\
&\le \frac{5}{n} + \frac{5\sqrt{C}}{n} \sum_{v=2}^{n-1} \frac{1}{v(v-1)^{1/4}} =: V_1(n) = \mathcal{O}(n^{-1}).
\end{aligned}
$$

$$B_n = \sum_{v=n}^{\infty} \frac{1}{v} \zeta_{1,v} \leq C \sum_{v=n}^{\infty} \frac{1}{v^{1+1/2}} =: B_1(n) = \mathcal{O}(n^{-1}).$$

Letting $M_1(n) = V_1(n) + B_1^2(n) = \mathcal{O}(n^{-1})$ establishes the desired result.

For Part 2, if Condition 3.3 holds, then

$$\begin{aligned} M_n &= \frac{5}{n} \zeta_0^{1/2} + \frac{5}{n} \sum_{v=2}^{n-1} \frac{1}{v} \zeta_{v-1}^{1/2} \\ &\leq \frac{5}{n} + \frac{5\sqrt{C}}{n} \sum_{v=2}^{n-1} \frac{1}{v(v-1)^{\delta/2}} =: V_2(n) = \mathcal{O}(n^{-1}). \end{aligned}$$

On the other hand,

$$B_n = \sum_{v=n}^{\infty} \frac{1}{v} \zeta_1 \leq C \sum_{v=n}^{\infty} \frac{1}{v^{1+\delta}} =: B_2(n) = \mathcal{O}(n^{-\delta}).$$

Since in this case $2\delta < 1$, letting $M_2(n) = V_2(n) + B_2^2(n) = \mathcal{O}(n^{-2\delta})$ establishes the desired result. □

The statements of Theorem 3.13 give the convergence rates of upper bounds in mean-squared errors for various types of distribution. Statement 1 says that, for all distributions with fast decaying tails, the bias of $\hat{H}_z$ decays sufficiently fast so that $|B_n|^2$ is dominated by $\mathrm{Var}(\hat{H}_z)$, which converges at a rate of $\mathcal{O}(1/n)$. It may be interesting to note that the so-called fast decaying distributions here include those with power decaying tails down to a threshold $\delta = 1/2$. Statement 2 says that, for each of the thick-tailed distributions, the squared bias $|B_n|^2$ dominates the convergence in mean-squared errors.

Example 3.8 *Suppose $p_k = Ck^{-\lambda}$, where $C > 0$ and $\lambda > 1$ are constants, for all $k \geq k_0$ where k_0 is some positive integer. It can be verified that $n^\delta \sum_k p_k(1-p_k)^n \to L > 0$ where $\delta = 1 - 1/\lambda$ for some constant L (see Exercise 13). By Theorem 3.13,*

1) *if $\lambda \geq 2$, then $\{p_k\} \in \mathcal{P}_2$ and therefore*

$$\mathrm{E}(\hat{H}_z - H)^2 \leq \mathcal{O}\,(n^{-1})\,;$$

and

2) *if $1 < \lambda < 2$, then $\{p_k\} \in \mathcal{P}_3$ and therefore*

$$\mathrm{E}(\hat{H}_z - H)^2 = \mathcal{O}\left(n^{-\frac{2(\lambda-1)}{\lambda}}\right).$$

Example 3.9 *Suppose $p_k = Ce^{-\lambda k}$, where $C > 0$ and $\lambda > 0$ are constants, for all $k \geq k_0$ where k_0 is some positive integer. It can be verified that $n\zeta_n < U$ for some constant $U > 0$. By Theorem 3.13, $\mathrm{E}(\hat{H}_z - H)^2 = \mathcal{O}\,(n^{-1})$.*

Furthermore, a sufficient condition may be established for a normal law of $\sqrt{n}(\hat{H}_z - H)$. Recall $H_2 = \sum_{k\geq1} p_k \ln^2 p_k$, $w(v+1,n)$ of (3.29), and C_v of (3.30). Let

$$\hat{H}_{2z} = \sum_{v=1}^{n-1} \left\{ C_v\, w(v+1,n) \sum_{k\geq1} \left[\hat{p}_k \prod_{j=0}^{v-1} \left(1 - \hat{p}_k - \frac{j}{n}\right) \right] \right\}. \tag{3.35}$$

Theorem 3.14 *For a nonuniform distribution $\{p_k; k \geq 1\} \in \mathscr{P}$ satisfying $H_2 < \infty$, if there exists an integer-valued function $K(n)$ such that, as $n \to \infty$,*

1) $K(n) \to \infty$,
2) $K(n) = o(\sqrt{n}/\ln n)$, and
3) $\sqrt{n}\sum_{k\geq K(n)} p_k \ln p_k \to 0$,

then

$$\sqrt{n}\,(\hat{H}_z - H) \xrightarrow{L} N(0,\sigma^2)$$

where $\sigma^2 = H_2 - H^2$.

Corollary 3.6 *Under the conditions of Theorem 3.14,*

$$\sqrt{n}\left(\frac{\hat{H}_z - H}{\sqrt{\hat{H}_{2z} - \hat{H}_z^2}}\right) \xrightarrow{L} N(0,1).$$

Toward proving Theorem 3.14 and Corollary 3.6, $Z_{1,v}$, $\hat{H}_z$, and $\hat{H}_{2z}$ are re-expressed as follows:

$$\begin{aligned}
Z_{1,v} &= \frac{n^{v+1}[n-(v+1)]!}{n!} \sum_{k\geq1} \left[\hat{p}_k \prod_{j=0}^{v-1} \left(1 - \hat{p}_k - \frac{j}{n}\right) \right] \\
&= \sum_{k\geq1} \left[\frac{n^{v+1}[n-(v+1)]!}{n!} \left(\hat{p}_k \prod_{j=0}^{v-1} \frac{n - Y_k - j}{n} \right) \right] \\
&= \sum_{k\geq1} \left(\hat{p}_k \prod_{j=0}^{v-1} \frac{n - Y_k - j}{n - j - 1} \right) \\
&= \sum_{k\geq1} \left[\hat{p}_k \prod_{j=1}^{v} \left(1 - \frac{Y_k - 1}{n - j}\right) \right],
\end{aligned}$$

and therefore,

$$\begin{aligned}
\hat{H}_z &= \sum_{v=1}^{n-1} \frac{1}{v} Z_{1,v} = \sum_{v=1}^{n-1} \frac{1}{v} \sum_k \hat{p}_k \prod_{j=1}^{v} \left(1 - \frac{Y_k - 1}{n - j}\right) \\
&= \sum_k \hat{p}_k \sum_{v=1}^{n-1} \frac{1}{v} \prod_{j=1}^{v} \left(1 - \frac{Y_k - 1}{n - j}\right) \\
&= \sum_k \hat{p}_k \sum_{v=1}^{n-Y_k} \frac{1}{v} \prod_{j=1}^{v} \left(1 - \frac{Y_k - 1}{n - j}\right)
\end{aligned}$$

and

$$\hat{H}_{2z} = \sum_{v=1}^{n-1} C_v Z_{1,v} = \sum_k \hat{p}_k \sum_{v=1}^{n-Y_k} C_v \prod_{j=1}^{v} \left(1 - \frac{Y_k - 1}{n - j}\right) .$$

Consider a modified probability distribution $\{p_{k,n}; k = 1, \dots, K(n)\}$ of $\{p_k\}$ as follows.

Let

$$p_{k,n} = \begin{cases} p_k, & \text{for} \quad 1 \le k \le K(n) - 1 \\ \sum k \ge K(n) p_k =: p_{K(n),n}, & \text{for} \quad k = K(n), \end{cases}$$

and

$$Y_{k,n} = \begin{cases} Y_k, & \text{for} \quad 1 \le k \le K(n) - 1 \\ \sum k \ge K(n) Y_k =: Y_{K(n),n}, & \text{for} \quad k = K(n), \end{cases}$$

and $\hat{p}_{k,n} = Y_{k,n}/n$ for $1 \le k \le K(n)$. Consequently, the following quantities are also modified accordingly,

$$\begin{aligned}
\zeta_v^* &= \sum_{k=1}^{K(n)} p_{k,n} (1 - p_{k,n})^v, \\
Z_{1,v}^* &= \frac{n^{1+v}[n - (1+v)]!}{n!} \sum_{k=1}^{K(n)} \left[\hat{p}_{k,n} \prod_{j=0}^{v-1} \left(1 - \hat{p}_{k,n} - \frac{j}{n}\right) \right], \\
H^* &= \sum_{k=1}^{K(n)} (-p_{k,n} \ln p_{k,n}) = \sum_{v=1}^{\infty} \frac{1}{v} \zeta_v^*,
\end{aligned}$$

$$\hat{H}_z^* = \sum_{v=1}^{n-1} \frac{1}{v} Z_{1,v}^* = \sum_{k=1}^{K(n)} \left[\hat{p}_{k,n} \sum_{v=1}^{n-Y_{k,n}} \frac{1}{v} \prod_{j=1}^{v} \left(1 - \frac{Y_{k,n}-1}{n-j}\right) \right],$$

$$\hat{H}^* = \sum_{k=1}^{K(n)} (-\hat{p}_{k,n} \ln \hat{p}_{k,n}),$$

$$H_2^* = \sum_{v=1}^{\infty} C_v \, \zeta_v^*,$$

$$\hat{H}_{2z}^* = \sum_{v=1}^{n-1} C_v Z_{1,v}^* = \sum_{k=1}^{K(n)} \left[\hat{p}_{k,n} \sum_{v=1}^{n-Y_{k,n}} C_v \prod_{j=1}^{v} \left(1 - \frac{Y_{k,n}-1}{n-j}\right) \right],$$

$$\hat{H}_2^* = \sum_{k=1}^{K(n)} \hat{p}_{k,n} \ln^2 \hat{p}_{k,n}. \tag{3.36}$$

The following two facts are needed throughout the proofs:

1) $\mathrm{E}(Z_{1,v}) = \zeta_v$ and $\mathrm{E}(Z_{1,v}^*) = \zeta_v^*$; and
2)

$$C_v = \sum_{i=1}^{v-1} \frac{1}{i(v-i)} = \frac{1}{v} \sum_{i=1}^{v-1} \left(\frac{1}{i} + \frac{1}{v-i} \right) \le \frac{2(\ln v + 1)}{v}.$$

The first fact is due to Zhang and Zhou (2010), and the second is easily verified (see Exercise 14).

Lemmas 3.8 and 3.9 are due to Zhang and Zhang (2012).

Lemma 3.8 *For any nonuniform distribution $\{p_k; k \ge 1\}$ satisfying $H_2 < \infty$, if there exists an integer-valued function $K(n)$ such that, as $n \to \infty$,*

1) $K(n) \to \infty$,
2) $K(n) = o(\sqrt{n})$, and
3) $\sqrt{n}\sum_{k \ge K(n)} p_k \ln p_k \to 0$,

then

a)

$$\sqrt{n} \left(\frac{\hat{H} - H}{\sigma} \right) \xrightarrow{L} N(0, 1),$$

b)

$$\sqrt{n} \left(\frac{\hat{H}^* - H^*}{\sigma} \right) \xrightarrow{L} N(0, 1).$$

where $\hat{H}$ and $\hat{H}^*$ are the plug-in estimators of H and H^*, respectively, and $\sigma^2 = H_2 - H^2$.

Lemma 3.9 *For a probability distribution* $\{p_k; k \geq 1\}$, *if there exists an integer-valued function* $K(n)$ *such that as* $n \to \infty$, $K(n) \to \infty$, *and* $\sqrt{n}\sum_{k\geq K(n)} p_k \ln p_k \to 0$, *then*

a) $(\sqrt{n} \ln n) p_{K(n),n} \to 0$, *and*
b) $-\sqrt{n} p_{K(n),n} \ln p_{K(n),n} \to 0$.

Five more lemmas, Lemmas 3.10 through 3.14, are also needed.

Lemma 3.10 *Under the conditions of Theorem 3.14,* $\sqrt{n}(\hat{H}_z - \hat{H}_z^*) = o_p(1)$.

Proof: Noting that for any $k \geq K(n)$, $Y_{k,n} \leq Y_{K(n),n}$ and

$$\begin{aligned} 0 &\leq \sum_{v=1}^{n-Y_{k,n}} \frac{1}{v} \prod_{j=1}^{v} \left(1 - \frac{Y_{k,n} - 1}{n-j}\right) - \sum_{v=1}^{n-Y_{K(n),n}} \frac{1}{v} \prod_{j=1}^{v} \left(1 - \frac{Y_{K(n),n} - 1}{n-j}\right) \\ &\leq \sum_{v=1}^{n-Y_{k,n}} \frac{1}{v} \prod_{j=1}^{v} \left(1 - \frac{Y_{k,n} - 1}{n-j}\right) \\ &\leq \sum_{v=1}^{n-Y_{k,n}} \frac{1}{v} \leq \ln n + 1, \end{aligned}$$

therefore,

$$\begin{aligned} 0 &\leq \sqrt{n}(\hat{H}_z - \hat{H}_z^*) \\ &= \sqrt{n} \sum_{k\geq K(n)} \hat{p}_k \sum_{v=1}^{n-Y_k} \frac{1}{v} \prod_{j=1}^{v} \left(1 - \frac{Y_k - 1}{n-j}\right) \\ &\quad - \sqrt{n}\hat{p}_{K(n),n} \sum_{v=1}^{n-Y_{K(n),n}} \frac{1}{v} \prod_{j=1}^{v} \left(1 - \frac{Y_{K(n),n} - 1}{n-j}\right) \\ &= \sqrt{n} \sum_{k\geq K(n)} \hat{p}_k \left[\sum_{v=1}^{n-Y_k} \frac{1}{v} \prod_{j=1}^{v} \left(1 - \frac{Y_k - 1}{n-j}\right) \right. \\ &\quad \left. - \sum_{v=1}^{n-Y_{K(n),n}} \frac{1}{v} \prod_{j=1}^{v} \left(1 - \frac{Y_{K(n),n} - 1}{n-j}\right) \right] \\ &\leq \sqrt{n}(\ln n + 1) \sum_{k\geq K(n)} \hat{p}_k. \end{aligned}$$

By Lemma 3.9,

$$\sqrt{n}(\ln n + 1)\mathrm{E}\left(\sum_{k\geq K(n)} \hat{p}_{k,n}\right) = \sqrt{n}(\ln n + 1) \sum_{k\geq K(n)} p_{k,n} \to 0,$$

hence $\sqrt{n}\left(\hat{H}_z - \hat{H}_z^*\right) = o_p(1)$ follows Chebyshev's inequality. □

Lemma 3.11 *As $n \to \infty$, under the conditions of Theorem 3.14, $\sqrt{n}(\mathrm{E}(\hat{H}_z^*) - H^*) \to 0$.*

Proof: The lemma follows the fact below.

$$
\begin{aligned}
0 &\le \sqrt{n}(H_n^* - \mathrm{E}(\hat{H}_z^*)) \\
&= \sqrt{n}\sum_{v=1}^{\infty}\frac{1}{v}\,\zeta_v^* - \sqrt{n}\sum_{v=1}^{n-1}\frac{1}{v}\,\zeta_v^* \\
&= \sqrt{n}\sum_{v=n}^{\infty}\frac{1}{v}\,\zeta_v^* \\
&= \sqrt{n}\sum_{v=n}^{\infty}\frac{1}{v}\sum_{k=1}^{K(n)} p_{k,n}(1-p_{k,n})^v \\
&= \sqrt{n}\sum_{k=1}^{K(n)} p_{k,n}\sum_{v=n}^{\infty}\frac{1}{v}(1-p_{k,n})^v \\
&\le \frac{1}{\sqrt{n}}\sum_{k=1}^{K(n)} p_{k,n}\sum_{v=n}^{\infty}(1-p_{k,n})^v \\
&\le \frac{1}{\sqrt{n}}\sum_{k=1}^{K(n)} p_{k,n}\frac{(1-p_{k,n})^n}{p_{k,n}} \\
&= \frac{1}{\sqrt{n}}\sum_{k=1}^{K(n)} (1-p_{k,n})^n \\
&\le \frac{K(n)}{\sqrt{n}} \to 0.
\end{aligned}
$$

□

Lemma 3.12 *As $n \to \infty$, under the conditions of Theorem 3.14, $\sqrt{n}(\mathrm{E}(\hat{H}^*) - H^*) \to 0$.*

Proof: Since $f(x) = -x\ln(x)$ is a concave function for $x > 0$, by Jensen's inequality,

$$
\sqrt{n}\sum_{k=1}^{K(n)} \mathrm{E}(-\hat{p}_{k,n}\ln\hat{p}_{k,n} + p_{k,n}\ln p_{k,n}) \le 0.
$$

Also by (3.3),

$$
\begin{aligned}
&\sqrt{n}\sum_{k=1}^{K(n)}(\mathrm{E}(\hat{H}_n^*) - H_n^*) \\
&= \sqrt{n}\sum_{k=1}^{K(n)} \mathrm{E}(-\hat{p}_{k,n}\ln\hat{p}_{k,n} + p_{k,n}\ln p_{k,n})1[p_{k,n} \ge 1/n]
\end{aligned}
$$

$$+\sqrt{n}\sum_{k=1}^{K(n)}\mathrm{E}(-\hat{p}_{k,n}\ln\hat{p}_{k,n}+p_{k,n}\ln p_{k,n})1[p_{k,n}<1/n]$$

$$\geq\sqrt{n}\left[-\frac{K(n)-1}{2n}+\frac{1}{12n^2}\left(1-\sum_{k=1}^{K(n)}\frac{1}{p_{k,n}}1[p_{k,n}\geq 1/n]\right)+O(n^{-3})\right]$$

$$+\sqrt{n}\sum_{k=1}^{K(n)}(p_{k,n}\ln p_{k,n})1[p_{k,n}<1/n]$$

$$\geq-\frac{\sqrt{n}K(n)}{2n}-\frac{\sqrt{n}K(n)n}{12n^2}-\frac{\sqrt{n}K(n)\ln n}{n}\to 0.$$

Therefore, $\sqrt{n}(\mathrm{E}(\hat{H}^*)-H^*)\to 0$. □

Lemma 3.13 *If a and b are such that $0<a<b<1$, then for any integer $m\geq 0$,*

$$b^m-a^m\leq mb^{m-1}(b-a).$$

Proof: Noting that $f(x)=x^m$ is convex on interval (0, 1) and

$$\frac{df(b)}{db}=mb^{m-1},$$

the result follows immediately. □

Lemma 3.14 *Under the conditions of Theorem 3.14, $\sqrt{n}(\hat{H}_z^*-\hat{H}^*)=o_p(1)$.*

A proof of Lemma 3.14 is given in the appendix of this chapter.

Proof of Theorem 3.14: Note that

$$\begin{aligned}\sqrt{n}(\hat{H}_z-H)-\sqrt{n}(\hat{H}^*-H^*)&=\sqrt{n}(\hat{H}_z-\hat{H}^*)-\sqrt{n}(H-H^*)\\&=\sqrt{n}(\hat{H}_z-\hat{H}_z^*)+\sqrt{n}(\hat{H}_z^*-\hat{H}^*)-\sqrt{n}(H-H^*)\\&=\sqrt{n}(\hat{H}_z-\hat{H}_z^*)+\sqrt{n}(\hat{H}_z^*-\hat{H}^*)\\&\quad+\sqrt{n}\sum_{k\geq K(n)}(p_k\ln p_k)-\sqrt{n}\quad p_{K(n),n}\ln p_{K(n),n}.\end{aligned}\tag{3.37}$$

Since $\sqrt{n}(\hat{H}^*-H^*)\xrightarrow{L}N(0,\sigma^2)$ by Part (b) of Lemma 3.8, it suffices to show that each of the four terms in the right-hand side of (3.37) is $o_p(1)$.

The first two terms are $o_p(1)$ by Lemmas 3.10 and 3.14, respectively, the third term goes to 0 by the conditions of Theorem 3.14, and the fourth term goes to 0 by Lemma 3.9. Therefore, the statement of the theorem follows Slutsky's lemma. □

To prove Corollary 3.6, a few more lemmas are needed.

Lemma 3.15 *Under the conditions of Theorem 3.14,* $\hat{H}_2 - \hat{H}_2^* = o_p(1)$.

Proof:

$$
\begin{aligned}
0 &\le \hat{H}_2 - \hat{H}_2^* \\
&= \sum_{k\ge K(n)} \hat{p}_{k,n}\ln^2\hat{p}_{k,n} - \hat{p}_{K(n),n}\ln^2\hat{p}_{K(n),n} \\
&= \sum_{k\ge K(n)} \hat{p}_{k,n}\ln^2\hat{p}_{k,n} - \sum_{k\ge K(n)} \hat{p}_{k,n}\ln^2\hat{p}_{K(n),n} \\
&= \sum_{k\ge K(n)} \hat{p}_{k,n}(\ln^2\hat{p}_{k,n} - \ln^2\hat{p}_{K(n),n}) \\
&\le \sum_{k\ge K(n)} \hat{p}_{k,n}\ln^2\hat{p}_{k,n} \\
&\le \ln^2 n \sum_{k\ge K(n)} \hat{p}_{k,n}.
\end{aligned}
$$

By Lemma 3.9, $\ln^2 n\mathrm{E}\left(\sum_{k\ge K(n)}\hat{p}_{k,n}\right) = (\ln^2 n)p_{K(n),n} \to 0$, $\hat{H}_2 - \hat{H}_2^* = o_p(1)$ follows Chebyshev's inequality. □

Lemma 3.16 *Under the conditions of Theorem 3.14,* $\hat{H}_{2z} - \hat{H}_{2z}^* = o_p(1)$.

Proof: Noting that for any $k \ge K(n)$, $Y_{k,n} \le Y_{K(n),n}$ and

$$
\begin{aligned}
0 &\le \sum_{v=1}^{n-Y_{k,n}} C_v \prod_{j=1}^{v}\left(1 - \frac{Y_{k,n}-1}{n-j}\right) - \sum_{v=1}^{n-Y_{K(n),n}} C_v \prod_{j=1}^{v}\left(1 - \frac{Y_{K(n),n}-1}{n-j}\right) \\
&\le \sum_{v=1}^{n-Y_{k,n}} C_v \prod_{j=1}^{v}\left(1 - \frac{Y_{k,n}-1}{n-j}\right) \\
&\le \sum_{v=1}^{n-Y_{k,n}} C_v \\
&\le \sum_{v=1}^{n} \frac{2(\ln v + 1)}{v} \\
&\le 2(\ln n + 1)^2,
\end{aligned}
$$

therefore

$$
\begin{aligned}
0 \le& \hat{H}_{2z} - \hat{H}_{2z}^* \\
&= \sum_{k\ge K(n)} \hat{p}_{k,n} \sum_{v=1}^{n-Y_{k,n}} C_v \prod_{j=1}^{v}\left(1 - \frac{Y_{k,n}-1}{n-j}\right)
\end{aligned}
$$

$$
\begin{aligned}
& - \hat{p}_{K(n),n} \sum_{v=1}^{n-Y_{K(n),n}} C_v \prod_{j=1}^{v} \left(1 - \frac{Y_{K(n),n} - 1}{n-j}\right) \\
&= \sum_{k \ge K(n)} \hat{p}_{k,n} \left[\sum_{v=1}^{n-Y_{k,n}} C_v \prod_{j=1}^{v} \left(1 - \frac{Y_{k,n} - 1}{n-j}\right) \right. \\
&\quad \left. - \sum_{v=1}^{n-Y_{K(n),n}} C_v \prod_{j=1}^{v} \left(1 - \frac{Y_{K(n),n} - 1}{n-j}\right) \right] \\
&\le 2(\ln n + 1)^2 \sum_{k \ge K(n)} \hat{p}_{k,n}.
\end{aligned}
$$

By Lemma 3.9,

$$
(\ln n + 1)^2 \mathrm{E} \left(\sum_{k \ge K(n)} \hat{p}_{k,n} \right) = (\ln n + 1)^2 \sum_{k \ge K(n)} p_{k,n} \to 0,
$$

therefore, $\hat{H}_{2z} - \hat{H}^*_{2z} = o_p(1)$ follows Chebyshev's inequality. □

Lemma 3.17 *As $n \to \infty$, under the conditions of Theorem 3.14,*

$$
\mathrm{E}(\hat{H}^*_{2z}) - H^*_2 \to 0.
$$

Proof:

$$
\begin{aligned}
0 &\le H^*_2 - E(\hat{H}^*_{2z}) \\
&= \sum_{v=1}^{\infty} C_v \zeta^*_v - \sum_{v=1}^{n-1} C_v \zeta^*_v \\
&= \sum_{v=n}^{\infty} C_v \zeta^*_v \\
&= \sum_{v=n}^{\infty} C_v \sum_{k=1}^{K(n)} p_{k,n} (1 - p_{k,n})^v \\
&= \sum_{k=1}^{K(n)} p_{k,n} \sum_{v=n}^{\infty} C_v (1 - p_{k,n})^v \\
&\le \frac{2(\ln n + 1)}{n} \sum_{k=1}^{K(n)} p_{k,n} \sum_{v=n}^{\infty} (1 - p_{k,n})^v \\
&\le \frac{2(\ln n + 1)}{n} \sum_{k=1}^{K(n)} p_{k,n} \frac{(1 - p_{k,n})^n}{p_{k,n}}
\end{aligned}
$$

$$= \frac{2(\ln n + 1)}{n} \sum_{k=1}^{K(n)} (1 - p_{k,n})^n$$
$$\leq \frac{2(\ln n + 1)K(n)}{n} \to 0.$$

□

Lemma 3.18 *Under the conditions of Theorem 3.14,* $\hat{H}_{2z}^* - \hat{H}_2^* = o_p(1)$.

A proof of Lemma 3.18 is given in the appendix of this chapter.

Proof of Corollary 3.6: Note that

$$\hat{H}_{2z} - H_2 = (\hat{H}_{2z} - \hat{H}_{2z}^*) + (\hat{H}_z^* - \hat{H}_2^*) + (\hat{H}_2^* - \hat{H}_2) + (\hat{H}_2 - H_2).$$

Each of the first three terms in the above-mentioned equation is $o_p(1)$ by Lemmas 3.16, 3.18, and 3.15, respectively. Also, it is shown, in the proof of Corollary 1 of Zhang and Zhang (2012), that $\hat{H}_2 - H_2 = o_p(1)$. Therefore, $\hat{H}_{2z} - H_2 = o_p(1)$.

By Theorem 3.14, and the fact that $\hat{H}_z^2 \xrightarrow{p} H^2$,

$$\hat{H}_{2z} - \hat{H}_z^2 \xrightarrow{p} H_2 - H^2 = \sigma^2.$$

Finally, the corollary follows Slutsky's lemma. □

Remark 3.3 *The sufficient condition (Part 2) given in Theorem 3.14 for the normality of $\hat{H}_z$ is slightly more restrictive than that of the plug-in estimator $\hat{H}$ as stated in Theorem 1 of Zhang and Zhang (2012), and consequently supports a smaller class of distributions. It can be shown that the sufficient conditions of Theorem 3.14 still holds for $p_k = C_\lambda k^{-\lambda}$ where $\lambda > 2$, but not for $p_k = C/(k^2 \ln^2 k)$, which satisfies the sufficient conditions of Theorem 1 of Zhang and Zhang (2012). However, simulation results indicate that the asymptotic normality of $\hat{H}_z$ in Theorem 3.14 and Corollary 3.6 may still hold for $p_k = C/(k^2 \ln^2 k)$ for $k \geq 1$ though not covered by the sufficient condition, which leads to a conjecture that the sufficient condition of Theorem 3.14 could be further relaxed.*

3.4 Appendix

3.4.1 Proof of Lemma 3.2

Proof of Part 1: As the notation in $g_n(p)$ suggests, the range for v is from 1 to $\min\{n-1, n(1-p)+1\}$. Noting

$$0 \leq \frac{1 - j/[n(1-p)]}{1 - j/n} \leq 1$$

subject to $j \le n(1-p)$, by Lemma 3.1,

$$
\begin{aligned}
&\left| w(v+1,n)\prod_{j=0}^{v-1}\left(1-p-\frac{j}{n}\right) - (1-p)^v \right| \\
&= (1-p)^v \left| \prod_{j=0}^{v-1}\left(\frac{1-\frac{j}{n(1-p)}}{1-\frac{j+1}{n}}\right) - 1 \right| \\
&= (1-p)^v \left| \left(\frac{n}{n-v}\right)\prod_{j=0}^{v-1}\left(\frac{1-\frac{j}{n(1-p)}}{1-\frac{j}{n}}\right) - 1 \right| \\
&= (1-p)^v \left| \left(1+\frac{v}{n-v}\right)\prod_{j=0}^{v-1}\left(\frac{1-\frac{j}{n(1-p)}}{1-\frac{j}{n}}\right) - 1 \right| \\
&\le (1-p)^v \left(\frac{v}{n-v}\right) + (1-p)^v \left| \prod_{j=0}^{v-1}\left(\frac{1-\frac{j}{n(1-p)}}{1-\frac{j}{n}}\right) - 1 \right| \\
&\le (1-p)^v \left(\frac{v}{n-v}\right) + (1-p)^v \sum_{j=0}^{v-1}\left| \left(\frac{1-\frac{j}{n(1-p)}}{1-\frac{j}{n}}\right) - 1 \right| \\
&= (1-p)^v \left(\frac{v}{n-v}\right) + (1-p)^v \frac{p}{1-p}\sum_{j=1}^{v-1}\frac{j}{n-j} \\
&\le (1-p)^{v-1} \left(\frac{v}{n-v}\right) + (1-p)^{v-1}\sum_{j=1}^{v-1}\frac{j}{n-j} \\
&= (1-p)^{v-1}\sum_{j=1}^{v}\frac{j}{n-j} \le (1-p)^{v-1}\frac{v^2}{n-v}.
\end{aligned}
$$

For a sufficiently large n, let $V_n = \lfloor n^{1/8} \rfloor$ where $\lfloor \cdot \rfloor$ is the floor of a real value.

$$
\begin{aligned}
&\sqrt{n}|h_n(p) - h(p)| \\
&\le \sqrt{n}p \sum_{v=1}^{\lfloor n(1-p)+1 \rfloor} \frac{1}{v}\left| w(v+1,n)\prod_{j=0}^{v-1}\left(1-p-\frac{j}{n}\right) - (1-p)^v \right| \\
&\quad + \sqrt{n}p \sum_{v=\lfloor n(1-p)+2 \rfloor}^{\infty} \frac{1}{v}(1-p)^v \\
&= \sqrt{n}p \sum_{v=1}^{V_n} \frac{1}{v}\left| w(v+1,n)\prod_{j=0}^{v-1}\left(1-p-\frac{j}{n}\right) - (1-p)^v \right|
\end{aligned}
$$

$$
\begin{aligned}
&+ \sqrt{n}p \sum_{v=V_n+1}^{\lfloor n(1-p)+1 \rfloor} \frac{1}{v} \left| w(v+1,n) \prod_{j=0}^{v-1} \left(1 - p - \frac{j}{n}\right) - (1-p)^v \right| \\
&+ \sqrt{n}p \sum_{v=\lfloor n(1-p)+2 \rfloor}^{\infty} \frac{1}{v}(1-p)^v \\
&=: \Delta_1 + \Delta_2 + \Delta_3.
\end{aligned}
$$

$$
\begin{aligned}
\Delta_1 &\le \sqrt{n}p \sum_{v=1}^{V_n} \frac{v}{n-v}(1-p)^{v-1} \\
&\le \frac{n^{5/8}}{n - n^{1/8}} \to 0. \\
\Delta_2 &\le p\sqrt{n} \sum_{v=V_n+1}^{\lfloor n(1-p)+1 \rfloor} \frac{v}{n-v}(1-p)^{v-1} \\
&\le \frac{p\sqrt{n}[n(1-p)+1]}{np-1} \sum_{v=V_n+1}^{\lfloor n(1-p)+1 \rfloor} (1-p)^{v-1} \\
&\le \frac{\sqrt{n}[n(1-p)+1]}{np-1}(1-p)^{\lfloor n^{1/8} \rfloor} \\
&\le \frac{\sqrt{n}[n(1-c)+1]}{nc-1}(1-c)^{\lfloor n^{1/8} \rfloor} \to 0. \\
\Delta_3 &\le \frac{\sqrt{n}}{n(1-p)}(1-p)^{\lfloor n(1-p)+2 \rfloor} \\
&= \frac{1}{\sqrt{n}}(1-p)^{\lfloor n(1-p)+1 \rfloor} \\
&\le \frac{1}{\sqrt{n}} \to 0.
\end{aligned}
$$

Hence, $\sup_{p\in(c,1)} \sqrt{n}|h_n(p) - h(p)| \to 0.$ □

Proof of Part 2: The proof is identical to that of the above-mentioned Part 1 until the expression $\Delta_1 + \Delta_2 + \Delta_3$ where each term is to be evaluated on the interval $[1/n, c]$. It is clear that $\Delta_1 \le O(n^{-3/8})$. For Δ_2, since $n(1-p)+1$ at $p = 1/n$ is $n > n-1$, it follows that

$$
\begin{aligned}
\Delta_2 &\le p\sqrt{n} \sum_{v=V_n+1}^{\min\{n-1, \lfloor n(1-p)+1 \rfloor\}} \frac{v}{n-v}(1-p)^{v-1} \\
&\le p\sqrt{n} \sum_{v=V_n+1}^{\min\{n-1, \lfloor n(1-1/n)+1 \rfloor\}} \frac{v}{n-v}(1-p)^{v-1}
\end{aligned}
$$

$$
\begin{aligned}
&= p\sqrt{n} \sum_{v=V_n+1}^{n-1} \frac{v}{n-v}(1-p)^{v-1} \\
&< p\sqrt{n}(n-1) \sum_{v=V_n+1}^{n-1} (1-p)^{v-1} \\
&< \sqrt{n}(n-1)(1-p)^{V_n} < \sqrt{n}(n-1) = O(n^{3/2}).
\end{aligned}
$$

$$
\begin{aligned}
\Delta_3 &= p\sqrt{n} \sum_{v=\min\{n-1,\lfloor n(1-p)+1\rfloor\}+1}^{\infty} \frac{1}{v}(1-p)^v \\
&< p\sqrt{n} \sum_{v=1}^{\infty} \frac{1}{v}(1-p)^v < \sqrt{n} = O(n^{1/2}).
\end{aligned}
$$

Therefore, $\Delta_1 + \Delta_2 + \Delta_3 = O(n^{3/2})$. □

Proof of Part 3: Let Z and $\phi(z)$ be a standard normal random variable and its density function, respectively, and let "$\simeq$" denote asymptotic equality. Since $\sqrt{n}(\hat{p} - p) \xrightarrow{L} N(0, p(1-p))$,

$$
\begin{aligned}
P(\hat{p} \le c) &\simeq \int_{-\infty}^{\sqrt{n}(c-p)/\sqrt{p(1-p)}} \phi(z)dz \\
&= \int_{\sqrt{n}(p-c)/\sqrt{p(1-p)}}^{\infty} \phi(z)dz \\
&< \int_{\sqrt{n}(p-c)/\sqrt{p(1-p)}}^{\infty} e^{-z\,[\sqrt{n}(p-c)/\sqrt{p(1-p)}]}\, dz \\
&= \frac{\sqrt{p(1-p)}}{\sqrt{n}(p-c)} \int_{[\sqrt{n}(p-c)/\sqrt{p(1-p)}]^2}^{\infty} e^{-x} dx \\
&= \frac{\sqrt{p(1-p)}}{\sqrt{n}(p-c)} \exp\{-\,[\sqrt{n}(p-c)/\sqrt{p(1-p)}]^2\} \\
&= n^{-1/2} \frac{\sqrt{p(1-p)}}{(p-c)} \exp\left\{-\frac{n(p-c)^2}{p(1-p)}\right\}.
\end{aligned}
$$

□

3.4.2 Proof of Lemma 3.5

Proof: Let $m = v + 1$. By Lemmas 3.3 and 3.4, and identity

$$
\binom{n}{m}^{-1} \sum_{c=1}^{m} c \binom{m}{c} \binom{n-m}{m-c} = \frac{m^2}{n},
$$

$$
\mathrm{Var}(Z_{1,v}) \le \binom{n}{m}^{-1} \sum_{c=1}^{m} c \binom{m}{c} \binom{n-m}{m-c} \frac{\sigma_m^2}{m} = \frac{m}{n}\sigma_m^2. \tag{3.38}
$$

Consider

$$\begin{aligned}\sigma_m^2 &= \text{Var}(\psi(X_1,\ldots,X_m))\\ &= \text{E}(\psi(X_1,\ldots,X_m))^2 - \left[\sum_{k=1}^{K} p_k(1-p_k)^{m-1}\right]^2.\end{aligned}$$

Let $Y_k^{(m)}$ denote the frequency of the kth letter in the sample of size m.

$$\begin{aligned}\sigma_m^2 &\le \text{E}(\psi(X_1,\ldots,X_m))^2\\ &= \frac{1}{m^2}\text{E}\left[\left(\sum_{k=1}^{K} 1[Y_k=1]\right)\left(\sum_{k'=1}^{K} 1[Y_{k'}=1]\right)\right]\\ &= \frac{1}{m^2}\text{E}\left(\sum_{k=1}^{K} 1[Y_k=1] + 2\sum_{1\le k<k'\le K} 1[Y_k=1]1[Y_{k'}=1]\right)\\ &= \frac{1}{m}\sum_{k=1}^{K} p_k(1-p_k)^{m-1} + \frac{2(m-1)}{m}\sum_{1\le k<k'\le K} p_k p_{k'}(1-p_k-p_{k'})^{m-2}\\ &\le \frac{1}{m}\sum_{k=1}^{K} p_k(1-p_k)^{m-1} + 2\sum_{1\le k<k'\le K} p_k p_{k'}(1-p_k-p_{k'}+p_k p_{k'})^{m-2}\\ &= \frac{1}{m}\sum_{k=1}^{K} p_k(1-p_k)^{m-1} + 2\sum_{1\le k<k'\le K} [p_k(1-p_k)^{m-2}p_{k'}(1-p_{k'})^{m-2}]\\ &\le \frac{1}{m}\sum_{k=1}^{K} p_k(1-p_k)^{m-1} + \left[\sum_{k=1}^{K} p_k(1-p_k)^{m-2}\right]^2\\ &= \frac{1}{m}\,\zeta_{m-1} + \zeta_{m-2}^2.\end{aligned}$$

By (3.38), $\text{Var}(Z_{1,v}) \le \frac{1}{n}\,\zeta_v + \frac{v+1}{n}\,\zeta_{v-1}^2$. □

3.4.3 Proof of Corollary 3.5

Proof: Noting $\zeta_v \ge \zeta_w$ if $v \le w$,

$$\begin{aligned}M_n &= \frac{5}{n}\left(\sum_{v=1}^{n-1}\frac{1}{v}\,\zeta_{v-1}^{1/2}\right)^2\\ &= \frac{5}{n}\left(\sum_{v=1}^{n-1}\frac{1}{v}\,\zeta_{v-1}^{1/2}\right)\left(\sum_{w=1}^{n-1}\frac{1}{w}\,\zeta_{w-1}^{1/2}\right)\end{aligned}$$

$$
\begin{aligned}
&= \frac{5}{n}\left(\sum_{v=1}^{n-1}\frac{1}{v^2}\,\zeta_{v-1} + 2\sum_{1\le v<w\le n-1}\frac{1}{vw}\,\zeta_{1,v-1}^{1/2}\,\zeta_{1,w-1}^{1/2}\right)\\
&\le \frac{5}{n}\left(\sum_{v=1}^{n-1}\frac{1}{v^2}\,\zeta_{v-1} + 2\sum_{1\le v<w\le n-1}\frac{1}{vw}\,\zeta_{v-1}\right)\\
&= \frac{5}{n}\left(\sum_{w=1}^{n-1}\frac{1}{w}\sum_{v=1}^{n-1}\frac{1}{v}\,\zeta_{v-1}\right)\\
&= \frac{5}{n}\left(\sum_{w=1}^{n-1}\frac{1}{w}\right)\left(\sum_{v=1}^{n-1}\frac{1}{v}\,\zeta_{v-1}\right).
\end{aligned}
$$

The expression in the first pair of parentheses above is the harmonic series, which has a well-known upper bound $1 + \ln(n-1)$. Consider the expression in the second pair of parentheses.

$$
\begin{aligned}
\sum_{v-1}^{n-1}\frac{1}{v}\,\zeta_{v-1} &= 1 + \sum_{v=2}^{n-1}\frac{1}{v}\,\zeta_{v-1} < 1 + \sum_{v=2}^{n-1}\frac{1}{v-1}\,\zeta_{v-1}\\
&= 1 + \sum_{v=1}^{n}\frac{1}{v}\,\zeta_{v}\\
&< 1 + \sum_{v=1}^{\infty}\frac{1}{v}\,\zeta_{v}\\
&= 1 + H.
\end{aligned}
$$

Therefore,

$$
M_n < 5(1+H)\left(\frac{1+\ln(n-1)}{n}\right) = \mathcal{O}(\ln n/n).
$$

□

3.4.4 Proof of Lemma 3.14

Proof:

$$
\begin{aligned}
\sqrt{n}(\hat{H}_z^* - \hat{H}^*) &= \sqrt{n}\left(\hat{H}_z^* + \sum_{k=1}^{K(n)}\hat{p}_{k,n}\ln\hat{p}_{k,n}\right)\\
&= \left\{\sqrt{n}\left[\hat{H}_z^* - \sum_{k=1}^{K(n)}\sum_{v=1}^{n-Y_{k,n}}\frac{1}{v}\,\hat{p}_{k,n}(1-\hat{p}_{k,n})^v\right]\right\}\\
&\quad - \left\{\sqrt{n}\sum_{k=1}^{K(n)}\sum_{v=n-Y_{k,n}+1}^{\infty}\frac{1}{v}\,\hat{p}_{k,n}(1-\hat{p}_{k,n})^v\right\}\\
&=: \{\mathcal{A}_1\} - \{\mathcal{A}_2\}.
\end{aligned}
$$

Since

$$
\begin{aligned}
0 \le \mathcal{A}_2 &= \sqrt{n} \sum_{k=1}^{K(n)} \sum_{v=n-Y_{k,n}+1}^{\infty} \frac{1}{v}\, \hat{p}_{k,n}(1-\hat{p}_{k,n})^{v} \\
&\le \sum_{k=1}^{K(n)} \frac{\sqrt{n}}{n-Y_{k,n}+1} \hat{p}_{k,n} \sum_{v=n-Y_{k,n}+1}^{\infty} (1-\hat{p}_{k,n})^{v} \\
&= \sum_{k=1}^{K(n)} \frac{\sqrt{n}}{n-Y_{k,n}+1} (1-\hat{p}_{k,n})^{n-Y_{k,n}+1} \\
&\le \frac{1}{\sqrt{n}} \sum_{k=1}^{K(n)} \frac{1}{1-\hat{p}_{k,n}+1/n} (1-\hat{p}_{k,n})^{n-Y_{k,n}+1} \\
&= \frac{1}{\sqrt{n}} \sum_{k=1}^{K(n)} \frac{1}{1-\hat{p}_{k,n}+1/n} (1-\hat{p}_{k,n})^{n-Y_{k,n}+1} 1[\hat{p}_{k,n} < 1] \\
&\le \frac{1}{\sqrt{n}} \sum_{k=1}^{K(n)} \frac{1}{1-\hat{p}_{k,n}} (1-\hat{p}_{k,n})^{n-Y_{k,n}+1} 1[\hat{p}_{k,n} < 1] \\
&= \frac{1}{\sqrt{n}} \sum_{k=1}^{K(n)} (1-\hat{p}_{k,n})^{n-Y_{k,n}} 1[\hat{p}_{k,n} < 1] \le \frac{K(n)}{\sqrt{n}},
\end{aligned}
$$

$\mathcal{A}_2 \xrightarrow{a.s.} 0$ and therefore $\mathcal{A}_2 \xrightarrow{p} 0$.

Before considering $\mathcal{A}_1$, following are several facts that are to be noted. First

$$
\left(1 - \frac{Y_{k,n}-1}{n-j}\right) \ge \left(1 - \frac{Y_{k,n}}{n}\right) = (1-\hat{p}_{k,n})
$$

if and only if

$$
0 \le j \le \frac{n}{Y_{k,n} + 1[Y_{k,n}=0]} =: \frac{1}{\hat{p}^*_{k,n}}; \tag{3.39}
$$

and second, after a few algebraic steps, $Z^*_{1,v}$ may be expressed as

$$
\begin{aligned}
Z^*_{1,v} &= \sum_{k=1}^{K(n)} \hat{p}_{k,n} \prod_{j=1}^{v} \left(1 - \frac{Y_{k,n}-1}{n-j}\right) \\
&= \sum_{k=1}^{K(n)} \hat{p}_{k,n} \prod_{j=1}^{J_k \wedge v} \left(1 - \frac{Y_{k,n}-1}{n-j}\right) \prod_{j=J_k\wedge v+1}^{v} \left(1 - \frac{Y_{k,n}-1}{n-j}\right)
\end{aligned} \tag{3.40}
$$

where $J_k = \lfloor 1/\hat{p}^*_{k,n} \rfloor$ and $\prod_{v=a}^{b}(\cdot) = 1$ if $a > b$.

$$
\begin{aligned}
\mathcal{A}_1 &= \sqrt{n}\left[\hat{H}_z^* - \sum_{k=1}^{K(n)} \sum_{v=1}^{n-Y_{k,n}} \frac{1}{v}\, \hat{p}_{k,n}(1-\hat{p}_{k,n})^v\right] \\
&= \sqrt{n} \sum_{k=1}^{K(n)} \hat{p}_{k,n} \sum_{v=1}^{n-Y_{k,n}} \frac{1}{v} \left[\prod_{j=1}^{J_k \wedge v} \left(1 - \frac{Y_{k,n}-1}{n-j}\right) \prod_{j=J_k\wedge v+1}^{v} \left(1 - \frac{Y_{k,n}-1}{n-j}\right)\right. \\
&= \left\{ \sqrt{n} \sum_{k=1}^{K(n)} \hat{p}_{k,n} \sum_{v=1}^{n-Y_{k,n}} \frac{1}{v} \left[\prod_{j=1}^{J_k \wedge v} \left(1 - \frac{Y_{k,n}-1}{n-j}\right)\right.\right. \\
&\quad \times \prod_{j=J_k\wedge v+1}^{v} \left(1 - \frac{Y_{k,n}-1}{n-j}\right) \\
&\quad \left.\left. - \prod_{j=1}^{J_k \wedge v} \left(1 - \frac{Y_{k,n}-1}{n-j}\right) (1-\hat{p}_{k,n})^{0\vee(v-J_k)}\right]\right\} \\
&\quad + \left\{ \sqrt{n} \sum_{k=1}^{K(n)} \hat{p}_{k,n} \sum_{v=1}^{n-Y_{k,n}} \frac{1}{v} \left[\prod_{j=1}^{J_k \wedge v} \left(1 - \frac{Y_{k,n}-1}{n-j}\right) (1-\hat{p}_{k,n})^{0\vee(v-J_k)}\right.\right. \\
&=: \{\mathcal{A}_{1,1}\} + \{\mathcal{A}_{1,2}\}\,.
\end{aligned}
$$

By (3.39), $\mathcal{A}_{1,1} \leq 0$ and $\mathcal{A}_{1,2} \geq 0$. It suffices to show that $\mathrm{E}(\mathcal{A}_{1,1}) \to 0$ and $\mathrm{E}(\mathcal{A}_{1,2}) \to 0$, respectively.

$$
\begin{aligned}
\mathcal{A}_{1,1} &= \sqrt{n}\,(\hat{H}_z^* - H^*) \\
&\quad - \sqrt{n} \left[\sum_{k=1}^{K(n)} \hat{p}_{k,n} \sum_{v=1}^{n-Y_{k,n}} \frac{1}{v} \prod_{j=1}^{J_k\wedge v} \left(1 - \frac{Y_{k,n}-1}{n-j}\right)(1-\hat{p}_{k,n})^{0\vee(v-J_k)}\right. \\
&\quad \left. - H^* \right] \\
&= \left\{\sqrt{n}\left(\hat{H}_z^* - H^*\right)\right\} \\
&\quad - \left\{ \sqrt{n} \left[\sum_{k=1}^{K(n)} \hat{p}_{k,n} \sum_{v=1}^{n-Y_{k,n}} \frac{1}{v} \prod_{j=1}^{J_k\wedge v} \left(1 - \frac{Y_{k,n}-1}{n-j}\right)(1-\hat{p}_{k,n})^{0\vee(v-J_k)}\right.\right. \\
&\quad \left.\left. - \sum_{k=1}^{K(n)} \hat{p}_{k,n} \sum_{v=1}^{n-Y_{k,n}} \frac{1}{v}\,(1-\hat{p}_{k,n})^v \right]\right\} \\
&\quad - \left\{ \sqrt{n} \left[\sum_{k=1}^{K(n)} \hat{p}_{k,n} \sum_{v=1}^{n-Y_{k,n}} \frac{1}{v}\,(1-\hat{p}_{k,n})^v - H^*\right]\right\} \\
&=: \{\mathcal{A}_{1,1,1}\} - \{\mathcal{A}_{1,1,2}\} - \{\mathcal{A}_{1,1,3}\}\,.
\end{aligned}
$$

$E(\mathcal{A}_{1,1,1}) \to 0$ follows Lemma 3.11. Then,

$$\begin{aligned}
\mathcal{A}_{1,1,3} &= \sqrt{n}\left[\sum_{k=1}^{K(n)} \hat{p}_{k,n} \sum_{v=1}^{n-Y_{k,n}} \frac{1}{v}\,(1-\hat{p}_{k,n})^{v} - H^*\right] \\
&= \sqrt{n}\left[\sum_{k=1}^{K(n)} \hat{p}_{k,n} \sum_{v=1}^{\infty} \frac{1}{v}\,(1-\hat{p}_{k,n})^{v} - H^*\right] \\
&\quad - \sqrt{n}\sum_{k=1}^{K(n)} \hat{p}_{k,n} \sum_{v=n-Y_{k,n}+1}^{\infty} \frac{1}{v}\,(1-\hat{p}_{k,n})^{v} \\
&= \{\sqrt{n}\,(\hat{H}^* - H^*)\} - \left\{\sqrt{n}\sum_{k=1}^{K(n)} \hat{p}_{k,n} \sum_{v=n-Y_{k,n}+1}^{\infty} \frac{1}{v}\,(1-\hat{p}_{k,n})^{v}\right\} \\
&=: \{\mathcal{A}_{1,1,3,1}\} - \{\mathcal{A}_2\}\,.
\end{aligned}$$

$E(\mathcal{A}_{1,1,3,1}) \to 0$ by Lemma 3.12, $E(\mathcal{A}_2) \to 0$ is established above, and therefore $E(\mathcal{A}_{1,1,3}) \to 0$.

$$\begin{aligned}
\mathcal{A}_{1,1,2} &= \sqrt{n}\left[\sum_{k=1}^{K(n)} \hat{p}_{k,n} \sum_{v=1}^{n-Y_{k,n}} \frac{1}{v} \prod_{j=1}^{J_k \wedge v} \left(1 - \frac{Y_{k,n}-1}{n-j}\right) (1-\hat{p}_{k,n})^{0 \vee (v-J_k)} \right.\\
&\quad \left. - \sum_{k=1}^{K(n)} \hat{p}_{k,n} \sum_{v=1}^{n-Y_{k,n}} \frac{1}{v}\,(1-\hat{p}_{k,n})^{\,v}\right] \\
&\le \sqrt{n}\left[\sum_{k=1}^{K(n)} \hat{p}_{k,n} \sum_{v=1}^{n-Y_{k,n}} \frac{1}{v} \prod_{j=1}^{J_k \wedge v} \left(1 - \frac{Y_{k,n}-1}{n-1}\right) (1-\hat{p}_{k,n})^{0 \vee (v-J_k)} \right.\\
&\quad \left. - \sum_{k=1}^{K(n)} \hat{p}_{k,n} \sum_{v=1}^{n-Y_{k,n}} \frac{1}{v}\,(1-\hat{p}_{k,n})^{\,v}\right] \\
&= \sqrt{n}\left[\sum_{k=1}^{K(n)} \hat{p}_{k,n} \sum_{v=1}^{n-Y_{k,n}} \frac{1}{v} \left(1 - \frac{Y_{k,n}-1}{n-1}\right)^{J_k \wedge v} (1-\hat{p}_{k,n})^{0 \vee (v-J_k)} \right.\\
&\quad \left. - \sum_{k=1}^{K(n)} \hat{p}_{k,n} \sum_{v=1}^{n-Y_{k,n}} \frac{1}{v}\,(1-\hat{p}_{k,n})^{J_k \wedge v}(1-\hat{p}_{k,n})^{0 \vee (v-J_k)}\right] \\
&= \sqrt{n}\sum_{k=1}^{K(n)} \hat{p}_{k,n} \sum_{v=1}^{n-Y_{k,n}} \frac{1}{v}\left[\left(1 - \frac{Y_{k,n}-1}{n-1}\right)^{J_k \wedge v} - (1-\hat{p}_{k,n})^{J_k \wedge v}\right] \\
&\quad \times (1-\hat{p}_{k,n})^{0 \vee (v-J_k)}
\end{aligned}$$

$$\le \sqrt{n}\sum_{k=1}^{K(n)}\hat{p}_{k,n}\sum_{v=1}^{n-Y_{k,n}}\frac{1}{v}\left[\left(1-\frac{Y_{k,n}-1}{n-1}\right)^{J_k\wedge v}-(1-\hat{p}_{k,n})^{J_k\wedge v}\right]$$

$$\le \sqrt{n}\sum_{k=1}^{K(n)}\hat{p}_{k,n}\sum_{v=1}^{n-Y_{k,n}}\frac{1}{v}\left[(J_k\wedge v)\left(1-\frac{Y_{k,n}-1}{n-1}\right)^{J_k\wedge v-1}\frac{n-Y_{k,n}}{n(n-1)}\right]$$

(by Lemma 3.13)

$$\le \frac{\sqrt{n}}{n-1}\sum_{k=1}^{K(n)}\hat{p}_{k,n}(1-\hat{p}_{k,n})\left[\sum_{v=1}^{n-Y_{k,n}}\frac{1}{v}(J_k\wedge v)\right]$$

$$= \frac{\sqrt{n}}{n-1}\sum_{k=1}^{K(n)}\hat{p}_{k,n}(1-\hat{p}_{k,n})\left[\sum_{v=1}^{J_k}\frac{1}{v}(J_k\wedge v)+\sum_{v=J_k+1}^{n-Y_{k,n}}\frac{1}{v}(J_k\wedge v)\right]$$

$$= \frac{\sqrt{n}}{n-1}\sum_{k=1}^{K(n)}\hat{p}_{k,n}(1-\hat{p}_{k,n})\left(J_k+J_k\sum_{v=J_k+1}^{n-Y_{k,n}}\frac{1}{v}\right)$$

$$\le \frac{\sqrt{n}}{n-1}\sum_{k=1}^{K(n)}\hat{p}_{k,n}\left(J_k+J_k\sum_{v=1}^{n}\frac{1}{v}\right)$$

$$\le \frac{\sqrt{n}}{n-1}\sum_{k=1}^{K(n)}\frac{Y_{k,n}}{n}\;\frac{n}{Y_{k,n}+1[Y_{k,n}=0]}\;(\ln n+2)$$

$$\le \frac{\sqrt{n}K(n)(\ln n+2)}{n-1}.$$

Therefore,

$$E(\mathcal{A}_{1,1,2})\le \mathcal{O}\left(\frac{\sqrt{n}K(n)\ln n}{n}\right)\to 0.$$

Finally, $E(\mathcal{A}_{1,2})=E(\mathcal{A}_{1,1,2})\to 0$. It follows that $\sqrt{n}(\hat{H}_z^*-\hat{H}^*)=o_p(1)$. □

3.4.5 Proof of Lemma 3.18

Proof:

$$\begin{aligned}\hat{H}_{2z}^*-\hat{H}_2^* &= \hat{H}_{2z}^*-\sum_{k=1}^{K(n)}\hat{p}_{k,n}\ln^2\hat{p}_{k,n}\\ &= \left\{\hat{H}_{2z}^*-\sum_{k=1}^{K(n)}\sum_{v=1}^{n-Y_{k,n}}C_v\hat{p}_{k,n}(1-\hat{p}_{k,n})^v\right\}\\ &\quad-\left\{\sum_{k=1}^{K(n)}\sum_{v=n-Y_{k,n}+1}^{\infty}C_v\hat{p}_{k,n}(1-\hat{p}_{k,n})^v\right\}\\ &=: \{\mathcal{B}_1\}\;-\;\{\mathcal{B}_2\}\,.\end{aligned}$$

Since

$$
\begin{aligned}
0 \le \mathcal{B}_2 &= \sum_{k=1}^{K(n)} \sum_{v=n-Y_{k,n}+1}^{\infty} C_v \hat{p}_{k,n} (1-\hat{p}_{k,n})^v \\
&\le \sum_{k=1}^{K(n)} \frac{2(\ln n + 1)}{n - Y_{k,n} + 1}\, \hat{p}_{k,n} \sum_{v=n-Y_{k,n}+1}^{\infty} (1-\hat{p}_{k,n})^v \\
&= \sum_{k=1}^{K(n)} \frac{2(\ln n + 1)}{n - Y_{k,n} + 1} (1-\hat{p}_{k,n})^{n-Y_{k,n}+1} \\
&\le \frac{2(\ln n + 1)}{n} \sum_{k=1}^{K(n)} \frac{1}{1-\hat{p}_{k,n}+1/n} (1-\hat{p}_{k,n})^{n-Y_{k,n}+1} \\
&= \frac{2(\ln n + 1)}{n} \sum_{k=1}^{K(n)} \frac{1}{1-\hat{p}_{k,n}+1/n} (1-\hat{p}_{k,n})^{n-Y_{k,n}+1} 1[\hat{p}_{k,n} < 1] \\
&\le \frac{2(\ln n + 1)}{n} \sum_{k=1}^{K(n)} \frac{1}{1-\hat{p}_{k,n}} (1-\hat{p}_{k,n})^{n-Y_{k,n}+1} 1[\hat{p}_{k,n} < 1] \\
&= \frac{2(\ln n + 1)}{n} \sum_{k=1}^{K(n)} (1-\hat{p}_{k,n})^{n-Y_{k,n}} 1[\hat{p}_{k,n} < 1] \\
&\le \frac{2(\ln n + 1)K(n)}{n},
\end{aligned}
$$

$\mathcal{B}_2 \xrightarrow{a.s.} 0$ and therefore $\mathcal{B}_2 \xrightarrow{p} 0$.

Next,

$$
\begin{aligned}
\mathcal{B}_1 &= \hat{H}_{2z}^* - \sum_{k=1}^{K(n)} \sum_{v=1}^{n-Y_{k,n}} C_v \hat{p}_{k,n} (1-\hat{p}_{k,n})^v \\
&= \sum_{k=1}^{K(n)} \hat{p}_{k,n} \sum_{v=1}^{n-Y_{k,n}} C_v \left[\prod_{j=1}^{J_k \wedge v} \left(1 - \frac{Y_{k,n}-1}{n-j}\right) \prod_{j=J_k\wedge v+1}^{v} \left(1 - \frac{Y_{k,n}-1}{n-j}\right) - (1-\hat{p}_{k,n})^v \right] \\
&= \left\{ \sum_{k=1}^{K(n)} \hat{p}_{k,n} \sum_{v=1}^{n-Y_{k,n}} C_v \left[\prod_{j=1}^{J_k \wedge v} \left(1 - \frac{Y_{k,n}-1}{n-j}\right) \prod_{j=J_k\wedge v+1}^{v} \left(1 - \frac{Y_{k,n}-1}{n-j}\right) \right.\right. \\
&\qquad \left.\left. - \prod_{j=1}^{J_k \wedge v} \left(1 - \frac{Y_{k,n}-1}{n-j}\right) (1-\hat{p}_{k,n})^{0 \vee (v-J_k)} \right] \right\} \\
&\quad + \left\{ \sum_{k=1}^{K(n)} \hat{p}_{k,n} \sum_{v=1}^{n-Y_{k,n}} C_v \left[\prod_{j=1}^{J_k \wedge v} \left(1 - \frac{Y_{k,n}-1}{n-j}\right) (1-\hat{p}_{k,n})^{0 \vee (v-J_k)} (1-\hat{p}_{k,n})^v \right] \right\} \\
&=: \{\mathcal{B}_{1,1}\} + \{\mathcal{B}_{1,2}\}\,.
\end{aligned}
$$

By (3.39), $\mathcal{B}_{1,1} \leq 0$ and $\mathcal{B}_{1,2} \geq 0$. It suffices to show that $E(\mathcal{B}_{1,1}) \to 0$ and $E(\mathcal{B}_{1,2}) \to 0$, respectively.

$$\begin{aligned}
\mathcal{B}_{1,1} &= (\hat{H}^*_{2z} - H^*_2) \\
&\quad - \left[\sum_{k=1}^{K(n)} \hat{p}_{k,n} \sum_{v=1}^{n-Y_{k,n}} C_v \prod_{j=1}^{J_k \wedge v} \left(1 - \frac{Y_{k,n}-1}{n-j}\right) (1-\hat{p}_{k,n})^{0 \vee (v-J_k)} \right. \\
&\quad \left. - H^*_2 \right] \\
&= \left\{ \hat{H}^*_{2z} - H^*_2 \right\} \\
&\quad - \left\{ \sum_{k=1}^{K(n)} \hat{p}_{k,n} \sum_{v=1}^{n-Y_{k,n}} C_v \prod_{j=1}^{J_k \wedge v} \left(1 - \frac{Y_{k,n}-1}{n-j}\right) (1-\hat{p}_{k,n})^{0 \vee (v-J_k)} \right. \\
&\quad \left. - \sum_{k=1}^{K(n)} \hat{p}_{k,n} \sum_{v=1}^{n-Y_{k,n}} C_v (1-\hat{p}_{k,n})^v \right\} \\
&\quad - \left\{ \sum_{k=1}^{K(n)} \hat{p}_{k,n} \sum_{v=1}^{n-Y_{k,n}} C_v (1-\hat{p}_{k,n})^v - H^*_2 \right\} \\
&=: \{\mathcal{B}_{1,1,1}\} - \{\mathcal{B}_{1,1,2}\} - \{\mathcal{B}_{1,1,3}\} .
\end{aligned}$$

$E(\mathcal{B}_{1,1,1}) \to 0$ follows Lemma 3.17.

Next,

$$\begin{aligned}
\mathcal{B}_{1,1,3} &= \sum_{k=1}^{K(n)} \hat{p}_{k,n} \sum_{v=1}^{n-Y_{k,n}} C_v (1-\hat{p}_{k,n})^v - H^*_2 \\
&= \left[\sum_{k=1}^{K(n)} \hat{p}_{k,n} \sum_{v=1}^{\infty} C_v (1-\hat{p}_{k,n})^v - H^*_2 \right] \\
&\quad - \sum_{k=1}^{K(n)} \hat{p}_{k,n} \sum_{v=n-Y_{k,n}+1}^{\infty} C_v (1-\hat{p}_{k,n})^v \\
&= \{\hat{H}^*_2 - H^*_2\} - \left\{ \sum_{k=1}^{K(n)} \hat{p}_{k,n} \sum_{v=n-Y_{k,n}+1}^{\infty} C_v (1-\hat{p}_{k,n})^v \right\} \\
&=: \{\mathcal{B}_{1,1,3,1}\} - \{\mathcal{B}_2\} .
\end{aligned}$$

The proof of $E(\mathcal{B}_{1,1,3,1}) \to 0$ is implied by the proof of Corollary 1 of Zhang and Zhang (2012). Also, $E(\mathcal{B}_2) \to 0$ is established above, and therefore, $E(\mathcal{B}_{1,1,3}) \to 0$.

Next,

$$\begin{aligned}
\mathcal{B}_{1,1,2} &= \sum_{k=1}^{K(n)} \hat{p}_{k,n} \sum_{v=1}^{n-Y_{k,n}} C_v \prod_{j=1}^{J_k\wedge v} \left(1 - \frac{Y_{k,n}-1}{n-j}\right) (1-\hat{p}_{k,n})^{0\vee(v-J_k)} \\
&\quad - \sum_{k=1}^{K(n)} \hat{p}_{k,n} \sum_{v=1}^{n-Y_{k,n}} C_v\, (1-\hat{p}_{k,n})^{v} \\
&\le \sum_{k=1}^{K(n)} \hat{p}_{k,n} \sum_{v=1}^{n-Y_{k,n}} C_v \prod_{j=1}^{J_k\wedge v} \left(1 - \frac{Y_{k,n}-1}{n-1}\right) (1-\hat{p}_{k,n})^{0\vee(v-J_k)} \\
&\quad - \sum_{k=1}^{K(n)} \hat{p}_{k,n} \sum_{v=1}^{n-Y_{k,n}} C_v\, (1-\hat{p}_{k,n})^{v} \\
&= \left[\sum_{k=1}^{K(n)} \hat{p}_{k,n} \sum_{v=1}^{n-Y_{k,n}} C_v \left(1 - \frac{Y_{k,n}-1}{n-1}\right)^{J_k\wedge v} (1-\hat{p}_{k,n})^{0\vee(v-J_k)}\right. \\
&\quad \left. - \sum_{k=1}^{K(n)} \hat{p}_{k,n} \sum_{v=1}^{n-Y_{k,n}} C_v\, (1-\hat{p}_{k,n})^{J_k\wedge v}(1-\hat{p}_{k,n})^{0\vee(v-J_k)}\right] \\
&= \sum_{k=1}^{K(n)} \hat{p}_{k,n} \sum_{v=1}^{n-Y_{k,n}} C_v \left[\left(1 - \frac{Y_{k,n}-1}{n-1}\right)^{J_k\wedge v}\right. \\
&\quad \left. - (1-\hat{p}_{k,n})^{J_k\wedge v}\right] (1-\hat{p}_{k,n})^{0\vee(v-J_k)} \\
&\le \sum_{k=1}^{K(n)} \hat{p}_{k,n} \sum_{v=1}^{n-Y_{k,n}} C_v \left[\left(1 - \frac{Y_{k,n}-1}{n-1}\right)^{J_k\wedge v} - (1-\hat{p}_{k,n})^{J_k\wedge v}\right] \\
&\le \sum_{k=1}^{K(n)} \hat{p}_{k,n} \sum_{v=1}^{n-Y_{k,n}} C_v \left[(J_k\wedge v) \left(1 - \frac{Y_{k,n}-1}{n-1}\right)^{J_k\wedge v-1} \frac{n-Y_{k,n}}{n(n-1)}\right] \\
&\quad \text{(by Lemma 3.13)} \\
&\le \frac{1}{n-1} \sum_{k=1}^{K(n)} \hat{p}_{k,n}(1-\hat{p}_{k,n}) \left[\sum_{v=1}^{n-Y_{k,n}} C_v (J_k\wedge v)\right] \\
&\le \frac{1}{n-1} \sum_{k=1}^{K(n)} \hat{p}_{k,n}(1-\hat{p}_{k,n}) J_k \sum_{v=1}^{n-Y_{k,n}} C_v \\
&\le \frac{1}{n-1} \sum_{k=1}^{K(n)} \hat{p}_{k,n}(1-\hat{p}_{k,n}) J_k \sum_{v=1}^{n} \frac{2(\ln n+1)}{v} \\
&\le \frac{2(\ln n+1)}{n-1} \sum_{k=1}^{K(n)} \hat{p}_{k,n} J_k (\ln n+1)
\end{aligned}$$

$$\le \frac{2(\ln n + 1)^2}{n-1} \sum_{k=1}^{K(n)} \frac{Y_{k,n}}{n} \frac{n}{Y_{k,n} + 1[Y_{k,n} = 0]}$$
$$\le \frac{2(\ln n + 1)^2 K(n)}{n-1}.$$

Therefore,

$$\mathrm{E}(\mathcal{B}_{1,1,2}) \le \mathcal{O}\left(\frac{K(n)\ln^2 n}{n}\right) \to 0.$$

Finally, $\mathrm{E}(\mathcal{B}_{1,2}) = \mathrm{E}(\mathcal{B}_{1,1,2}) \to 0$. It follows that $\hat{H}_{2z}^* - \hat{H}_2^* = o_p(1)$. □

3.5 Remarks

The existing literature on entropy estimation seems to suggest a heavy focus on bias reduction. Although bias reduction is undoubtedly important, an argument could be made to support a shift in research focus to other worthy directions. For example, the distributional characteristics of proposed estimators are, with a few exceptions, largely ignored in the literature. The lack of results in weak convergence of proposed entropy estimators is perhaps partially due to the inherent difficulty in establishing such results in theory. Nevertheless, one must have such results in order to make reasonable statistical inference about the underlying entropy.

As evidenced by each and every one of the many normal laws introduced in this chapter, the asymptotic variance always contains the second entropic moment $H_2 = \sum_{k\ge1} p_k \ln^2 p_k$. To make inference about the underlying entropy H, one must have not only a good estimator for H but also a good estimator for H_2. Future research on nonparametric estimation of H_2 would be beneficial.

Another issue needing more investigation is the sample size consideration in estimating entropy. How large should sample size n be in order to produce a reasonably reliable estimate of H? Answers to this question would be interesting both in theory and in practice. Many papers in the existing literature discuss bias reduction in cases of small samples with sizes as small as $n = 10$ or $n = 20$. Given the fact that entropy is meant to be a measure of chaos in a complex system, it would seem quite forced to estimate entropy based on such small samples. In the existing literature, there does not seem to exist a guideline on how large a sample must be to make entropy estimation realistically meaningful. On the other hand, there is a natural perspective based on Turing's formula. Simply put, Turing's formula T_1 estimates the sample noncoverage of the alphabet, and therefore $1 - T_1$ estimates the sample coverage of the alphabet. Given a sample data set, if $1 - T_1$ is small, then perhaps one should not force an estimate of entropy. For example, if a data set contains only singletons, that is, $N_1 = n$, then $1 - T_1 = 0$, suggesting that the sample is perhaps not large enough to have any meaningful coverage.

3.6 Exercises

1 Let X be the number of heads observed when a fair coin is tossed four times. Find the entropy of X.

2 A fair coin is repeatedly tossed until the first head is observed. Let X be the number of tosses required to observe the first head. Find the entropy of X.

3 Verify the bias (3.3) for the plug-in estimator $\hat{H}$ in (3.2).

4 Verify the bias (3.6) for the jackknife estimator $\hat{H}_{JK}$ in (3.5).

5 Show the following.
a) The Dirichlet prior of (3.7) leads to the Dirichlet posterior of (3.8).
b) For each k, the mean of Dirichlet posterior of (3.8) is

$$E(p_k|\mathbf{y}) = \frac{y_k + \alpha_k}{n + \sum_{k=1}^{K} \alpha_k}.$$

6 Show that if $\{p_k\}$ is uniformly distributed, that is, $p_k = 1/K$ for each $k = 1, \dots, K$, then σ^2 (3.11) is zero.

7 Prove Theorem 3.2.

8 Prove Theorem 3.3.

9 Verify that $H < \infty$ under the distribution $\{p_k; k \geq 1\}$ where $p_k = c_\lambda k^{-\lambda}$ and λ is any real value satisfying $\lambda > 1$.

10 In Example 3.4, show

$$\sqrt{n} \int_{K(n)}^{\infty} \frac{C_\lambda}{x^\lambda} \ln\left(\frac{C_\lambda}{x^\lambda}\right) dx = \frac{C_\lambda \lambda}{\lambda - 1} \frac{\sqrt{n} \ln K(n)}{(K(n))^{\lambda-1}} + \left[\frac{C_\lambda \lambda}{(\lambda-1)^2} - \frac{C_\lambda \ln C_\lambda}{\lambda - 1}\right] \frac{\sqrt{n}}{(K(n))^{\lambda-1}}.$$

11 Let a_j, $j = 1, \dots, n$, be real numbers satisfying $|a_j| \leq 1$ for every j. Show that $\left|\prod_{j=1}^{n} a_j - 1\right| \leq \sum_{j=1}^{n} |a_j - 1|$.

12 For any given distribution $\{p_k\}$ with finite K, $|\mathrm{E}(\hat{H}_{z2} - H_2)|$ decays exponentially in n.

13 Suppose $p_k = Ck^{-\lambda}$, where $C > 0$ and $\lambda > 1$ are constants, for all $k \geq k_0$ where k_0 is some positive integer. Show that

$$\lim_{n\to\infty} n^{\delta} \sum_k p_k(1-p_k)^n,$$

where $\delta = 1 - 1/\lambda$, exists. (Hint: Use Euler–Maclaurin lemma.)

14 Show that

$$C_\nu = \sum_{i=1}^{\nu-1} \frac{1}{i(\nu-i)} = \frac{1}{\nu}\sum_{i=1}^{\nu-1}\left(\frac{1}{i} + \frac{1}{\nu-i}\right) \leq \frac{2(\ln \nu + 1)}{\nu}.$$

15 Show
a) $\lim_{p\to 0} p \ln p = 0$; and
b) $\lim_{p\to 0} p\ln^2 p = 0$.

16 Show that for any $p \in (0,1)$ and any nonnegative integer j less or equal to $n(1-p)$,

$$0 \leq \frac{1 - j/[n(1-p)]}{1 - j/n} \leq 1.$$

17 Show that $Z_{1,\nu}$ in (3.26) may be re-expressed as

$$Z_{1,\nu} = \sum_{k\geq 1}\left[\hat{p}_k \prod_{j=1}^{\nu}\left(1 - \frac{Y_k - 1}{n-j}\right)\right].$$

18 Show that

$$\sum_{k\geq 1} \frac{\ln\ln(k+1)}{(k+1)\ln^2(k+1)} < \infty.$$

19 Let X be a discrete random variables defined on the set of natural numbers $\mathbb{N}$, with probability distribution $\{p_x; x \in \mathbb{N}\}$. Show that if $\mathrm{E}(X^\lambda) < \infty$ for some constant $\lambda > 0$, then the entropy of X exists, that is, $H(X) = -\sum_{x=1}^{\infty} p_x \ln p_x < \infty$.

20 Let X_1 and X_2 be two random elements on the same alphabet $\mathscr{X} = \{\ell_1, \ell_2\}$ with probability distributions $\mathbf{p}_1 = \{p_1, 1-p_1\}$ and $\mathbf{p}_2 = \{p_2, 1-p_2\}$, respectively. Let θ be a bivariate random variable, independent of X_1 and X_2, such that $\mathrm{P}(\theta = 1) = q$ and $\mathrm{P}(\theta = 2) = 1 - q$. Let

$$X_\theta = X_1 1[\theta = 1] + X_2 1[\theta = 2].$$

Show that

$$H(X_\theta) \geq qH(X_1) + (1 - q)H(X_2),$$

with equality if and only if $H(\theta) = 0$.

4

Estimation of Diversity Indices

Diversity is a general concept pertaining to the nature of assortment in a population with multiple species. Originated in ecology where the diversity of species in an ecosystem is of great interest, the concept of diversity has become increasingly relevant in many other fields of study in modern sciences, for example, of genetic diversity within a biological species, of word diversity of an author, of diversity of an investment portfolio, and so on. More generally, in information science, one is interested in the diversity of letters of some alphabet. While the meaning of the word, diversity, is quite clear, it is not always obvious how it may be committed to a quantitative measure. As a central theme in ecology, a large volume of research can be found on how a diversity index should be defined and how it could be estimated. Proposed diversity indices and their estimators are quite numerous. Shannon's entropy introduced in Shannon (1948) and Simpson's index introduced in Simpson (1949) are among the earliest diversity indices found in the literature. A set of popular diversity indices include Emlen's index, the Gini – Simpson index, Hill's diversity number, Rényi's entropy, and Tsallis entropy. These are, respectively, named after the authors of Emlen (1973), Gini (1912), Hill (1973), Rényi's (1961), and Tsallis (1988). For a comprehensive and in-depth introduction of the subject, readers may wish to refer to Krebs (1999), Magurran (2004), and Marcon (2014). In summary, while the ideas of what constitute diversity indices and how to estimate them are quite diverse, the implied consensus in the current literature seems to be that diversity is a multifaceted concept and therefore no single mathematical index should be expected to capture it entirely.

This chapter describes a unified perspective for all of the above-mentioned diversity indices based on a re-parameterization and a general nonparametric estimation procedure for these indices.

Statistical Implications of Turing's Formula, First Edition. Zhiyi Zhang.

4.1 A Unified Perspective on Diversity Indices

Consider a population with countably many species, $\mathscr{X} = \{\ell_k; k \geq 1\}$, where each letter ℓ_k stands for one distinct species. Let $p_k = P(X = \ell_k)$ and $\mathbf{p} = \{p_k; k \geq 1\}$ where X is a random element drawn from the population. Denote the cardinality of $\mathscr{X}$ by $K = \sum_{k\geq 1} 1[p_k > 0]$.

Let a general index of diversity be defined as a function $\theta = \theta(\mathscr{X}, \mathbf{p})$, which assigns a real value to every given pair $\{\mathscr{X}, \mathbf{p}\}$. Axiom 4.1 is commonly accepted as minimal constraints on a diversity index θ:

Axiom 4.1 *A diversity index $\theta = \theta(\mathscr{X}, \mathbf{p})$ satisfies*

$\mathcal{A}_{01}$: $\theta = \theta(\mathscr{X}, \mathbf{p}) = \theta(\mathbf{p})$ *and*

$\mathcal{A}_{02}$: $\theta(\mathbf{p}) = \theta(\mathbf{p}^*)$,

where $\mathbf{p}^*$ is any rearrangement of $\mathbf{p}$.

Axiom $\mathcal{A}_{01}$ implies that the value of a diversity index only depends on the probability distribution on the alphabet and not on the alphabet $\mathscr{X}$ itself. In other words, a diversity index is determined only by the species proportions in the population regardless of what the species are.

Axiom A_{02} implies that a diversity index θ does not alter its value if any two indices, k_1 and k_2, exchange their integer values, that is to say, θ assumes a value regardless of the order in which the species are arranged in the population. Axiom A_{02} was first formally introduced by Rényi (1961) as the symmetry axiom, and it was also known as the permutation invariant property of diversity indices.

Many more reasonable axiomatic conditions have been discussed in the diversity literature, but Axion 4.1 encircles the realm of discussion in this chapter. The following is a list of several commonly used diversity indices satisfying Axion 4.1.

Example 4.1 *Simpson's index* $\lambda = \sum_{k\geq 1} p_k^2$.

Example 4.2 *The Gini–Simpson index* $1 - \lambda = \sum_{k\geq 1} p_k(1 - p_k)$.

The Gini–Simpson index is one of the earliest diversity indices discussed in the literature. It is popular for many reasons. In addition to its simplicity, it has a probabilistic interpretation with regard to the intuitive notion of diversity. Consider two independently and identically distributed random elements from $\mathscr{X}$, say X_1 and X_2, under a probability distribution, $\{p_k; k \geq 1\}$. A more (or less) diverse population could be partially characterized by a higher (or a lower) value of $P(X_1 \neq X_2)$, that is, the probability that X_1 and X_2 are not equal. However,

$$P(X_1 \neq X_2) = 1 - P(X_1 = X_2) = 1 - \sum_{k\geq 1} P(X_1 = X_2 = \ell_k)$$
$$= 1 - \sum_{k\geq 1} p_k^2 = 1 - \lambda.$$

Example 4.3 *Shannon's entropy* $H = -\sum_{k\geq 1} p_k \ln p_k$.

Example 4.4 *Rényi's entropy* $H_\alpha = (1-\alpha)^{-1} \ln\left(\sum_{k\geq 1} p_k^\alpha\right)$ *for any* $\alpha > 0$, $\alpha \neq 1$.

Example 4.5 *Tsallis' entropy* $T_\alpha = (1-\alpha)^{-1}\left(\sum_{k\geq 1} p_k^\alpha - 1\right)$ *for any* $\alpha > 0$, $\alpha \neq 1$.

Example 4.6 *Hill's diversity number* $N_\alpha = \left(\sum_{k\geq 1} p_k^\alpha\right)^{1/(1-\alpha)}$ *for any* $\alpha > 0$, $\alpha \neq 1$.

Example 4.7 *Emlen's index* $D = \sum_k p_k e^{-p_k}$.

Example 4.8 *The richness index* $K = \sum_{k\geq 1} 1[p_k > 0]$.

To better facilitate an investigation into the commonalities of these diversity indices, a notion of equivalence between two diversity indices was introduced by Zhang and Grabchak (2016) and is given in Definition 4.1 below.

Definition 4.1 *Two diversity indices,* θ_1 *and* θ_2*, are said to be equivalent if there exists a strictly increasing continuous function* $g(\cdot)$ *such that* $\theta_1 = g(\theta_2)$*. The equivalence is denoted by* $\theta_1 \Leftrightarrow \theta_2$.

Consider two populations, P_A and P_B, in diversity comparison. Suppose P_A is more diverse than P_B by means of an index θ_2, that is, $\theta_2(P_A) > \theta_2(P_B)$. Then since $g(\cdot)$ is strictly increasing,

$$\theta_1(P_A) = g(\theta_2(P_A)) > g(\theta_2(P_B)) = \theta_1(P_B).$$

The reverse of the above-mentioned argument is also true because the inverse function of g, g^{-1}, exists and it is also strictly increasing (see Exercise 3). This argument suggests that the equivalence of Definition 4.1 is an order-preserving relationship. That said however, it is to be noted that this relationship does not preserve the incremental difference in diversity between P_A and P_B, that is, it is not guaranteed that $\theta_1(P_A) - \theta_1(P_B) = \theta_2(P_A) - \theta_2(P_B)$.

Example 4.9 *Rényi's entropy H_α, Tsallis' entropy T_α, and Hill's diversity number N_α are equivalent to each other for $\alpha \in (0,1)$. To verify this claim, it suffices (why?) to show that all three indices are equivalent to a common core index*

$$h_\alpha = \sum_{k\geq 1} p_k^\alpha, \tag{4.1}$$

where α is a constant satisfying $\alpha \in (0,1)$. Toward that end, consider the following three continuous functions as the function $g(x)$ in Definition 4.1 for Rényi's, Tsallis,' and Hill's indices, respectively,

$$g_1(x) = \frac{\ln x}{(1-\alpha)}, \quad g_2(x) = \frac{x-1}{(1-\alpha)}, \quad and \quad g_3(x) = x^{1/(1-\alpha)},$$

where α is a fixed constant in $(0,1)$. By the fact that each of these functions is continuous and strictly increasing for $x \in (0,\infty)$ and by Definition 4.1, the desired result follows.

Recall, in Chapter 2, a special subclass of the generalized Simpson's indices is denoted as

$$\zeta_1 = \left\{ \zeta_v = \sum_{k\geq 1} p_k(1-p_k)^v; v \geq 0 \right\}. \tag{4.2}$$

Also recall, in Chapter 3, a finite Shannon's entropy has a representation

$$H = \sum_{v\geq 1} \frac{1}{v}\,\zeta_v,$$

which is a linear combination of all the elements in $\boldsymbol{\zeta}_1$. Finally, recall Theorem 2.2, which states that a distribution $\{p_k; k \geq 1\}$ on $\mathscr{X}$ and its associated panel of generalized Simpson's indices $\{\zeta_v; v \geq 1\}$ uniquely determine each other up to a permutation on the index set $\{k; k \geq\}$.

In light of the above-mentioned recalled facts, $\boldsymbol{\zeta} := \boldsymbol{\zeta}_1 = \{\zeta_v; v \geq 1\}$ deserves a name for future reference.

Definition 4.2 *For every probability sequence $\{p_k; k \geq 1\}$ on $\mathscr{X}$, $\boldsymbol{\zeta} = \{\zeta_v; v \geq 1\}$ is said to be the entropic basis of $\mathbf{p} = \{p_k; k \geq 1\}$.*

The following theorem provides a unifying perspective on all diversity indices satisfying Axioms 4.1.

Theorem 4.1 *Given an alphabet $\mathscr{X} = \{\ell_k; k \geq 1\}$ and an associated probability distribution $\mathbf{p} = \{p_k; k \geq 1\}$, a diversity index θ satisfying Axioms 4.1 is a function of the entropic basis $\zeta = \{\zeta_v; v \geq 0\}$.*

Proof: By Theorem 2.2, $\boldsymbol{\zeta}$ determines $\mathbf{p}$ up to a permutation and, in turn, determines any diversity index θ satisfying Axioms 4.1. □

The statement of Theorem 4.1 holds for any probability distribution $\mathbf{p}$ regardless of whether K is finite or infinite. Theorem 4.1 essentially offers a re-parameterization of $\mathbf{p}$ (up to a permutation) in terms of ζ. This re-parameterization is not just an arbitrary one, it has several statistical implications. First of all, every element of ζ contains information about the entire distribution and not just one frequency p_k. This helps to deal with the problem of estimating probabilities of unobserved species. Second, for a random sample of size n, there are good estimators of ζ_v for $v = 1, 2, \ldots, n-1$. These estimators are introduced in Chapter 2 and are briefly recalled in the following section.

While a general diversity index can be any function of the entropic basis, most commonly used diversity indices in practice are either linear functions or equivalent to linear functions of the entropic basis.

Definition 4.3 *A diversity index θ is said to be a linear diversity index if it is a linear combination of the members of the entropic basis, that is,*

$$\theta = \theta(\mathbf{p}) = \sum_{v=0}^{\infty} w_v \zeta_v = \sum_{v=0}^{\infty} w_v \sum_{k\geq 1} p_k(1-p_k)^v \tag{4.3}$$

where w_v is a function of v such that $\sum_{v=0}^{n} w_v \zeta_v < \infty$.

Definition 4.3 identifies a subclass of indices among all functions of ζ, that is, all diversity indices satisfying Axioms 4.1. Every member of this subclass is a weighted linear form of $\{\zeta_v\}$. While there are no fundamental reasons why a search of a good diversity index should be restricted to this subclass, it happens to cover all of the popular indices in the literature, up to the equivalence relationship given in Definition 4.1. The following examples demonstrate the linearity of several popular diversity indices.

Example 4.10 *Simpson's index.*

$$\lambda = \sum_{k\geq 1} p_k^2 = \zeta_0 - \zeta_1.$$

(See Exercise 6.)

Example 4.11 *The Gini–Simpson index.*

$$1 - \lambda = \sum_{k\geq 1} p_k(1-p_k) = \zeta_1.$$

Example 4.12 *Shannon's entropy.*

$$H = -\sum_{k\geq 1} p_k \ln p_k = \sum_{v=1}^{\infty} \frac{1}{v}\, \zeta_v,$$

as shown in Chapter 3.

Example 4.13 *The index in (4.1),*

$$h_\alpha = \sum_{k\geq 1} p_k^\alpha = \zeta_0 + \sum_{v=1}^{\infty} \left[\prod_{i=1}^{v} \left(\frac{i-\alpha}{i} \right) \zeta_v \right],$$

where $\alpha > 0$ and $\alpha \neq 1$. (See Exercise 7.)

Example 4.14 *Emlen's index.*

$$D = \sum_{k\geq 1} p_k e^{-p_k} = \sum_{v=0}^{\infty} \frac{e^{-1}}{v!} \zeta_v$$

(see Exercise 8).

Example 4.15 *The richness index.*

$$\begin{aligned} K &= \sum_{k\geq 1} 1[p_k > 0] \\ &= \sum_{k\geq 1} 1[p_k > 0] \frac{p_k}{1-(1-p_k)} \\ &= \sum_{k\geq 1} 1[p_k > 0] p_k \sum_{v=0}^{\infty} (1-p_k)^v \\ &= \sum_{v=0}^{\infty} \sum_{k\geq 1} 1[p_k > 0] p_k (1-p_k)^v = \sum_{v=0}^{\infty} \zeta_v. \end{aligned}$$

The richness indices, K, and the evenness indices, for example, Gini–Simpson's $1 - \lambda$, are generally thought as two qualitatively very different types of indices. Many arguments for such a difference can be found in the existing literature (see, for example, Peet (1974); Heip, Herman, and Soetaert (1998), and Purvis and Hector (2000)). However, Example 4.15 demonstrates that, in the perspective of entropic basis, they both are linear diversity indices and merely differ in the weighting scheme w_v in Definition 4.3.

Example 4.16 *The generalized Simpson's indices.*

$$\begin{aligned} \zeta_{u,m} &= \sum_{k\geq 1} p_k^u (1-p_k)^m \\ &= \sum_{k\geq 1} p_k [1-(1-p_k)]^{u-1} (1-p_k)^m \\ &= \sum_{k\geq 1} p_k \sum_{v=0}^{u-1} (-1)^{u-1-v} \binom{u-1}{v} (1-p_k)^{u-1-v} (1-p_k)^m \\ &= \sum_{v=0}^{u-1} (-1)^v \binom{u-1}{v} \zeta_{m+v}. \end{aligned}$$

Linear diversity indices may also be derived from a general form of

$$\theta = \sum_{k\geq 1} p_k h(p_k) \tag{4.4}$$

where $h(p)$ is an analytic function on an open interval containing $(0, 1]$. More specifically, (4.4) may be re-expressed as, provided that $0 \leq \theta < \infty$,

$$\begin{aligned}\theta &= \sum_{k\geq 1}\sum_{v\geq 0} \frac{h^{(v)}(1)}{v!} p_k (p_k - 1)^v \\ &= \sum_{k\geq 1}\sum_{v\geq 0} \frac{(-1)^v h^{(v)}(1)}{v!} p_k (1 - p_k)^v \\ &= \sum_{v\geq 0} \frac{(-1)^v h^{(v)}(1)}{v!} \zeta_v \end{aligned} \tag{4.5}$$

where $h^{(v)}(1)$ is the vth derivative of $h(p)$ evaluated at $p = 1$. The exchange of the two summations in the above-mentioned expression is supported by Fubini's lemma, assuming θ is finite. (4.5) conforms with (4.3).

4.2 Estimation of Linear Diversity Indices

In this section, two general approaches to nonparametric statistical estimation of linear diversity indices are described. Let $X_1, X_2, \dots, X_n$ be independent and identically distributed (*iid*) random observations from $\mathcal{X}$ according to $\mathbf{p}$. Let $\{Y_k = \sum_{i=1}^n 1[X_i = \ell_k]\}$ be the sequence of observed counts in an *iid* sample of size n and let $\hat{\mathbf{p}} = \{\hat{p}_k = Y_k/n\}$ be the observed sample proportions. Suppose it is of interest to estimate θ of (4.3) where $\{w_v; v \geq 0\}$ is pre-chosen but $\mathbf{p} = \{p_k; k \geq 1\}$ is unknown.

Replacing each p_k in (4.3) with $\hat{p}_k$, the resulting estimator,

$$\hat{\theta} = \sum_{v\geq 0} w_v \sum_{k\geq 1} \hat{p}_k (1 - \hat{p}_k)^v, \tag{4.6}$$

is often referred to as the plug-in estimator. The plug-in estimator is an intuitive estimator and has many desirable properties asymptotically. However, it is known to have a slowly decaying bias, particularly when the sample size n is relatively small and the alphabet $\mathcal{X}$ is relatively large. For this reason, the plug-in estimator is not always the first one chosen in practice. Nevertheless, since its statistical properties are better known than those of other estimators, it is often used as a reference estimator.

On the other hand, by $Z_{1,v}$ in (3.26) and the fact $\mathrm{E}(Z_{1,v}) = \zeta_v$ for every v satisfying $1 \leq v \leq n - 1$, the following estimator in Turing's perspective is readily available,

$$\hat{\theta}^\sharp = \sum_{v=0}^{n-1} w_v Z_{1,v}, \tag{4.7}$$

and is the main object of this chapter.

Several statistical properties of $\hat{\theta}^{\sharp}$ may be described under the following condition.

Condition 4.1 *A probability distribution* $\mathbf{p} = \{p_k; k \geq 1\}$ *and an associated linear diversity index in the form of (4.3) are such that*

1) *the effective cardinality of* $\mathcal{X}$ *is finite, that is,* $K = \sum_{k\geq 1} 1[p_k > 0] < \infty$; *and*
2) *the weights,* w_v, *for all* $v \geq 0$, *are bounded, that is, there exists an* $M > 0$ *such that* $|w_v| \leq M$ *for all* $v \geq 0$.

Condition 4.1 guarantees that the summation in (4.3) always converges (see Exercise 10). It can be verified that the assumption that $|w_n| \leq M$ is satisfied by all of the linear diversity indices discussed in Examples 4.10 through 4.16 (see Exercise 11). Condition 4.1 also guarantees that estimator $\hat{\theta}^{\sharp}$ has a bias that decays exponentially fast in sample size n.

Theorem 4.2 *Under Condition 4.1,*

$$|\mathrm{E}(\hat{\theta}^{\sharp}) - \theta| \leq MK(1-p_{\wedge})^n \tag{4.8}$$

where $p_{\wedge} = \min\{p_k, \dots, p_K\}$.

Proof: The U-statistics construction of $Z_{1,v}$ in Chapter 3 establishes the fact that $\mathrm{E}(Z_{1,v}) = \zeta_v$ for v satisfying $v = 1, \dots, n-1$. Therefore,

$$\begin{aligned}
|\mathrm{E}(\hat{\theta}^{\sharp}) - \theta| &= \left| \sum_{v=1}^{n-1} w_v \mathrm{E}(Z_{1,v}) - \sum_{v=0}^{\infty} w_v \zeta_v \right| = \left| \sum_{v=n}^{\infty} w_v \zeta_v \right| \\
&\leq |w_v| \sum_{v=n}^{\infty} \sum_{k=1}^{K} p_k (1-p_k)^v = |w_v| \sum_{k=1}^{K} (1-p_k)^n \\
&\leq MK(1-p_{\wedge})^n.
\end{aligned}$$

□

Remark 4.1 *While decaying rapidly, the bias of* $\hat{\theta}^{\sharp}$ *could still be sizable when* M, K *are relatively large and* n *and* $p_{\wedge}$ *are relatively small. Practitioners often find bias reduction desirable in many realistic situations. In the existing literature, there are many proposed ways to reduce bias and to enhance accuracy in estimation of diversity indices. For a recent update of this research space, readers may wish to refer to Chao and Jost (2012) and Chao and Jost (2015). Zhang and Grabchak (2013) proposed a bias reduction methodology that is particularly suitable for general estimators in the form of (4.7).*

As in any sound statistical practice, reliability of estimated diversity indices must be supported by distributional characteristics of the estimators. Toward that end, the asymptotic normality of $\hat{\theta}^{\sharp}$ is established in the following. The approach here consists of two parts. The first part is to establish asymptotic normality of the plug-in estimator $\hat{\theta}$ in (4.6), and then the second part is to show that $\hat{\theta}^{\sharp}$ and $\hat{\theta}$ are sufficiently close to warrant the same asymptotic distributional behavior.

Let

$$\mathbf{v} = (p_1, p_2, \dots, p_{K-1})^{\tau} \quad \text{and} \quad \hat{\mathbf{v}} = (\hat{p}_1, \hat{p}_2, \dots, \hat{p}_{K-1})^{\tau}. \tag{4.9}$$

Lemma 4.1 is a well-known multivariate normal approximation to the multinomial distribution (see Exercise 17).

Lemma 4.1 *Suppose the probability distribution* $\mathbf{p} = \{p_k; k = 1, \dots, K\}$ *is such that* $p_k > 0$ *for each* k*:*

$$\sqrt{n}(\hat{\mathbf{v}} - \mathbf{v}) \xrightarrow{L} MVN(\emptyset, \Sigma(\mathbf{v})), \tag{4.10}$$

where $\emptyset$ *is a* $(K-1)$*-dimensional zero vector and* $\Sigma(\mathbf{v})$ *is the* $(K-1) \times (K-1)$ *covariance matrix given by*

$$\Sigma(\mathbf{v}) = \begin{pmatrix} p_1(1-p_1) & -p_1p_2 & \cdots & -p_1p_{K-1} \\ -p_2p_1 & p_2(1-p_2) & \cdots & -p_2p_{K-1} \\ \vdots & \vdots & \cdots & \vdots \\ -p_{K-1}p_1 & -p_{K-1}p_2 & \cdots & p_{K-1}(1-p_{K-1}) \end{pmatrix}. \tag{4.11}$$

Let $h(p) = \sum_{v\geq 0} w_v(1-p)^v$, write the index θ of (4.3) as a function of $(p_1, \dots, p_{K-1})$, and denote the resulting function as

$$G(\mathbf{v}) = \sum_{k=1}^{K} p_k h(p_k) = \sum_{k=1}^{K-1} p_k h(p_k) + \left(1 - \sum_{k=1}^{K-1} p_k\right) h\left(1 - \sum_{k=1}^{K-1} p_k\right).$$

Let the gradient of $G(\mathbf{v})$ be denoted as

$$g(\mathbf{v}) = \nabla G(\mathbf{v}) = \left(\frac{\partial}{\partial p_1} G(\mathbf{v}), \dots, \frac{\partial}{\partial p_{K-1}} G(\mathbf{v})\right)^{\tau} =: (g_1(\mathbf{v}), \dots, g_{K-1}(\mathbf{v}))^{\tau}. \tag{4.12}$$

For each j, $j = 1, \dots, K-1$,

$$\begin{aligned} g_j(\mathbf{v}) = \frac{\partial}{\partial p_j} G(\mathbf{v}) &= h(p_j) + p_j h'(p_j) - h\left(1 - \sum_{k=1}^{K-1} p_k\right) \\ &\quad - \left(1 - \sum_{k=1}^{K-1} p_k\right) h'\left(1 - \sum_{k=1}^{K-1} p_k\right) \\ &= \sum_{v\geq 0} w_v(1-p_j)^v - p_j \sum_{v\geq 0} w_v v(1-p_j)^{v-1} \\ &\quad - \sum_{v\geq 0} w_v(1-p_K)^v + p_K \sum_{v\geq 0} w_v v(1-p_K)^{v-1} \end{aligned}$$

$$= \sum_{v \geq 0} w_v[(1-p_j)^v - (1-p_K)^v]$$
$$- \sum_{v \geq 0} w_v v[p_j(1-p_j)^{v-1} - p_K(1-p_K)^{v-1}]. \tag{4.13}$$

An application of the delta method gives the following result.

Theorem 4.3 *Suppose Condition 4.1 holds and $g^\tau(\mathbf{v})\Sigma(\mathbf{v})g(\mathbf{v}) > 0$, then*

$$\sqrt{n}(\hat{\theta} - \theta)(g^\tau(\mathbf{v})\Sigma(\mathbf{v})g(\mathbf{v}))^{-\frac{1}{2}} \xrightarrow{L} N(0,1). \tag{4.14}$$

Tanabe and Sagae (1992) showed that, provided that $K \geq 2$ and that $p_k > 0$ for each and every k, $k = 1, \dots, K$, the covariance matrix $\Sigma(\mathbf{v})$ is a positive definite matrix. Therefore, $g^\tau(\mathbf{v})\Sigma(\mathbf{v})g(\mathbf{v}) > 0$ if and only if $g(\mathbf{v}) \neq 0$. However, for some combinations of $h(p)$ and $\mathbf{p}$, $g(\mathbf{v}) = 0$ and the condition of Theorem 4.3 does not hold.

Example 4.17 *Shannon's entropy. Let $\theta = \sum_{k=1}^K p_k \ln(1/p_k)$. Then $g^\tau(\mathbf{v})\Sigma(\mathbf{v})$ $g(\mathbf{v}) = 0$ if and only if $\mathbf{p} = \{p_k; k = 1, \dots, K\}$ is uniform.*

Proof: In this case, $h(p) = -\ln p$. For each j, $j = 1, \dots, K-1$,

$$g_j(\mathbf{v}) = \frac{\partial}{\partial p_j}\theta(\mathbf{p}) = \frac{\partial}{\partial p_j}\sum_{k=1}^K p_k h(p_k)$$
$$= -\ln(p_j) - 1 + \ln p_K + 1 = \ln(p_K/p_j).$$

If $p_j = 1/K$ for every j, then $g_j(\mathbf{v}) = 0$. On the other hand, if $\ln(p_K/p_j) = 0$ for each and every j, then $p_j = p_K = 1/K$. □

Example 4.18 *Emlen's index. Let $\theta = \sum_{k=1}^K p_k e^{p_k}$. Then $g^\tau(\mathbf{v})\Sigma(\mathbf{v})g(\mathbf{v}) = 0$ if and only if $\mathbf{p} = \{p_k; k = 1, \dots, K\}$ is uniform.*

Proof: In this case, $h(p) = e^{-p}$. For each j, $j = 1, \dots, K-1$,

$$g_j(\mathbf{v}) = \frac{\partial}{\partial p_j}\theta(\mathbf{p}) = e^{-p_j}(1-p_j) - e^{-p_K}(1-p_K).$$

If $p_j = 1/K$ for every j, then $g_j(\mathbf{v}) = 0$. Suppose for each and every j $g_j(\mathbf{v}) = 0$, then

$$e^{-p_j}(1-p_j) = e^{-p_K}(1-p_K). \tag{4.15}$$

Noting $f(p) = e^{-p}(1-p)$ is a strictly monotone function on interval $(0,1)$, (4.15) implies $p_j = p_K$ for every j (see Exercise 19). □

As evidenced by the last expression of (4.3), $g(\mathbf{v})$ is a continuous function of $\mathbf{v}$ for $\mathbf{v} \in (0,1]^{K-1}$. The said continuity and the fact $\hat{\mathbf{v}} \xrightarrow{p} \mathbf{v}$ imply that $g(\hat{\mathbf{v}}) \xrightarrow{p} g(\mathbf{v})$ by the continuous mapping theorem of Mann and Wald (1943). The following corollary follows immediately from Slutsky's theorem.

Corollary 4.1 *Under the condition of Theorem 4.3,*

$$\sqrt{n}(\hat{\theta} - \theta)(g^{\tau}(\hat{\mathbf{v}})\Sigma(\hat{\mathbf{v}})g(\hat{\mathbf{v}}))^{-\frac{1}{2}} \xrightarrow{L} N(0,1).$$

Corollary 4.1 provides a means of statistical inference. It is to be noted however that in practice it is often the case that $\hat{p}_k = 0$ for some index values of k, that is, the letters in the alphabet that are not covered by a sample of size n. For simplicity, suppose $\hat{p}_K > 0$ (if not, a switch of two indices will suffice). Two issues exist in such a situation. First, noting $g_j(\mathbf{v})$ for each j is only defined for $p_j > 0$, some components of $g(\hat{\mathbf{v}}) = (g_1(\hat{\mathbf{v}}), \dots, g_{K-1}(\hat{\mathbf{v}}))^{\tau}$ may not be well defined if some components of $\hat{\mathbf{v}}$ are zeros, say $g_j(\hat{\mathbf{v}})$. Estimating $g_j(\mathbf{v})$ by a not well-defined statistic $g_j(\hat{\mathbf{v}})$ is problematic. This issue can be resolved by setting $g_j(\hat{\mathbf{v}}) = 0$, or any fixed value for that matter, since $g_j(\hat{\mathbf{v}})$ converges to the true value of $g_j(\mathbf{v})$ and therefore is bounded away from the defined value in probability. Second, if $\hat{p}_k = 0$ for some index values of k, though $p_k > 0$ and therefore $\Sigma(\mathbf{v})$ is positive definite, $\Sigma(\hat{\mathbf{v}})$ is not. This raises the question on the positivity of $g^{\tau}(\hat{\mathbf{v}})\Sigma(\hat{\mathbf{v}})g(\hat{\mathbf{v}})$. However, it may be shown that $g^{\tau}(\hat{\mathbf{v}})\Sigma(\hat{\mathbf{v}})g(\hat{\mathbf{v}})$, after redefining $g_j(\hat{\mathbf{v}})$ for all $\hat{p}_j = 0$, is positive (see Exercise 18).

Another interesting feature of Corollary 4.1 is that it does not distinguish a species that does not exist in the population, that is, $p_k = 0$, from a species that does exist but is not observed in the sample, that is, $p_k > 0$ but $\hat{p}_k = 0$. This is so because both $\Sigma(\hat{\mathbf{v}})$ and the redefined $g(\hat{\mathbf{v}})$ have zeros in locations that would make the difference inconsequential in computation. For this reason, Corollary 4.1 may be used regardless of whether K is known *a priori*.

The asymptotic normal laws established in Theorem 4.3 and Corollary 4.1 for the plug-in estimator may be extended to that for the estimator of Turing's perspective, defined in (4.7).

Theorem 4.4 *Let $\hat{\theta}^{\sharp}$ be as in (4.7). Suppose Condition 4.1 holds and $g^{\tau}(\mathbf{v})\Sigma(\mathbf{v})g(\mathbf{v}) > 0$. Then*

$$\sqrt{n}\left(\hat{\theta}^{\sharp} - \theta\right)(g^{\tau}(\mathbf{v})\Sigma(\mathbf{v})g(\mathbf{v}))^{-\frac{1}{2}} \xrightarrow{L} N(0,1).$$

Corollary 4.2 *Let $\hat{\theta}^{\sharp}$ be as in (4.7). Suppose Condition 4.1 holds and $g^{\tau}(\mathbf{v})\Sigma(\mathbf{v})g(\mathbf{v}) > 0$. Then*

$$\sqrt{n}\left(\hat{\theta}^{\sharp} - \theta\right)(g^{\tau}(\hat{\mathbf{v}})\Sigma(\hat{\mathbf{v}})g(\hat{\mathbf{v}}))^{-\frac{1}{2}} \xrightarrow{L} N(0,1). \tag{4.16}$$

To prove Theorem 4.4, two lemmas, Lemmas 4.2 and 4.3, are needed. Both lemmas are stated in the following without proofs. A proof of Lemma 4.2 can be found in Zhang and Grabchak (2016), and a proof of Lemma 4.3 can be found in Hoeffding (1963).

For $p \in [0,1]$ and $n \in \mathbb{N} = \{1,2, \dots\}$, consider the following two functions:

$$f_n(p) = p \sum_{v=1}^{\lfloor n(1-p)+1 \rfloor} w_v \prod_{j=1}^{v} \left(1 - \frac{np-1}{n-j}\right) \quad \text{and}$$

$$f(p) = p \sum_{v=1}^{\infty} w_v (1-p)^v.$$

Lemma 4.2 *Suppose Condition 4.1 holds.*

1) If $0 < c < d < 1$, then

$$\lim_{n\to\infty} \sup_{p\in[c,d]} \sqrt{n}|f_n(p) - f(p)| = 0.$$

2) Let $p_n \in [0,1]$ be such that $np_n \in \{0,1,2,\dots,n\}$, then

$$\sqrt{n}|f_n(p_n) - f(p_n)| \le \sqrt{n}(n+1) \le 2n^{3/2}.$$

Lemma 4.3 *(Hoeffding's inequality) Let Y be a binomial random variable with parameters n and $p > 0$ and $\hat{p} = Y/n$. For any $\varepsilon > 0$,*

1) $\mathrm{P}(\hat{p} - p \le -\varepsilon) \le e^{-2\varepsilon^2 n}$ and
2) $\mathrm{P}(\hat{p} - p \ge \varepsilon) \le e^{-2\varepsilon^2 n}$.

Recall the result of Exercise 17 of Chapter 3 that $Z_{1,v}$ in (3.26) may be re-expressed as

$$Z_{1,v} = \sum_{k\ge1} \left[\hat{p}_k \prod_{j=1}^{v} \left(1 - \frac{Y_k - 1}{n-j}\right)\right]. \tag{4.17}$$

As a result of (4.17), $\hat{\theta}^\sharp$ in (4.7) may be re-expressed as

$$\hat{\theta}^\sharp = w_0 + \sum_{v=1}^{n-1} w_v \sum_{k=1}^{K} \hat{p}_k \prod_{j=1}^{v} \left(1 - \frac{Y_k - 1}{n-j}\right) \tag{4.18}$$

$$= w_0 + \sum_{k=1}^{K} \hat{p}_k \sum_{v=1}^{n-Y_k} w_v \prod_{j=1}^{v} \left(1 - \frac{Y_k - 1}{n-j}\right) \tag{4.19}$$

$$= w_0 + \sum_{k=1}^{K} \hat{\theta}_k^\sharp, \tag{4.20}$$

where

$$\hat{\theta}_k^\sharp = \hat{p}_k \sum_{v=1}^{n-Y_k} w_v \prod_{j=1}^{v} \left(1 - \frac{Y_k - 1}{n-j}\right) = \hat{p}_k \sum_{v=1}^{\infty} w_v \prod_{j=1}^{v} \left(1 - \frac{Y_k - 1}{n-j}\right). \quad (4.21)$$

Also recall from (4.6)

$$\hat{\theta} = w_0 + \sum_{k=1}^{K} \hat{\theta}_k, \quad (4.22)$$

where $\hat{\theta}_k = \hat{p}_k \sum_{v=1}^{\infty} w_v (1-\hat{p}_k)^v$.

Proof of Theorem 4.4: Since

$$\sqrt{n}(\hat{\theta}^\sharp - \theta) = \sqrt{n}(\hat{\theta}^\sharp - \hat{\theta}) + \sqrt{n}(\hat{\theta} - \theta),$$

by Theorem 4.3 and Slutsky's theorem it suffices to show that

$$\sqrt{n}(\hat{\theta}^\sharp - \hat{\theta}) \xrightarrow{p} 0.$$

However, from (4.20) and (4.22),

$$\sqrt{n}(\hat{\theta}^\sharp - \hat{\theta}) = \sum_{k=1}^{K} \sqrt{n}(\hat{\theta}_k^\sharp - \hat{\theta}_k),$$

and therefore it suffices to show that, for each k,

$$\sqrt{n}(\hat{\theta}_k^\sharp - \hat{\theta}_k) \xrightarrow{p} 0.$$

Toward that end,

$$\begin{aligned}\sqrt{n}(\hat{\theta}_k^\sharp - \hat{\theta}_k) &= \left\{ \sqrt{n}\left(\hat{\theta}_k^\sharp - \hat{\theta}_k\right) 1[\hat{p}_k \le p_k/2] \right.\\ &\quad \left. + \sqrt{n}\left(\hat{\theta}_k^\sharp - \hat{\theta}_k\right) 1[\hat{p}_k \ge (1+p_k)/2] \right\}\\ &\quad + \left\{ \sqrt{n}\left(\hat{\theta}_k^\sharp - \hat{\theta}_k\right) 1[p_k/2 < \hat{p}_k < (1+p_k)/2] \right\}\\ &=: \{\mathcal{A}_1\} + \{\mathcal{A}_2\}.\end{aligned}$$

By Part 2 of Lemmas 4.2 and 4.3,

$$\begin{aligned}E|\mathcal{A}_1| &\le 2n^{3/2}\left(\mathrm{P}\left(\hat{p}_k > \frac{1+p_k}{2}\right) + \mathrm{P}\left(\hat{p}_k \le \frac{p_k}{2}\right)\right)\\ &= 2n^{3/2}\left(\mathrm{P}\left(\hat{p}_k - p_k \frac{1-p_k}{2}\right) + \mathrm{P}\left(\hat{p}_k - p_k \le -\frac{p_k}{2}\right)\right)\\ &\le 2n^{3/2}[e^{-n(1-p_k)^2/2} + e^{-np_k^2/2}] \to 0.\end{aligned}$$

Thus, it follows that $\mathcal{A}_1 \xrightarrow{p} 0$. By Part 1 of Lemma 4.2, it follows that $\mathcal{A}_2 \xrightarrow{p} 0$. □

Finally, the same justification for Corollary 4.1 also applies to Corollary 4.2.

4.3 Estimation of Rényi's Entropy

In the last section, a general nonparametric estimation approach to linear diversity indices is introduced. To extend the derived results further to diversity indices that are not linear but only equivalent to a linear kernel, $\theta_{LK} = \sum_{v=0}^{\infty} w_v\ \zeta_v$, consider $\theta = \psi(\theta_{LK})$ where $\psi(\cdot)$ is a function differentiable at θ_{LK}. Let $\psi'(\cdot)$ be the first derivative of $\psi(\cdot)$ and

$$\tilde{\theta}^\sharp = \psi(\hat{\theta}^\sharp) \tag{4.23}$$

where $\hat{\theta}^\sharp$, as in (4.7), is an estimator of $\theta = \psi(\theta_{LK})$. An application of the delta method immediately gives the following theorem and corollary.

Theorem 4.5 *Suppose Condition 4.1 holds and* $g^\tau(\mathbf{v})\Sigma(\mathbf{v})g(\mathbf{v}) > 0$. *Then*

$$\sqrt{n}\left(\psi(\hat{\theta}^\sharp) - \theta\right)\left(g^\tau(\mathbf{v})\Sigma(\mathbf{v})g(\mathbf{v})\psi'(\theta_{LK})\right)^{-\frac{1}{2}} \xrightarrow{L} N(0,1).$$

Corollary 4.3 *Suppose Condition 4.1 holds and* $g^\tau(\mathbf{v})\Sigma(\mathbf{v})g(\mathbf{v}) > 0$. *Then*

$$\sqrt{n}\left(\psi(\hat{\theta}^\sharp) - \theta\right)\left(g^\tau(\hat{\mathbf{v}})\Sigma(\hat{\mathbf{v}})g(\hat{\mathbf{v}})\psi'(\hat{\theta}^\sharp)\right)^{-\frac{1}{2}} \xrightarrow{L} N(0,1).$$

Furthermore, if $|\psi'(t)| < D$ for some $D > 0$ on the interval of t between θ_{LK} and $\inf_*\{\hat{\theta}^\sharp\}$ and the interval between θ_{LK} and $\sup_*\{\hat{\theta}^\sharp\}$, where both $\inf_*$ and $\sup_*$ are taken over all possible samples of all sizes, then the bias of $\psi(\hat{\theta}^\sharp)$ decays at least as fast as the bias of $\hat{\theta}^\sharp$, which is exponential in n under Condition 4.1. To see this, it suffices to consider the first-order Taylor expansion for $\psi(\hat{\theta}^\sharp)$:

$$\psi(\hat{\theta}^\sharp) - \psi(\theta_{LK}) = \psi'(\xi)(\hat{\theta}^\sharp - \theta_{LK}) \tag{4.24}$$

where the random variable ξ is between θ_{LK} and $\hat{\theta}^\sharp \in \left(\inf_*\{\hat{\theta}^\sharp\}, \sup_*\{\hat{\theta}^\sharp\}\right)$. Taking expectation on both sides of (4.24) gives

$$\left|E(\psi(\hat{\theta}^\sharp) - \psi(\theta_{LK}))\right| \le D\left|E(\hat{\theta}^\sharp - \theta_{LK})\right| \le DMK(1-p_\wedge)^n \tag{4.25}$$

where the last inequality is due to (4.8).

Rényi's entropy is one of the most frequently used diversity indices by ecologists, and it is one of the two diversity indices among the many discussed in this chapter that are not linear in entropic basis, the other one being Hill's diversity number that is a logarithmic transformation of Rényi's entropy. Theorem 4.5 and Corollary 4.3 may be applied to Rényi's entropy.

For a fixed $\alpha > 0$ such that $\alpha \ne 1$, let $\psi(t) = (1-\alpha)^{-1}\ln t$. Note that

$$H_\alpha = \psi(h_\alpha) = \frac{\ln h_\alpha}{1-\alpha},$$

where H_α is Rényi's entropy and $h_\alpha = \sum_{k=1}^{K} p_k^\alpha$ is Rényi's equivalent entropy, as in Example 4.13. Let

$$\hat{h}_\alpha^\sharp = 1 + \sum_{k=1}^{K} \hat{p}_k \sum_{v=1}^{n-Y_k} w_v \prod_{j=1}^{v} \left(1 - \frac{Y_k - 1}{n - j}\right), \tag{4.26}$$

where for $v \geq 1$

$$w_v = \prod_{i=1}^{v} \left(\frac{i - \alpha}{i}\right).$$

$\hat{h}_\alpha^\sharp$ is the estimator of h_α in the form of (4.7). Let

$$\hat{H}_\alpha^\sharp = \frac{\ln \hat{h}_\alpha^\sharp}{1 - \alpha} = \psi(\hat{h}_\alpha^\sharp).$$

Since

$$\psi'(t) = \frac{1}{(1 - \alpha)t}, \tag{4.27}$$

Theorem 4.5 and the fact that $h_\alpha > 0$ imply the following.

Theorem 4.6 *Provided that* $g^\tau(\mathbf{v})\Sigma(\mathbf{v})g(\mathbf{v}) > 0$,

$$\sqrt{n}\left(\hat{H}_\alpha^\sharp - H_\alpha\right) |1 - \alpha| h_\alpha [g^\tau(\mathbf{v})\Sigma(\mathbf{v})g(\mathbf{v})]^{-\frac{1}{2}} \xrightarrow{L} N(0, 1).$$

Corollary 4.4 *Provided that* $g^\tau(\mathbf{v})\Sigma(\mathbf{v})g(\mathbf{v}) > 0$,

$$\sqrt{n}(\hat{H}_\alpha^\sharp - H_\alpha)|1 - \alpha| \hat{h}_\alpha^\sharp [g^\tau(\hat{\mathbf{v}})\Sigma(\hat{\mathbf{v}})g(\hat{\mathbf{v}})]^{-\frac{1}{2}} \xrightarrow{L} N(0, 1).$$

Next it is to establish that when $\alpha \in (0, 1)$, the bias of $\hat{H}_\alpha^\sharp$ decays exponentially in the sample size n. This fact follows the following three observations.

1) First, observe that the linear kernel of Rényi's entropy, or Rényi's equivalent index, as in Example 4.13, satisfies

   $$\theta_{LK} = h_\alpha = \sum_{k \geq 1} p_k^\alpha > 0$$

   for any $\alpha > 0$. In particular, for $\alpha \in (0, 1)$,

   $$\theta_{LK} = h_\alpha = \sum_{k \geq 1} p_k^\alpha \geq 1.$$

2) Second, consider $\hat{h}_\alpha^\sharp$ of (4.26). Since $\alpha \in (0, 1)$, $(i - \alpha) > 0$ for all $i = 1, \dots,$ $w_v > 0$, and hence $\hat{\theta}^\sharp = \hat{h}_\alpha^\sharp \geq 1$, one has

   $$m = \min\left\{h_\alpha, \inf_{*}\left\{\hat{h}_\alpha^\sharp\right\}\right\} \geq 1.$$

3) Third, since $\psi'(t)$ of (4.27) satisfies

   $$\psi'(t) = \frac{1}{(1 - \alpha)t} \leq \frac{1}{(1 - \alpha)m} =: D$$

   for $t \in [m, +\infty)$, Equation (4.25) holds. Therefore, the bias of $\hat{H}_\alpha^\sharp$ decays exponentially in n.

Finally, a large-sample statistical testing procedure for a difference in the diversity measured by Rényi's entropy between two populations (possibly defined by two different locations or two different time points) is described in the following.

Suppose two independent *iid* samples of sizes n_1 and n_2 are to be taken from a same alphabet $\mathscr{X}$ under two different distributions: $\{p_{1,k}; k \geq 1\}$ and $\{p_{2,k}; k \geq 1\}$. The samples are summarized, respectively, by two independent sets of the observed frequencies

$$\mathbf{Y}_1 = \{Y_{1,1}, Y_{1,2}, \dots, Y_{1,k}, \dots\}$$
$$\mathbf{Y}_2 = \{Y_{2,1}, Y_{2,2}, \dots, Y_{2,k}, \dots\}$$

and their corresponding observed relative frequencies

$$\hat{\mathbf{p}}_1 = \{\hat{p}_{1,1}, \hat{p}_{1,2}, \dots, \hat{p}_{1,k}, \dots\} = \{Y_{1,1}/n_1, Y_{1,2}/n_1, \dots, Y_{1,k}/n_1, \dots\}$$
$$\hat{\mathbf{p}}_2 = \{\hat{p}_{2,1}, \hat{p}_{2,2}, \dots, \hat{p}_{2,k}, \dots\} = \{Y_{2,1}/n_2, Y_{2,2}/n_2, \dots, Y_{2,k}/n_2, \dots\}.$$

For a fixed $\alpha > 0$ and $\alpha \neq 1$, denote Rényi's entropies for the two populations by $H_\alpha(1)$ and $H_\alpha(2)$. Suppose it is of interest to test the hypothesis

$$H_0 : H_\alpha(1) = H_\alpha(2) \quad \text{versus} \quad H_a : H_\alpha(1) \neq H_\alpha(2)$$

at a level of significance $\beta \in (0, 1)$.

The testing procedure may be carried out in the following 10 steps.

1) To calculate, using (4.26),

$$\hat{h}^{\sharp}_\alpha(1) = 1 + \sum_{k=1}^{K} \hat{p}_{1,k} \sum_{v=1}^{n_1 - Y_{1,k}} w_v \prod_{j=1}^{v} \left(1 - \frac{Y_{1,k} - 1}{n_1 - j}\right) \quad \text{and}$$
$$\hat{h}^{\sharp}_\alpha(2) = 1 + \sum_{k=1}^{K} \hat{p}_{2,k} \sum_{v=1}^{n_2 - Y_{2,k}} w_v \prod_{j=1}^{v} \left(1 - \frac{Y_{2,k} - 1}{n_2 - j}\right),$$

where

$$w_v = \prod_{i=1}^{v} \left(\frac{i - \alpha}{i}\right).$$

2) To calculate

$$\hat{H}^{\sharp}_\alpha(1) = \frac{\ln \hat{h}^{\sharp}_\alpha(1)}{1 - \alpha} \quad \text{and} \quad \hat{H}^{\sharp}_\alpha(2) = \frac{\ln \hat{h}^{\sharp}_\alpha(2)}{1 - \alpha}.$$

3) To calculate vectors

$$g(\hat{\mathbf{v}}_1) = \left(\alpha\left(\hat{p}^{\alpha-1}_{1,1} - \hat{p}^{\alpha-1}_{1,K}\right), \dots, \alpha\left(\hat{p}^{\alpha-1}_{1,K-1} - \hat{p}^{\alpha-1}_{1,K}\right)\right)^\tau \quad \text{and}$$
$$g(\hat{\mathbf{v}}_2) = \left(\alpha\left(\hat{p}^{\alpha-1}_{2,1} - \hat{p}^{\alpha-1}_{2,K}\right), \dots, \alpha\left(\hat{p}^{\alpha-1}_{2,K-1} - \hat{p}^{\alpha-1}_{2,K}\right)\right)^\tau.$$

4) To make adjustment to $g(\hat{\mathbf{v}}_1)$ and $g(\hat{\mathbf{v}}_2)$ by setting the entries, when $\hat{p}_{i,j} = 0$, $i = 1, 2$, and $j = 1, 2, \dots, K-1$, to zero, that is,

$$\begin{aligned}\overline{g}(\hat{\mathbf{v}}_1) = \big(&\alpha(\hat{p}_{1,1}^{\alpha-1} - \hat{p}_{1,K}^{\alpha-1})1[\hat{p}_{1,1} > 0]\,, \\ &\dots, \alpha(\hat{p}_{1,K-1}^{\alpha-1} - \hat{p}_{1,K}^{\alpha-1})1[\hat{p}_{1,K-1} > 0]\big)^{\tau}, \text{and} \\ \overline{g}(\hat{\mathbf{v}}_2) = \big(&\alpha(\hat{p}_{2,1}^{\alpha-1} - \hat{p}_{2,K}^{\alpha-1})1[\hat{p}_{2,1} > 0]\,, \\ &\dots, \alpha(\hat{p}_{2,K-1}^{\alpha-1} - \hat{p}_{2,K}^{\alpha-1})1[\hat{p}_{2,K-1} > 0]\big)^{\tau}.\end{aligned}$$

5) To calculate the two $(K-1) \times (K-1)$ covariance matrices

$$\Sigma(\hat{\mathbf{v}}_1) = \begin{pmatrix} \hat{p}_{1,1}(1-\hat{p}_{1,1}) & -\hat{p}_{1,1}\hat{p}_{1,2} & \cdots & -\hat{p}_{1,1}\hat{p}_{1,K-1} \\ -\hat{p}_{1,2}\hat{p}_{1,1} & \hat{p}_{1,2}(1-\hat{p}_{1,2}) & \cdots & -\hat{p}_{1,2}\hat{p}_{1,K-1} \\ \vdots & \vdots & \cdots & \vdots \\ -\hat{p}_{1,K-1}\hat{p}_{1,1} & -\hat{p}_{1,K-1}\hat{p}_{1,2} & \cdots & \hat{p}_{1,K-1}(1-\hat{p}_{1,K-1}) \end{pmatrix}$$

and

$$\Sigma(\hat{\mathbf{v}}_2) = \begin{pmatrix} \hat{p}_{2,1}(1-\hat{p}_{2,1}) & -\hat{p}_{2,1}\hat{p}_{2,2} & \cdots & -\hat{p}_{2,1}\hat{p}_{2,K-1} \\ -\hat{p}_{2,2}\hat{p}_{2,1} & \hat{p}_{2,2}(1-\hat{p}_{2,2}) & \cdots & -\hat{p}_{2,2}\hat{p}_{2,K-1} \\ \vdots & \vdots & \cdots & \vdots \\ -\hat{p}_{2,K-1}\hat{p}_{2,1} & -\hat{p}_{2,K-1}\hat{p}_{2,2} & \cdots & \hat{p}_{2,K-1}(1-\hat{p}_{2,K-1}) \end{pmatrix}.$$

6) To calculate the two estimated variances

$$\hat{\sigma}_1^2 = (1-\alpha)^2\left(\hat{h}_{\alpha}^{\sharp}(1)\right)^2 (\overline{g}^{\tau}(\hat{\mathbf{v}}_1)\Sigma(\hat{\mathbf{v}}_1)\overline{g}(\hat{\mathbf{v}}_1))$$

and

$$\hat{\sigma}_2^2 = (1-\alpha)^2\left(\hat{h}_{\alpha}^{\sharp}(2)\right)^2 (\overline{g}^{\tau}(\hat{\mathbf{v}}_2)\Sigma(\hat{\mathbf{v}}_2)\overline{g}(\hat{\mathbf{v}}_2)).$$

7) To find the estimated combined variance

$$\hat{\sigma}^2 = \hat{\sigma}_1^2 + \frac{n_1}{n_2}\hat{\sigma}_2^2.$$

8) To calculate the test statistic

$$Z_{\alpha} = \frac{\sqrt{n_1}(\hat{H}_{\alpha}^{\sharp}(1) - \hat{H}_{\alpha}^{\sharp}(2))}{\hat{\sigma}}. \tag{4.28}$$

9) To make a decision to
   a) reject H_0 if $|Z_{\alpha}| \geq z_{\beta/2}$, or
   b) not to reject H_0 if $|Z_{\alpha}| < z_{\beta/2}$

   where $z_{\beta/2}$ is the $[100 \times (1-\beta/2)]$th percentile of the standard normal distribution.

10) To calculate a $[100 \times (1-\beta)]\%$ confidence interval for the difference in Rényi's entropy,

$$\left(\hat{H}_{\alpha}^{\sharp}(1) - \hat{H}_{\alpha}^{\sharp}(2)\right) \pm z_{\beta/2}\frac{\hat{\sigma}}{\sqrt{n_1}}. \tag{4.29}$$

Many ecologists like to compare Rényi's diversity profiles of two populations based on two samples. The estimated single-sample profiles are obtained by $\hat{H}_\alpha^\sharp(1)$ and $\hat{H}_\alpha^\sharp(2)$ as functions of α varying from a to b, satisfying $0 < a < b$. The estimated difference of the two profiles is obtained by $\hat{H}_\alpha^\sharp(1) - \hat{H}_\alpha^\sharp(2)$ as a function α. Letting α vary over (a, b) satisfying $0 < a < b$, (4.29) provides a $[100 \times (1 - \beta)]\%$ point-wise confidence band for the profile difference in Rényi's diversity entropy.

4.4 Remarks

The first two fundamental questions in biodiversity studies are as follows:

1) What is diversity?
2) How should it be mathematically measured?

Much discussion of diversity has ensued since (Fisher, Corbet, and Williams (1943), MacArthur (1955), Margalef (1958)), and it is still ongoing. One of the main objectives of the discussion has been to pin down one mathematical diversity index that would be universally accepted. However, after more than half of a century of intense discussion, it should be abundantly clear that any one diversity index would, at its best, be a summary parameter of the underlying distribution $\{p_k; k \geq 1\}$ to reflect a subjectively perceived notion of "diversity." A conceptual parallel of diversity may be found in Statistics, where the dispersion of a random variable is of interest. For example, the variance of a random variable is commonly used as a measure of dispersion associated with an underlying distribution. However, the popularity of variance has more to do with the mathematical convenience it entails than its appropriateness in measuring dispersion. In fact, the variance does not have any more or less intrinsic merit in measuring dispersion *per se* than, say, the expectation of absolute deviation from the mean or any even central moments of the random variable, when they exist. One should never lose sight of the ultimate end objective in Statistics: the knowledge of the underlying distribution. Moments are merely the means and not the end. Similarly, in diversity studies of ecological populations, or more generally $\mathscr{X}$ and its associated $\{p_k\}$, any diversity indices are merely the means but the distribution of all species is the end.

Much has been published in diversity literature, but rarely can a simpler and wiser statement be found than the opening paragraph of Purvis and Hector (2000):

> To proceed very far with the study of biodiversity, we need to pin the concept down. … However, any attempt to measure biodiversity quickly runs into the problem that it is a fundamentally multidimensional concept: it cannot be reduced sensibly to a single number.

That said however, it must be acknowledged that any efforts toward a systematic understanding of what axiomatic conditions would constitute diversity indices should very much be of value. Earlier in this chapter, it is established that under the two very basic axioms, $\mathcal{A}_{01}$ and $\mathcal{A}_{02}$, any diversity index is a function of $\{\zeta_\nu; \nu \geq 1\}$, which already offers a powerful unifying perspective. There are many possible additional axioms that could be contemplated. For example, the following maximization axiom, call it $\mathcal{A}_M$, is often mentioned in the literature.

Axiom 4.2 *Given an alphabet $\mathscr{X}$ with effective cardinality K and its associated probability distribution $\mathbf{p}$, a diversity index $\theta = \theta(\mathbf{p})$ must satisfy*

$$\mathcal{A}_M\colon\ \max\{\theta(\mathbf{p});\ \textit{all}\ \mathbf{p}\ \textit{on}\ \mathscr{X}\} = \theta(\mathbf{p}_u)$$

where $\mathbf{p}_u$ is the uniform distribution, that is, $p_k = 1/K$ for every $k = 1, \ldots, K$.

A stronger version of Axiom $\mathcal{A}_M$ is given in the following.

Axiom 4.3 *Given an alphabet $\mathscr{X}$ with effective cardinality K and its associated probability distribution $\mathbf{p}$, a diversity index $\theta = \theta(\mathbf{p})$ must satisfy*

$$\mathcal{A}_M^*\colon\ \theta(\mathbf{p})\ \textit{attains its maximum at, and only at,}\ \mathbf{p} = \mathbf{p}_u$$

where $\mathbf{p}_u$ is the uniform distribution, that is, $p_k = 1/K$ for every $k = 1, \ldots, K$, $\mathbf{p}$.

Axioms $\mathcal{A}_M$ and $\mathcal{A}_M^* i$ are heuristically reasonable if the ecological interest of all different species is equal, which often is not the case in practice. The question then becomes whether such an additional constraint should be included in the general axiomatic system for diversity indices.

The following is another axiom, also known as the minimization axiom. The minimization axiom is thought by many ecologists to be more fundamental than any other requirements on reasonable diversity indices.

Axiom 4.4 *A diversity index $\theta = \theta(\mathbf{p})$ must satisfy*

$$\mathcal{A}_m\colon\ \min\{\theta(\mathbf{p});\ \textit{all}\ \mathbf{p}\ \textit{on}\ \mathscr{X}\} = \theta(\mathbf{p}_o)$$

where $\mathbf{p}_o$ is a single-point-mass distribution, that is, $p_1 = 1$ and $K = 1$.

A stronger version of Axiom $\mathcal{A}_M$ is given in the following.

Axiom 4.5 *A diversity index $\theta = \theta(\mathbf{p})$ must satisfy*

$$\mathcal{A}_m^*\colon\ \theta(\mathbf{p})\ \textit{attains its minimum at, and only at,}\ \mathbf{p} = \mathbf{p}_o$$

where $\mathbf{p}_o$ is a single-point-mass distribution, that is, $p_1 = 1$ and $K = 1$.

Example 4.19 *The Gini–Simpson diversity index, $\theta = \sum_{k\geq 1} p_k(1 - p_k)$, satisfies both $\mathcal{A}_M$ and $\mathcal{A}_m$ (see Exercise 12).*

Grabchak, Marcon, Lang and Zhang (2016) showed that the generalized Simpson's indices also satisfy $\mathcal{A}_M$ and $\mathcal{A}_m$ under mild restrictions.

There is yet another interesting axiom, known as the replication principle. The replication principle requires that "if there are N equally large, equally diverse groups with no species in common, the diversity of the pooled groups must be N times the diversity of a single group." For a detailed discussion on the replication principle, readers may wish to refer to Jost (2006) and Chao, Chiu, and Jost (2010).

Remark 4.2 *The phrases "equally large" and "equally diverse" are subject to interpretation. However, they are interpreted here within as "equal in number of existing species" and "equal in diversity as measured by the underlying diversity index," respectively.*

The replication principle is a strong axiom, so strong it pins down essentially on one index under mild conditions.

Let $\theta(\mathbf{p})$ be a diversity index satisfying the replication principle. Suppose $\theta(\mathbf{p})$ also satisfies $\mathcal{A}_{01}$ and $\mathcal{A}_{02}$. Then $\theta_0 = \theta(\mathbf{p}_o)$ is uniquely defined, that is, the index assumes a same value, namely $\theta_0 = \theta(\mathbf{p}_o)$, for every group containing one species. This implies that any two singleton alphabets, $\{\ell_1\}$ and $\{\ell_2\}$, are equally diverse by the underlying index $\theta(\mathbf{p})$. Since any alphabet with finite effective cardinality $K = \sum_{k\geq 1} 1[p_k > 0] \geq 2$ may be viewed as a combination of K singleton alphabets, it follows that

$$\theta(\mathbf{p}) = \theta_0 K \tag{4.30}$$

for any distribution $\mathbf{p} = \{p_1, \dots, p_K\}$ such that $p_k > 0$ for each k.

Suppose $\theta(\mathbf{p})$ also satisfies $\mathcal{A}_m^*$. Then it follows that $\theta_0 > 0$. Or else, if $\theta_0 \leq 0$, then $\theta_0 K \leq \theta_0 \leq 0$ would violate $\mathcal{A}_m^*$. This argument establishes that: if a diversity index, $\theta(\mathbf{p})$, satisfies $\mathcal{A}_{01}$, $\mathcal{A}_{02}$, $\mathcal{A}_m^*$, and the replication principle, then it must be of the form in (4.30) with θ_0 being a positive constant. Up to a positive multiplicative factor, θ_0, there is essentially only one index in the form of (4.30), the richness index K given in Example 4.15. Pinning down on a unique index is a major advantage of the replication principle. However, as in every situation of viewing a high-dimensional object in a reduced dimensional space, some finer aspects of the object often are lost. For example, a common complaint about the richness index, K, is that it does not distinguish between $\{p_1 = 0.5000, p_2 = 0.5000\}$ and $\{p_1 = 0.9999, p_2 = 0.0001\}$ in diversity.

Of course even if one has pinned down a good diversity index, it is often not observable in practice, and therefore the issues of sample-based estimation come to play. For all practical purposes, a third fundamental question should be added to the first two given above in the beginning of this section. The question is: how to estimate a diversity index. The answer to this question is unfortunately not trivial at all. A glance at the popular diversity indices in the literature reveals that, as argued earlier in this chapter, most of the indices are of the form, or equivalent to,

$$\theta = \sum_{k\geq 1} p_k h(p_k)$$

where the weight $h(p)$ loads heavily on small values of p. Such is the nature of many diversity indices. However, estimation of such indices runs into the same problems as those of estimation of Shannon's entropy, that is, it is difficult to recover the values of $p_k h(p_k)$ for the letters, ℓ_k's, that are not observed in a sample. The plug-in estimator tends to have large biases for the same reason the plug-in estimator of Shannon's entropy does.

Similar to the literature of Shannon's entropy estimation, there is no lack of proposed estimators of diversity indices that have reduced biases. However, not enough attention is paid to the distributional characteristics of these estimators. While it is perhaps difficult to develop asymptotic distributions for many of the proposed estimators, they nevertheless are necessary knowledge in practice. Only when sufficient distributional characteristics are known about an estimator, can one make scientific inferences based on it about the underlying population. This chapter summarizes a few distributional results for estimators of various diversity indices, but much more research is needed in this direction.

4.5 Exercises

1 Suppose $X_1, X_2, \dots, X_{v+1}$ are $v + 1$ *iid* random elements drawn from $\mathscr{X} = \{\ell_k; k \geq 1\}$ with probability distribution $\{p_k; k \geq 1\}$. Find the probability that not all $v + 1$ random elements take a same letter. Discuss the relevance of the above probabilities to diversity.

2 Suppose X_1, X_2, X_3 are *iid* random elements drawn from $\mathscr{X} = \{\ell_k; k \geq 1\}$ with probability distribution $\{p_k; k \geq 1\}$.

a) Find the probability that the random elements take three different letters.
b) Find the probability that the three random elements take exactly two different letters.
c) Find the probability that the three random elements take at least two different letters and discuss the relevance of the probability to diversity.

3 Let $g(x)$ be a differentiable function with domain $(-\infty, +\infty)$. Suppose $g(x)$ is strictly increasing.

a) Show that the inverse function of $g(x)$, $g^{-1}(t)$, exists on some interval (a, b).
b) Show that $g^{-1}(t)$ is strictly increasing on (a, b).

4 Suppose $f(s)$ and $g(t)$ are both continuous and strictly increasing functions with domain $(-\infty, +\infty)$. Let $h(t) = f(g(t))$.
a) Show that $h(t)$ is continuous on $(-\infty, +\infty)$.
b) Show that $h(t)$ is strictly increasing on $(-\infty, +\infty)$.

5 Let θ, θ_1, and θ_2 be three diversity indices. Use Definition 4.1 to show that if θ_1 and θ_2 are both equivalent to θ, then θ_1 is equivalent to θ_2.

6 Show that, for Simpson's index, $\lambda = \sum_{k\geq 1} p_k^2 = \zeta_0 - \zeta_1$.

7 Show that, for Rényi's equivalent entropy in (4.1),

$$h_\alpha = \sum_{k\geq 1} p_k^r = \zeta_0 + \sum_{v=1}^{\infty} \left[\prod_{i=1}^{v} \left(\frac{i-\alpha}{i} \right) \zeta_v \right],$$

where $\alpha > 0$ and $\alpha \neq 1$.

8 Show that, for Emlen's index,

$$D = \sum_{k\geq 1} p_k e^{-p_k} = \sum_{v=0}^{\infty} \frac{e^{-1}}{v!} \zeta_v.$$

9 Show that the last equation of Example 4.16 holds, that is,

$$\sum_{k\geq 1} p_k \sum_{v=0}^{u-1} (-1)^{u-1-v} \binom{u-1}{v} (1-p_k)^{u-1-v} (1-p_k)^m = \sum_{v=0}^{u-1} (-1)^v \binom{u-1}{v} \zeta_{m+v}.$$

10 Show that, under Condition 4.1, θ in (4.3) is well defined, that is,

$$\lim_{m\to\infty} \sum_{v=0}^{m} w_v \sum_{k\geq 1} p_k (1-p_k)^v$$

exists.

11 Verify that, in each of Examples 4.10 through 4.16, the linear coefficients w_v satisfies $|w_v| \leq M$ for some $M > 0$.

12 Show that the Gini–Simpson diversity index, $\theta = \sum_{k=1}^{K} p_k(1-p_k)$, satisfies both $\mathcal{A}_M$ and $\mathcal{A}_m$.

13 Show that Shannon's entropy

$$H = -\sum_{k=1}^{K} p_k \ln p_k$$

satisfies both $\mathcal{A}_M$ and $\mathcal{A}_m$.

14 Show that Rényi's entropy

$$H_\alpha = \frac{1}{1-\alpha} \ln \left(\sum_{k=1}^{K} p_k^\alpha \right),$$

where $\alpha > 0$ and $\alpha \neq 1$, satisfies both $\mathcal{A}_M$ and $\mathcal{A}_m$.

15 For the generalized Simpson's indices,

$$\zeta_{u,v} = \sum_{k=1}^{K} p_k^u (1-p_k)^v,$$

a) show that $\mathcal{A}_m$ is satisfied for any pair $u \geq 1$ and $v \geq 1$;
b) show that $\mathcal{A}_M$ is satisfied for any pair $u = v \geq 1$; and
c) give an example of $\{p_k\}$, u and v, such that $\mathcal{A}_M$ is not satisfied.

16 For a distribution $\{p_k; k = 1, \ldots, K\}$ on a finite alphabet $\mathscr{X}$, let

$$\theta = \sum_{k=1}^{K} p_k \ln^2 p_k.$$

a) Show that $\mathcal{A}_m$ is satisfied.
b) Give an example of $\{p_k\}$ such that $\mathcal{A}_M$ is not satisfied.
c) Discuss sufficient conditions on $\{p_k\}$ such that $\mathcal{A}_M$ is satisfied.

17 Consider a multinomial distribution with $K = 3$ categories with probabilities, p_1, p_2, and $p_3 = 1 - p_1 - p_2$, respectively. Let Y_1, Y_2, and Y_3 be the observed frequencies of the three different categories in an *iid* sample of size n. Show that

$$\sqrt{n}\left(\begin{pmatrix} Y_1/n \\ Y_2/n \end{pmatrix} - \begin{pmatrix} p_1 \\ p_2 \end{pmatrix} \right) \xrightarrow{L} MVN(\emptyset, \Sigma)$$

where $\Sigma = \begin{pmatrix} p_1(1-p_1) & -p_1p_2 \\ -p_1p_2 & p_2(1-p_2) \end{pmatrix}$.

18 Let $\mathbf{v}$ and $\hat{\mathbf{v}}$ be defined as in (4.9). Let $\Sigma(\mathbf{v})$ be defined as in (4.11). For each $j, j = 1, \ldots, K-1$, let $g_j(\hat{\mathbf{v}})$ be defined as in (4.12) if $\hat{p}_j > 0$, but let $g_j(\hat{\mathbf{v}}) = 0$ if $\hat{p}_j = 0$. Suppose $\hat{p}_K > 0$. Show $g^\tau(\hat{\mathbf{v}})\Sigma(\hat{\mathbf{v}})g(\hat{\mathbf{v}}) > 0$. (Hint: Consider

$\hat{\mathbf{p}} = \{\hat{p}_1, \dots, \hat{p}_K\}$. Deleting all the zeros in $\hat{\mathbf{p}}$, reenumerating and renaming the remaining nonzero components give $\mathbf{q} = \{q_1, \dots, q_{K^*}\}$ where $K^* = \sum_{k=1} 1[\hat{p}_k \neq 0]$, which is a probability distribution.)

19 Suppose $f(x)$ is a strictly monotone function on interval $[a, b]$. Then for any $x_1 \in [a, b]$ and $x_2 \in [a, b]$, $f(x_1) = f(x_2)$ implies $x_1 = x_2$.

20 Let

$$\mathbf{p}_1 = \{0.5, 0.5\} \text{ and } \mathbf{p}_2 = \{0.9, 0.1\}. \tag{4.31}$$

For each of the two probability distributions, calculate the diversity indices of the following.

a) The Gini–Simpson index.
b) Shannon's entropy.
c) Rényi's entropy.
d) Tsallis' entropy.
e) Hill's diversity number.
f) Emlen's index.
g) The richness index.

21 Show that Rényi's entropy, H_α, approaches Shannon's entropy, H, as α approaches 1, that is,

$$\lim_{\alpha \to 1} \frac{\ln\left(\sum_{k\geq 1} p_k^\alpha\right)}{1-\alpha} = -\sum_{k\geq 1} p_k \ln p_k.$$

22 Sketch and compare $\theta_{\mathbf{p}} = \sum_{k\geq 1} p_k (1-p_k)^v$ over the interval from $v = 1$ to $v = 50$ for both probability distributions in (4.31) of Exercise 20.

23 Sketch and compare $\theta_{\mathbf{p}} = \ln\left(\sum_{k\geq 1} p_k^\alpha\right)/(1-\alpha)$ over the interval from $\alpha = 0$ to $\alpha = 5$ for both probability distributions in (4.31) of Exercise 20.

5

Estimation of Information

5.1 Introduction

Consider the following three countable alphabets

$$\begin{aligned}\mathscr{X} &= \{x_i; i \geq 1\},\\ \mathscr{Y} &= \{y_j; j \geq 1\},\\ \mathscr{X} \times \mathscr{Y} &= \{(x_i, y_j); i \geq 1, j \geq 1\}.\end{aligned}$$

Given a joint probability distribution

$$\mathbf{p}_{X,Y} = \{p_{i,j}; i \geq 1, j \geq 1\}$$

on $\mathscr{X} \times \mathscr{Y}$, consider the two marginal probability distributions on $\mathscr{X}$ and $\mathscr{Y}$, respectively,

$$\begin{aligned}\mathbf{p}_X &= \left\{ p_{i,\cdot} = \sum_{j\geq 1} p_{i,j}; i \geq 1 \right\}\\ \mathbf{p}_Y &= \left\{ p_{\cdot,j} = \sum_{i\geq 1} p_{i,j}; i \geq 1 \right\},\end{aligned}$$

and denote the two underlying random elements as X and Y. Furthermore, let

$$\mathbf{p}_{XY} = \{p_{i,\cdot}\, p_{\cdot,j}; i \geq 1, j \geq 1\}$$

(to be distinguished from $\mathbf{p}_{X,Y} = \{p_{i,j}; i \geq 1, j \geq 1\}$).

Consider also the conditional probability distributions, respectively, of X on $\mathscr{X}$ given $Y = y_j$ where $j \geq 1$ is a specific index value and of Y on $\mathscr{Y}$ given $X = x_i$ where $i \geq 1$ is a specific index value, that is,

$$\begin{aligned}\mathbf{p}_{X|y_j} &= \left\{ p_{x_i|y_j} = \frac{p_{i,j}}{\sum_{k\geq 1} p_{k,j}}; i \geq 1 \right\}\\ \mathbf{p}_{Y|x_i} &= \left\{ p_{y_j|x_i} = \frac{p_{i,j}}{\sum_{k\geq 1} p_{i,k}}; j \geq 1 \right\}.\end{aligned}$$

Statistical Implications of Turing's Formula, First Edition. Zhiyi Zhang.

By Definition 3.1, entropy may be defined for each of these probability distributions, that is, for $\mathbf{p}_{X,Y}$, $\mathbf{p}_X$, $\mathbf{p}_Y$, $\mathbf{p}_{X|y_j}$, and $\mathbf{p}_{Y|x_i}$, and the respective entropies are

$$\begin{aligned} H(X,Y) &= -\sum_{i\geq 1}\sum_{j\geq 1} p_{i,j} \ln p_{i,j} \\ H(X) &= -\sum_{i\geq 1} p_{i,\cdot} \ln p_{i,\cdot} \\ H(Y) &= -\sum_{j\geq 1} p_{\cdot,j} \ln p_{\cdot,j} \\ H(X|Y=y_j) &= -\sum_{i\geq 1} p_{x_i|y_j} \ln p_{x_i|y_j} \\ H(Y|X=x_i) &= -\sum_{j\geq 1} p_{y_j|x_i} \ln p_{y_j|x_i}. \end{aligned}$$

Definition 5.1 *Given a joint probability distribution* $\mathbf{p}_{X,Y} = \{p_{i,j}\}$ *on* $\mathscr{X} \times \mathscr{Y}$, *the (expected) conditional entropy of* Y *given* X *is*

$$H(Y|X) = \sum_{i\geq 1} p_{i,\cdot} H(Y|X=x_i). \tag{5.1}$$

An exchange of X and Y in Definition 5.1 gives the expected conditional entropy of X given Y is

$$H(X|Y) = \sum_{j\geq 1} p_{\cdot,j} H(X|Y=y_j).$$

In information theory, $H(Y|X)$ is often referred to as "the conditional entropy of Y given X." However, it is to be noted that it is not entropy of a probability distribution in the sense of Definition 3.1. $H(Y|X=x_i)$ is entropy of a conditional probability distributionand so is $H(X|Y=y_j)$. $H(Y|X)$ is a weighted average of conditional entropies $H(Y|X=x_i)$, $i \geq 1$, with respect to the marginal probability distribution of X, and $H(X|Y)$ is a weighted average of conditional entropies $H(X|Y=y_j)$, $j \geq 1$, with respect to the marginal probability distribution of Y.

Lemma 5.1 *Given a joint probability distribution* $\mathbf{p}_{X,Y} = \{p_{i,j}; i \geq 1, j \geq 1\}$ *on* $\mathscr{X} \times \mathscr{Y}$,

1) $H(X,Y) = H(X) + H(Y|X)$ *and*
2) $H(X,Y) = H(Y) + H(X|Y)$.

Proof: For Part 1, noting that, for each pair of indices (i,j), $p_{i,j} \leq p_{i,\cdot}$, $\ln p_{i,j} \leq \ln p_{i,\cdot}$, and hence $-\ln p_{i,j} \geq -\ln p_{i,\cdot} \geq 0$,

$$\begin{aligned} H(X,Y) &= -\sum_{i\geq 1}\sum_{j\geq 1} p_{i,j} \ln p_{i,j} \\ &= -\sum_{i\geq 1}\sum_{j\geq 1} p_{i,j} \ln(p_{y_j|x_i} p_{i,\cdot}) \end{aligned}$$

$$
\begin{aligned}
&= -\sum_{i\geq 1}\sum_{j\geq 1} p_{i,j} \ln p_{y_j|x_i} - \sum_{i\geq 1}\sum_{j\geq 1} p_{i,j} \ln p_{i,\cdot} \\
&= -\sum_{i\geq 1} p_{i,\cdot} \sum_{j\geq 1} p_{y_j|x_i} \ln p_{y_j|x_i} - \sum_{i\geq 1} p_{i,\cdot} \ln p_{i,\cdot} \\
&= H(Y|X) + H(X).
\end{aligned}
$$

Part 2 follows from the symmetry of $H(X, Y)$ with respect to X and Y. □

Corollary 5.1 *Given a joint probability distribution* $\mathbf{p}_{X,Y} = \{p_{i,j}; i \geq 1, j \geq 1\}$ *on* $\mathscr{X} \times \mathscr{Y}$*,*

1) $H(X) \leq H(X, Y)$,
2) $H(Y) \leq H(X, Y)$, *and*
3) $H(X) + H(Y) - H(X, Y) \leq H(X, Y)$.

Proof: Since, for each $i \geq 1$, the entropy of the conditional probability distribution $H(Y|X = x_i)$ is non-negative, the expected conditional entropy $H(Y|X) \geq 0$, being a positively weighted average of the conditional entropies, is also non-negative. For Part 1, by Lemma 5.1, the desired result follows from

$$H(X, Y) = H(X) + H(Y|X) \geq H(X).$$

Part 2 follows similarly. Part 3 is a consequence of Parts 1 and 2. □

By Corollary 5.1, the condition $H(X, Y) < \infty$ implies both $H(X) < \infty$ and $H(Y) < \infty$.

Consider two probability distributions

$$\mathbf{p} = \{p_k; k \geq 1\} \quad \text{and} \quad \mathbf{q} = \{q_k; k \geq 1\}$$

on a same alphabet $\mathscr{X}$.

Definition 5.2 *For two probability distributions* $\mathbf{p}$ *and* $\mathbf{q}$ *on a same alphabet* $\mathscr{X}$*, the relative entropy or the Kullback–Leibler divergence of* $\mathbf{p}$ *and* $\mathbf{q}$ *is*

$$D(\mathbf{p}\|\mathbf{q}) = \sum_{k\geq 1} p_k \ln\left(\frac{p_k}{q_k}\right), \tag{5.2}$$

observing the conventions that, for each summand $p \ln(p/q)$*,*

1) $p \ln(p/q) = 0$, *if* $p = 0$, *and*
2) $p \ln(p/q) = +\infty$, *if* $p > 0$ *and* $q = 0$.

By the above-mentioned definition, if there exists a k such that $p_k > 0$ and $q_k = 0$, then $D(\mathbf{p}\|\mathbf{q}) = +\infty$.

Suppose $\mathbf{p}$ and $\mathbf{q}$ have the same support, that is, for each k, $p_k > 0$ if and only if $q_k > 0$. Then letting $\{r_i; i \geq 1\}$ be the sequence of distinct values in

$$\{q_k/p_k \,:\, k \geq 1 \text{ and } p_k > 0\}$$

and

$$p_i^* = \sum_{k:q_k/p_k=r_i} p_k,$$

the following table defines a random variable R and its probability distribution.

$$\begin{array}{c|cccc} R & r_1 & \cdots & r_i & \cdots \\ \hline \mathrm{P}(r) & p_1^* & \cdots & p_i^* & \cdots \end{array}. \tag{5.3}$$

Theorem 5.1 *Let* $\mathbf{p}$ *and* $\mathbf{q}$ *be two probability distributions on a same alphabet* $\mathscr{X}$ *and with the same support. Suppose* $\mathrm{E}(R) < \infty$ *where* R *is as given in (5.3). Then*

$$D(\mathbf{p}||\mathbf{q}) \geq 0. \tag{5.4}$$

Moreover, the equality holds if and only if $\mathbf{p} = \mathbf{q}$.

To prove Theorem 5.1, Jensen's inequality is needed and is stated below as a lemma without proof.

Lemma 5.2 *(Jensen's inequality) If* $g(x)$ *is a convex function on an open interval* (a, b) *and* X *is a random variable with* $\mathrm{P}(X \in (a, b)) = 1$ *and finite expectation, then*

$$\mathrm{E}(g(X)) \geq g(\mathrm{E}(X)). \tag{5.5}$$

Moreover, if $g(x)$ *is strictly convex, then* $\mathrm{E}(g(X)) = g(\mathrm{E}(X))$ *if and only if* $\mathrm{P}(X = c) = 1$ *where* c *is some constant.*

Proof of Theorem 5.1: By Definition 5.2 and noting $g(t) = -\ln(t)$ is a strictly convex function on $(0, \infty)$,

$$\begin{aligned} D(\mathbf{p}||\mathbf{q}) &= \sum_{k\geq 1} p_k \ln\left(\frac{p_k}{q_k}\right) = \sum_{k:p_k>0} p_k \ln\left(\frac{p_k}{q_k}\right) = \sum_{i\geq 1} p_i^*(-\ln r_i) \\ &= \mathrm{E}(-\ln R) \geq -\ln \mathrm{E}(R) \qquad (5.6) \\ &= -\ln\left(\sum_{i\geq 1} p_i^* r_i\right) = -\ln\left(\sum_{i\geq 1}\left(\sum_{k:q_k/p_k=r_i} p_k\right) r_i\right) \\ &= -\ln\left(\sum_{i\geq 1}\left(\sum_{k:q_k/p_k=r_i} p_k \frac{q_k}{p_k}\right)\right) \\ &= -\ln\left(\sum_{i\geq 1}\left(\sum_{k:q_k/p_k=r_i} q_k\right)\right) = -\ln 1 = 0 \end{aligned}$$

where the inequality in (5.6) is due to Lemma 5.2.

If $D(\mathbf{p}||\mathbf{q}) = 0$, then the inequality in (5.6) is forced to be an equality, which, again by Lemma 5.2, implies that R is constant with probability one, that is, $q_k/p_k = r$ or $q_k = p_k r$ for every k for which $p_k > 0$. Since $\mathbf{p}$ and $\mathbf{q}$ have the same support, summing over all k gives $r = 1$, that is, $p_k = q_k$ for every k. This completes the proof. □

Theorem 5.1 gives fundamental support to the Kullback–Leibler divergence as a reasonable measure of the difference between two distributions on a common alphabet. However, it is to be noted that $D(\mathbf{p}||\mathbf{q})$ is not symmetric with respect to $\mathbf{p}$ and $\mathbf{q}$, and therefore not a distance measure. For this reason, the following symmetrized Kullback–Leibler divergence, as given in Kullback and Leibler (1951), is sometimes used.

$$D(\mathbf{p}, \mathbf{q}) = D(\mathbf{p}||\mathbf{q}) + D(\mathbf{q}||\mathbf{p}). \tag{5.7}$$

Definition 5.3 *The mutual information (MI) of random elements, $(X, Y) \in \mathcal{X} \times \mathcal{Y}$ with a joint probability distribution $\mathbf{p}_{X,Y}$, is*

$$MI = MI(X, Y) = D(\mathbf{p}_{X,Y}||\mathbf{p}_{XY}) = \sum_{i\geq 1}\sum_{j\geq 1} p_{i,j} \ln\left(\frac{p_{i,j}}{p_{i,\cdot}p_{\cdot,j}}\right). \tag{5.8}$$

Mutual information may be viewed as a degree of dependence between X and Y as measured by the Kullback–Leibler divergence between $\mathbf{p}_{X,Y}$ and $\mathbf{p}_{XY}$ on $\mathcal{X} \times \mathcal{Y}$. Definition 5.3 and Theorem 5.1 immediately give Theorem 5.2.

Theorem 5.2 *For any joint distribution $\mathbf{p}_{X,Y}$ on $\mathcal{X} \times \mathcal{Y}$,*

$$MI(X, Y) \geq 0. \tag{5.9}$$

Moreover, the equality holds if and only if X and Y are independent.

Theorem 5.3 *Suppose $H(X, Y) < \infty$ for a joint distribution $\mathbf{p}_{X,Y}$ on $\mathcal{X} \times \mathcal{Y}$. Then*

$$MI(X, Y) = H(X) + H(Y) - H(X, Y), \tag{5.10}$$

$$MI(X, Y) = H(Y) - H(Y|X), \tag{5.11}$$

$$MI(X, Y) = H(X) - H(X|Y), \tag{5.12}$$

$$MI(X, Y) = MI(Y, X), \tag{5.13}$$

$$MI(X, X) = H(X), \quad \textit{and} \tag{5.14}$$

$$MI(X, Y) \leq H(X, Y). \tag{5.15}$$

Proof: By Corollary 5.1, $H(X, Y) < \infty$ implies $H(X) < \infty$ and $H(Y) < \infty$, and it follows that

$$MI(X, Y) = \sum_{i\geq 1}\sum_{j\geq 1} p_{i,j} \ln\left(\frac{p_{i,j}}{p_{i,\cdot}p_{\cdot,j}}\right)$$

$$= \sum_{i \geq 1} \sum_{j \geq 1} p_{i,j} (\ln p_{i,j} - \ln p_{i,\cdot} - \ln p_{\cdot,j})$$

$$= H(X) + H(Y) - H(X, Y),$$

and therefore (5.10) holds.

By Parts 1 and 2 of Lemma 5.1,

$$MI(X, Y) = H(X) + H(Y) - H(X, Y) = H(Y) - H(Y|X),$$

$$MI(X, Y) = H(X) + H(Y) - H(X, Y) = H(X) - H(X|Y),$$

and therefore (5.11) and (5.12) hold.

Equation (5.13) is obvious by definition.

Noting $P(X = x, X = x) = P(X = x)$, $MI(X, X) = H(X)$ by definition, that is, (5.14) holds.

Equation (5.15) is a restatement of Part 3 of Corollary 5.1. □

In Theorem 5.3, MI is expressed in various linear forms of entropies. These forms are well defined if the joint distribution $\mathbf{p}_{X,Y}$ has a finite support in $\mathcal{X} \times \mathcal{Y}$. However, if the support is infinite, these linear forms of entropies only have well-defined meaning under the condition of $H(X, Y) < \infty$, which implies $H(X) < \infty$ and $H(Y) < \infty$.

Example 5.1 *Let, for* $i = 1, 2, \dots,$

$$p_i = \frac{1}{C(i+1)\ln^2(i+1)}$$

where C is a constant such that $\sum_{i \geq 1} p_i = 1$. *It was demonstrated in Example 3.3 that this distribution has an infinite entropy. Let the joint distribution* $\mathbf{p}_{X,Y}$ *be such that*

$$p_{i,j} = p_i \times p_j$$

for each pair of (i, j), *where* $i \geq 1$ *and* $j \geq 1$. *By Definition 5.3,* $MI = 0$. *However, by Corollary 5.1,* $H(X, Y) = \infty$, *and therefore, the condition of Theorem 5.3 does not hold.*

Entropy may be viewed as a dispersion measure on an alphabet, much like the variance of a random variable on the real line. It is perhaps interesting to note that the expression in (5.11) (and therefore also that in (5.12)) has an ANOVA-like intuitive interpretation as follows.

ANOVA-Like Interpretation of Mutual Information: *$H(Y)$ is the total amount of random dispersion in Y. $H(Y|X)$ is the (average) amount of dispersion in Y that is unexplained by X. The remainder, $H(Y) - H(Y|X)$, therefore may be viewed as the amount of dispersion in Y that is explained by*

X. The larger the value of $MI(X, Y) = H(Y) - H(Y|X)$, the higher the level of association is between X and Y.

To illustrate the characteristics of *MI*, consider the following three examples.

Example 5.2 *Let $\mathbf{p}_{X,Y} = \{p_{i,j}\}$ be such that*

$$p_{i,j} = \begin{cases} \frac{1}{33} & \text{if } 1 \le i \le 11 \text{ and } 1 \le j \le 3 \\ 0 & \text{otherwise} \end{cases}$$

where i and j are positive integers. Let $\mathcal{X}$ be a subset of the xy-plane containing the following 33 points, denoted by (x_i, y_j):

$$(x_i, y_j) = (0.1(i - 1), 0.6 - 0.1(j - 1))$$

where $i = 1, \ldots, 11$ and $j = 1, 2, 3$ (see Figure 5.1a).

Consider the pair of random variables (X, Y) such that

$$\mathrm{P}((X, Y) = (x_i, y_j)) = p_{i,j}.$$

Since X and Y are independent,

$$\rho = \frac{\mathrm{Cov}(X, Y)}{\sqrt{\mathrm{Var}(X) \times \mathrm{Var}Y}} = 0 \quad \text{and} \quad MI(X, Y) = 0. \tag{5.16}$$

Example 5.3 *Let $\mathbf{p}_{X,Y} = \{p_{i,j}\}$ be such that*

$$p_{i,j} = \begin{cases} \frac{1}{42} & \text{if } 1 \le i \le 21 \text{ and } 1 \le j \le 2 \\ 0 & \text{otherwise} \end{cases}$$

where i and j are positive integers. Let $\mathcal{X}$ be a subset of the xy-plane containing the following 42 points, denoted by (x_i, y_j):

$$(x_i, y_j) = (0.05(i - 1), 0.05(2 + i - 2j))$$

where $i = 1, \ldots, 21$ and $j = 1, 2$ (see Figure 5.1b).

Consider the pair of random variables (X, Y) such that

$$\mathrm{P}((X, Y) = (x_i, y_j)) = p_{i,j}.$$

$$H(X) = \ln 21, \tag{5.17}$$

$$H(Y) = \frac{4}{42} \ln 42 + \frac{19}{21} \ln 21, \tag{5.18}$$

$$H(X, Y) = \ln 42. \tag{5.19}$$

It may be verified that

$$\rho = \frac{\mathrm{Cov}(X, Y)}{\sqrt{\mathrm{Var}(X) \times \mathrm{Var}Y}} = 0.9866 \quad \text{and} \quad MI(X, Y) = 2.4274. \tag{5.20}$$

Example 5.4 *Let* $\mathbf{p}_{X,Y} = \{p_{i,j}\}$ *be such that*

$$p_{i,j} = \begin{cases} \frac{1}{80} & \text{if } 1 \le i \le 80 \text{ and } j = 1 \\ 0 & \text{otherwise} \end{cases}$$

where i and j are positive integers. Let $\mathcal{X}$ *be a subset of the xy-plane containing the following 80 points, denoted by* (x_i, y_j)*:*

$$(x_i, y_j) = \begin{cases} \left(-2\pi + (i-1)\frac{4\pi}{80}, \cos(x_i)\right) & \text{if } 1 \le i \le 40 \\ \left(-2\pi + i\frac{4\pi}{80}, \cos(x_i)\right) & \text{if } 41 \le i \le 80 \end{cases}$$

(see Figure 5.1c).

Consider the pair of random variables (X, Y) *such that*

$$P((X, Y) = (x_i, y_j)) = p_{i,j}.$$

$$H(X) = \ln 80, \tag{5.21}$$

$$H(Y) = 19\left(\frac{4}{80}\right)\ln\left(\frac{80}{4}\right) + 2\left(\frac{2}{80}\right)\ln\left(\frac{80}{2}\right), \tag{5.22}$$

$$H(X, Y) = \ln 80. \tag{5.23}$$

It may be verified that

$$\rho = \frac{\mathrm{Cov}(X, Y)}{\sqrt{\mathrm{Var}(X) \times \mathrm{Var}Y}} = 0 \quad \text{and} \quad MI(X, Y) = 3.0304. \tag{5.24}$$

Examples 5.2–5.4 demonstrate that MI has an advantage over the coefficient of correlation ρ in capturing a possibly nonlinear relationship between a pair of random variables (X, Y). In fact, an even stronger characteristic of MI is demonstrated by revisiting Example 5.4.

Suppose, in Example 5.4, the random pair (X, Y) is redefined in such a way that it amounts to an exchange of positions along the x-axis (see Figure 5.2a) or an exchange of positions along the y-axis (see Figure 5.2b). The values of mutual information $MI(X, Y)$ remain unchanged under such permutations on the index sets, $\{i; i \ge 1\}$ or $\{j; j \ge 1\}$, respectively. In fact, Figure 5.2c represents the set of (x_i, y_j) points in Example 5.4 but after a sequence of permutations, first on $\{i; i \ge 1\}$ and then on $\{j; j \ge 1\}$, respectively. The resulting data set, at a first glance, lacks any visible relationship between X and Y, linear or nonlinear. Yet, the mutual information $MI(X, Y)$ in Figure 5.2c remains unchanged from Figure 5.1c. The seemingly anti-intuitive phenomenon here is caused by the fact that one is often conditioned and, therefore, expects to see a continuous functional relationship between X and Y. Such a functional relationship does not necessarily exist on a general joint alphabet $\mathcal{X} \times \mathcal{Y}$, where the letters do not necessarily have any natural orders much less numerical meanings.

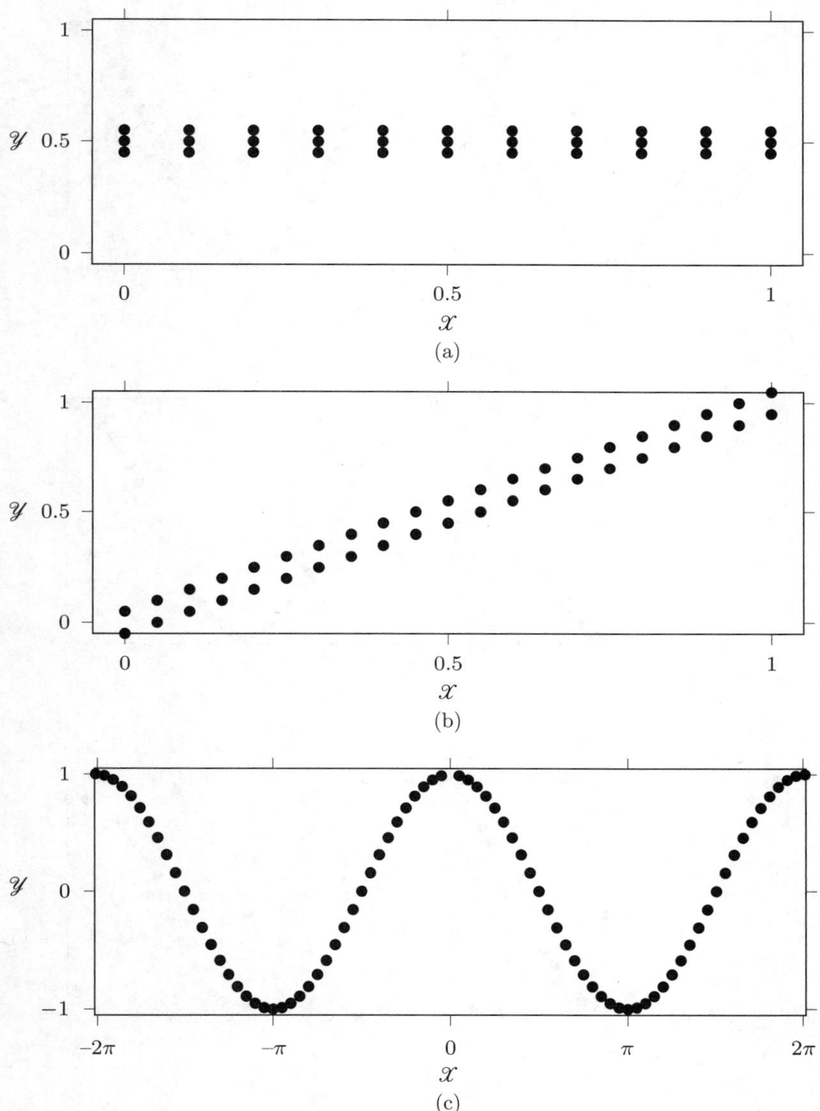

Figure 5.1 Letters with positive probabilities: (a) Example 5.2, (b) Example 5.3, and (c) Example 5.4

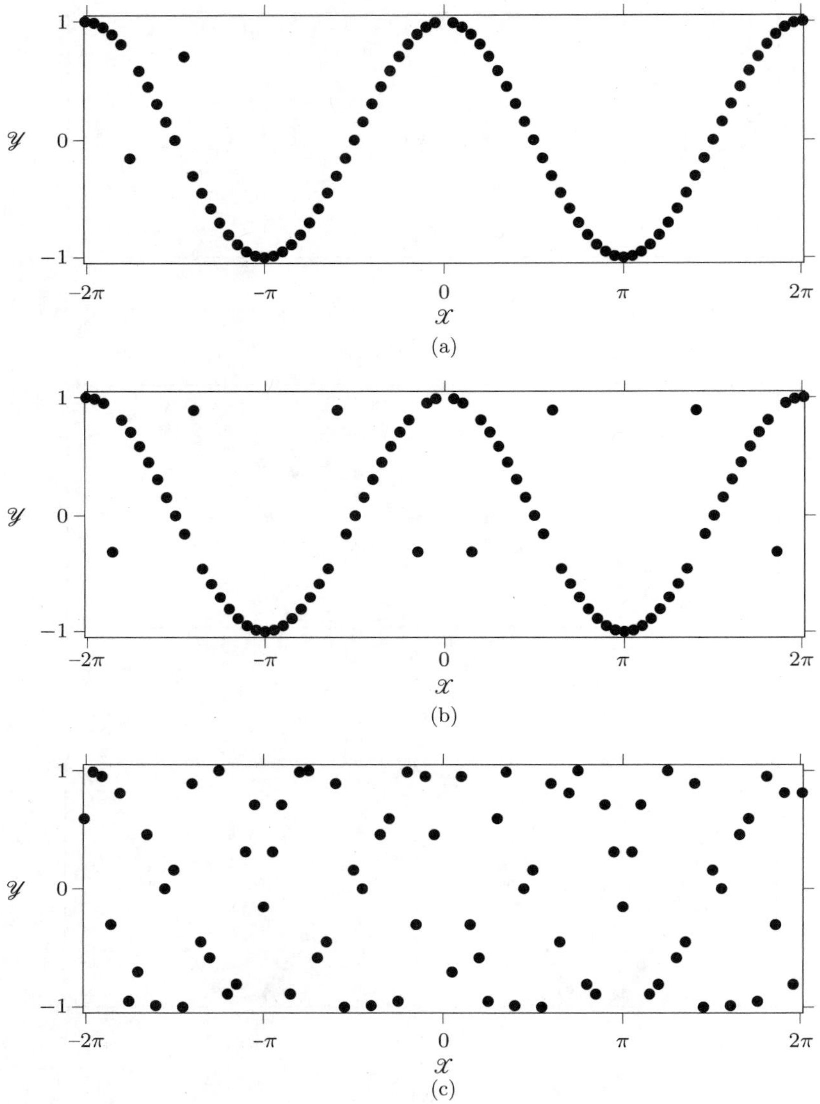

Figure 5.2 Invariance of *MI* under respective permutations by letters in $\mathscr{X}$ and $\mathscr{Y}$

The unchanged positive *MI* in Figure 5.2c from Figure 5.1c is explained by the fact that, for each given x_i in Figure 5.2c, there is one and only one y_j carrying positive probability. That is to say that the y-value is uniquely determined by an x-value, and hence a higher degree of association and *MI*. Since this relationship is not changed from Figure 5.1c, the *MI* remains unchanged.

While MI as in Definition 5.3 has been shown to have many desirable properties, it does not have a uniform upper bound since entropies can be arbitrarily large. The lack of such an upper bound for MI makes it difficult to convey an intuitive sense of how strong the association between X and Y is even when it is known to be positive. To alleviate this issue, the standardized mutual information (SMI), κ, between X and Y is given in Definition 5.4 below.

Definition 5.4 *Provided that $H(X,Y) < \infty$, the standardized MI is given by*

$$\kappa = \frac{MI}{H(X,Y)} = \frac{H(X)+H(Y)-H(X,Y)}{H(X,Y)} = \frac{H(X)+H(Y)}{H(X,Y)} - 1. \tag{5.25}$$

Definition 5.5 *Random elements $X \in \mathscr{X}$ and $Y \in \mathscr{Y}$ are said to have a one-to-one correspondence under a joint probability distribution $\mathbf{p}_{X,Y}$ on $\mathscr{X} \times \mathscr{Y}$ if*

1) *for every i satisfying $\mathrm{P}(X = x_i) > 0$, there exists a unique j such that $\mathrm{P}(Y = y_j | X = x_i) = 1$, and*
2) *for every j satisfying $\mathrm{P}(Y = y_j) > 0$, there exists a unique i such that $\mathrm{P}(X = x_i | Y = y_j) = 1$.*

Theorem 5.4 *Suppose $H(X,Y) < \infty$. Then*

$$0 \le \kappa \le 1. \tag{5.26}$$

Moreover,

1) *$\kappa = 0$ if and only if X and Y are independent, and*
2) *$\kappa = 1$ if and only if X and Y have a one-to-one correspondence.*

Proof: By Theorem 5.2 and Corollary 5.1,

$$0 \le MI \le H(X,Y).$$

Dividing all three parts above by $H(X,Y)$ gives (5.26).

Since $\kappa = 0$ if and only if $MI = 0$, $\kappa = 0$ if and only if X and Y are independent by Theorem 5.2.

If X and Y have a one-to-one correspondence, then for each i satisfying $\mathrm{P}(X = x_i) > 0$ let the unique corresponding j be denoted by j_i. Noting $p_{i,j_i} = p_{i,\cdot}$ and therefore $\ln(p_{i,j_i}/p_{i,\cdot}) = 0$,

$$\begin{aligned} H(X,Y) &= -\sum_{i\ge1,j\ge1} p_{i,j} \ln p_{i,j} \\ &= -\sum_{j\ge1}\sum_{i\ge1} p_{i,\cdot}\left(\frac{p_{i,j}}{p_{i,\cdot}}\right) \ln\left[p_{i,\cdot}\left(\frac{p_{i,j}}{p_{i,\cdot}}\right)\right] \end{aligned}$$

$$= -\sum_{j\geq 1}\sum_{i\geq 1} p_{i,\cdot}\left(\frac{p_{i,j}}{p_{i,\cdot}}\right)\ln p_{i,\cdot} - \sum_{j\geq 1}\sum_{i\geq 1} p_{i,\cdot}\left(\frac{p_{i,j}}{p_{i,\cdot}}\right)\ln\left(\frac{p_{i,j}}{p_{i,\cdot}}\right)$$

$$= -\sum_{i\geq 1} p_{i,\cdot}\ln p_{i,\cdot} - \sum_{i\geq 1} p_{i,\cdot}\left(\frac{p_{i,j_i}}{p_{i,\cdot}}\right)\ln\left(\frac{p_{i,j_i}}{p_{i,\cdot}}\right)$$

$$= H(X).$$

Similarly, $H(X,Y) = H(Y)$, and therefore $H(X) + H(Y) = 2H(X,Y)$, in turn, it implies

$$\kappa = \frac{H(X) + H(Y)}{H(X,Y)} - 1 = 2 - 1 = 1.$$

On the other hand, if $\kappa = 1$, then $H(X) + H(Y) = 2H(X,Y)$ and therefore

$$(H(X) - H(X,Y)) + (H(Y) - H(X,Y)) = 0.$$

However, by Corollary 5.1, both of the above-mentioned additive terms are nonpositive, which implies

$$H(X) = H(X,Y) \quad \text{and} \quad H(Y) = H(X,Y),$$

which in turn, by Lemma 5.1, imply

$$H(X|Y) = 0 \quad \text{and} \quad H(Y|X) = 0.$$

By Definition 5.1, $H(Y|X) = 0$ implies $H(Y|X = x_i) = 0$ for every i satisfying $\mathrm{P}(X = x_i) > 0$, which in turn implies that the conditional probability distribution of Y given $X = x_i$ puts a probability of mass one on a single point in $\mathscr{Y}$. By symmetry, $H(X|Y) = 0$ implies that the conditional probability distribution of X given $Y = y_j$ also puts a probability of mass one on a single point in $\mathscr{X}$. □

Example 5.5 *Revisiting Example 5.2 and noting* $H(X,Y) = \ln 33 > 0$,

$$\rho = 0,\ MI = 0,\ \textit{and}\ \kappa = 0.$$

Example 5.6 *Revisiting Example 5.3 and noting* $H(X,Y) = \ln 42$,

$$\rho = 0.9866,\ MI = 2.4174,\ \textit{and}\ \kappa = 0.6468.$$

Example 5.7 *Consider the following set of five points,* (x,y)*, on the xy-plane with the given associated probability distribution.*

(X,Y)	$(-2,2)$	$(-1,-1)$	$(0,0)$	$(1,1)$	$(2,-2)$
$\mathrm{P}(X = x, Y = y)$	0.05	0.2	0.5	0.2	0.05

See Figure 5.3.

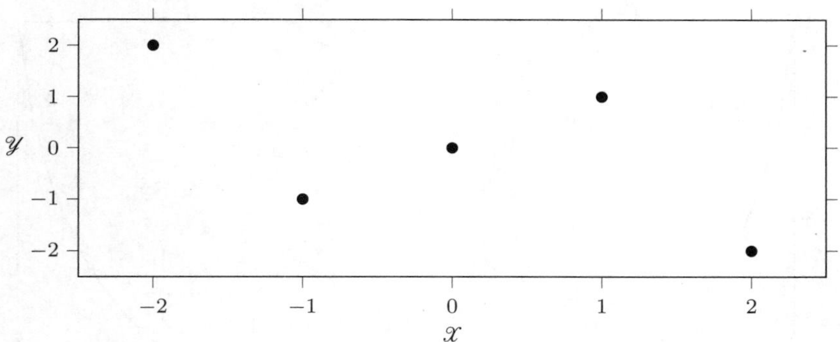

Figure 5.3 Points with positive probabilities in Example 5.7

It may be verified that

$$H(X) = H(Y) = H(X, Y) = 1.2899 \tag{5.27}$$

and that

$$\rho = 0, \quad MI(X, Y) = 1.2899\,, \quad \text{and} \quad \kappa = 1. \tag{5.28}$$

Note that the fact $\kappa = 1$ may be directly derived from Theorem 5.4.

Example 5.8 *Revisiting Example 5.4 and noting $H(X, Y) = \ln 80$,*

$$\rho = 0, \quad MI = 3.0304, \quad \text{and} \quad \kappa = 0.6916.$$

Note that $\kappa < 1$ in this case because, while X uniquely determines Y, Y does not uniquely determine X. Readers are reminded that the information being measured by MI, and hence κ, is "mutual," and it requires a one-to-one correspondence between X and Y for $\kappa = 1$.

The next example illustrates how the standardized *MI*, κ, changes as the joint probability distribution changes.

Example 5.9 *Consider the following family of joint probability distributions of (X, Y) on $\{0, 1\} \times \{0, 1\}$. For each m, $m = 0, 1, 2, \ldots, 50$,*

	$X = 0$	$X = 1$
$Y = 0$	$0.50 - 0.01\,m$	$0.01\,m$
$Y = 1$	$0.01\,m$	$0.50 - 0.01\,m$

or

$$\begin{array}{|c|c|}\hline 0.50 & 0.00 \\ \hline 0.00 & 0.50 \\ \hline \end{array}\,, \begin{array}{|c|c|}\hline 0.49 & 0.01 \\ \hline 0.01 & 0.49 \\ \hline \end{array}\,, \ldots, \begin{array}{|c|c|}\hline 0.25 & 0.25 \\ \hline 0.25 & 0.25 \\ \hline \end{array}\,, \ldots, \begin{array}{|c|c|}\hline 0.00 & 0.50 \\ \hline 0.50 & 0.00 \\ \hline \end{array}\,. \tag{5.29}$$

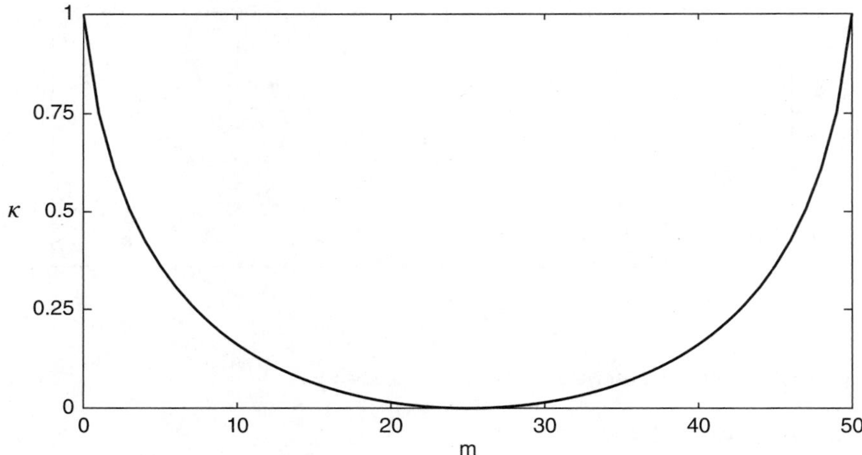

Figure 5.4 Plot of κ of (5.30) for $m = 0, \ldots, 50$.

When $m = 1$ or $m = 50$, $\kappa = 1$, by Theorem 5.4. For any other m,

$$\kappa = \frac{\ln 2 - [(m/50)\ \ln(50/m - 1) - \ln(1 - m/50)]}{\ln 2 + [(m/50)\ \ln(50/m - 1) - \ln(1 - m/50)]}. \tag{5.30}$$

See Exercise 20.

A plot of κ as a function of m is given in Figure 5.4. As m increases from $m = 0$, the distribution at the head of (5.29) moves away from a perfect one-to-one correspondence between X and Y in increments of 2% each. First, κ decreases rapidly from $\kappa = 1$, then changes slowly as it moves closer to $\kappa = 0$ at $m = 25$, which represents a state of independence between X and Y, and finally accelerates its ascend to $\kappa = 1$ again as it approaches $m = 50$.

The sharp drop of κ from the one observed in Example 5.9 as m moves away from zero is quite typical for the standardized MI, when a small perturbation is imposed on a joint distribution with a perfect one-to-one correspondence. Consequently, in most realistic data sets, the calculated standardized MI is often significantly below one. For a comprehensive discussion on some of the above-mentioned indices of information, as well as many others, readers may refer to Cover and Thomas (2006).

5.2 Estimation of Mutual Information

In this section, two estimators of MI are described, one is the plug-in estimator and the other is based on Turing's perspective. Both of them have asymptotic normality. Toward that end, let

$$\mathscr{X} = \{x_i; i = 1, \ldots, K_1\} \quad \text{and} \quad \mathscr{Y} = \{y_j; j = 1, \ldots, K_2\} \tag{5.31}$$

be two finite alphabets with cardinalities $K_1 < \infty$ and $K_2 < \infty$, respectively. Consider the Cartesian product $\mathscr{X} \times \mathscr{Y}$ with a joint probability distribution $\mathbf{p}_{X,Y} = \{p_{i,j}\}$. Assume that $p_{i,\cdot} > 0$ and $p_{\cdot,j} > 0$ for all $1 \leq i \leq K_1$ and $1 \leq j \leq K_2$. Let

$$K = \sum_{i,j} 1[p_{i,j} > 0] \tag{5.32}$$

be the number of positive joint probabilities in $\{p_{i,j}\}$.

For every pair of (i,j), let $f_{i,j}$ be the observed frequency of the random pair (X, Y) taking value (x_i, y_j), where $i = 1, \dots, K_1$ and $j = 1, \dots, K_2$, in an *iid* sample of size n from $\mathscr{X} \times \mathscr{Y}$; and let $\hat{p}_{i,j} = f_{i,j}/n$ be the corresponding relative frequency. Consequently, $\hat{\mathbf{p}}_{X,Y} = \{\hat{p}_{i,j}\}$, $\hat{\mathbf{p}}_X = \{\hat{p}_{i,\cdot}\}$, and $\hat{\mathbf{p}}_Y = \{\hat{p}_{\cdot,j}\}$ are the sets of observed joint and marginal relative frequencies.

5.2.1 The Plug-In Estimator

To be instructive, consider first the case of $K = K_1K_2$, that is, $p_{i,j} > 0$ for every pair (i,j) where $i = 1, \dots, K_1$ and $i = 1, \dots, K_2$. Let $\mathbf{p}$ be a specifically arranged $p_{i,j}$ as follows.

$$\mathbf{p} = (p_{1,1}, p_{1,2}, \dots, p_{1,K_2}, p_{2,1}, p_{2,2}, \dots, p_{2,K_2}, \dots, p_{K_1,1}, \dots, p_{K_1,K_2-1})^\tau.$$

Accordingly let

$$\hat{\mathbf{p}} = (\hat{p}_{1,1}, \hat{p}_{1,2}, \dots, \hat{p}_{1,K_2}, \hat{p}_{2,1}, \hat{p}_{2,2}, \dots, \hat{p}_{2,K_2}, \dots, \hat{p}_{K_1,1}, \dots, \hat{p}_{K_1,K_2-1})^\tau.$$

For notation convenience, the specifically rearranged $p_{i,j}$s and $\hat{p}_{i,j}$s in $\mathbf{p}$ and $\hat{\mathbf{p}}$ are re-enumerated by a single index k and are denoted by

$$\mathbf{v} = (p_1, \dots, p_{K-1})^\tau \quad \text{and} \quad \hat{\mathbf{v}} = (\hat{p}_1, \dots, \hat{p}_{K-1})^\tau \tag{5.33}$$

where $K = K_1K_2$. It is to be noted that for any pair of (i,j),

$$k = (i-1)K_2 + j.$$

The multivariate central limit theorem gives that

$$\sqrt{n}(\hat{\mathbf{v}} - \mathbf{v}) \xrightarrow{L} MVN(0, \Sigma) \tag{5.34}$$

where Σ is a $(K_1K_2 - 1) \times (K_1K_2 - 1)$ covariance matrix as given in the following.

$$\Sigma := \Sigma(v) := \begin{pmatrix} p_1(1-p_1) & -p_1p_2 & \cdots & -p_1p_{K-1} \\ -p_1p_2 & p_2(1-p_2) & \cdots & -p_2p_{K-1} \\ \cdots & \cdots & \cdots & \cdots \\ \cdots & \cdots & \cdots & \cdots \\ \cdots & \cdots & \cdots & \cdots \\ -p_{K-1}p_1 & -p_{K-1}p_2 & \cdots & p_{K-1}(1-p_{K-1}) \end{pmatrix}. \tag{5.35}$$

Shannon's entropies for $\mathscr{X}$, $\mathscr{Y}$, and $\mathscr{X} \times \mathscr{Y}$, with their respective distributions, are defined as

$$G_X(\mathbf{v}) = H(X) = -\sum_i p_{i,\cdot} \ln p_{i,\cdot}, \tag{5.36}$$

$$G_Y(\mathbf{v}) = H(Y) = -\sum_j p_{\cdot,j} \ln p_{\cdot,j}, \tag{5.37}$$

$$G_{XY}(\mathbf{v}) = H(X,Y) = -\sum_i \sum_j p_{i,j} \ln p_{i,j} = -\sum_k p_k \ln p_k. \tag{5.38}$$

Let $\hat{H}(X)$, $\hat{H}(Y)$, and $\hat{H}(X,Y)$ be the plug-in estimators of $H(X)$, $H(Y)$, and $H(X,Y)$, respectively, that is,

$$\hat{H}(X) = -\sum_i \hat{p}_{i,\cdot} \ln \hat{p}_{i,\cdot}, \tag{5.39}$$

$$\hat{H}(Y) = -\sum_j \hat{p}_{\cdot,j} \ln \hat{p}_{\cdot,j}, \tag{5.40}$$

$$\hat{H}(X,Y) = -\sum_i \sum_j \hat{p}_{i,j} \ln \hat{p}_{i,j} = -\sum_k \hat{p}_k \ln \hat{p}_k. \tag{5.41}$$

The respective gradients for $G_X(\mathbf{v})$, $G_Y(\mathbf{v})$, and $G_{XY}(\mathbf{v})$ are

$$g_X(\mathbf{v}) = \nabla G_X(\mathbf{v}) = \left(\frac{\partial}{\partial p_1} G_X(\mathbf{v}), \dots, \frac{\partial}{\partial p_{K-1}} G_X(\mathbf{v})\right)^\tau,$$

$$g_Y(\mathbf{v}) = \nabla G_Y(\mathbf{v}) = \left(\frac{\partial}{\partial p_1} G_Y(\mathbf{v}), \dots, \frac{\partial}{\partial p_{K-1}} G_Y(\mathbf{v})\right)^\tau,$$

$$g_{XY}(\mathbf{v}) = \nabla G_{XY}(\mathbf{v}) = \left(\frac{\partial}{\partial p_1} G_{XY}(\mathbf{v}), \dots, \frac{\partial}{\partial p_{K-1}} G_{XY}(\mathbf{v})\right)^\tau.$$

For every k, $k = 1, \dots, K-1$, such that $p_k = p_{i,j}$, the following facts can be verified (see Exercise 11):

$$\frac{\partial}{\partial p_k} G_X(\mathbf{v}) = \frac{\partial H(X)}{\partial p_k} = \ln p_{K_1,\cdot} - \ln p_{i,\cdot}, \tag{5.42}$$

$$\frac{\partial}{\partial p_k} G_Y(\mathbf{v}) = \frac{\partial H(Y)}{\partial p_k} = \ln p_{\cdot,K_2} - \ln p_{\cdot,j}, \tag{5.43}$$

$$\frac{\partial}{\partial p_k} G_{XY}(\mathbf{v}) = \frac{\partial H(X,Y)}{\partial p_k} = \ln p_K - \ln p_k. \tag{5.44}$$

Let

$$A := A(\mathbf{v}) : = \begin{pmatrix} a_1^\tau \\ a_2^\tau \\ a_3^\tau \end{pmatrix} := \begin{pmatrix} \frac{\partial}{\partial p_1} G_X(\mathbf{v}) & \cdots & \frac{\partial}{\partial p_{K-1}} G_X(\mathbf{v}) \\ \frac{\partial}{\partial p_1} G_Y(\mathbf{v}) & \cdots & \frac{\partial}{\partial p_{K-1}} G_Y(\mathbf{v}) \\ \frac{\partial}{\partial p_1} G_{XY}(\mathbf{v}) & \cdots & \frac{\partial}{\partial p_{K-1}} G_{XY}(v) \end{pmatrix}. \tag{5.45}$$

According to (5.42)–(5.44), the first row of A, written in $(K_1 - 1)$ blocks of size K_2 and 1 block of size $(K_2 - 1)$, is

$$
\begin{array}{llllll}
a_1^\tau = (& \ln(p_{K_1,\cdot}/p_{1,\cdot}), & \cdots & \ln(p_{K_1,\cdot}/p_{1,\cdot}), & \ln(p_{K_1,\cdot}/p_{1,\cdot}), \\
& \ln(p_{K_1,\cdot}/p_{2,\cdot}), & \cdots & \ln(p_{K_1,\cdot}/p_{2,\cdot}), & \ln(p_{K_1,\cdot}/p_{2,\cdot}), \\
& \vdots & \vdots\vdots\vdots & \vdots & \vdots \\
& \ln(p_{K_1,\cdot}/p_{K_1-1,\cdot}), & \cdots & \ln(p_{K_1,\cdot}/p_{K_1-1,\cdot}), & \ln(p_{K_1,\cdot}/p_{K_1-1,\cdot}), \\
& 0, & \cdots & 0 \ ); &
\end{array}
$$

the second row of A, written in $(K_1 - 1)$ blocks of size K_2 and 1 block of size $(K_2 - 1)$, is

$$
\begin{array}{llllll}
a_2^\tau = (& \ln(p_{\cdot,K_2}/p_{\cdot,1}), & \ln(p_{\cdot,K_2}/p_{\cdot,2}), & \cdots & \ln(p_{\cdot,K_2}/p_{\cdot,K_2-1}), & 0, \\
& \ln(p_{\cdot,K_2}/p_{\cdot,1}), & \ln(p_{\cdot,K_2}/p_{\cdot,2}), & \cdots & \ln(p_{\cdot,K_2}/p_{\cdot,K_2-1}), & 0, \\
& \vdots & \vdots & \vdots\vdots\vdots & \vdots & \vdots \\
& \ln(p_{\cdot,K_2}/p_{\cdot,1}), & \ln(p_{\cdot,K_2}/p_{\cdot,2}), & \cdots & \ln(p_{\cdot,K_2}/p_{\cdot,K_2-1}), & 0, \\
& \ln(p_{\cdot,K_2}/p_{\cdot,1}), & \ln(p_{\cdot,K_2}/p_{\cdot,2}), & \cdots & \ln(p_{\cdot,K_2}/p_{\cdot,K_2-1}) &);
\end{array}
$$

and the third row of A, written in $(K_1 - 1)$ blocks of size K_2 and 1 block of size $(K_2 - 1)$, is

$$
\begin{array}{llllll}
a_3^\tau = (& \ln(p_K/p_{1,1}), & \cdots & \ln(p_K/p_{1,K_2-1}), & \ln(p_K/p_{1,K_2}), \\
& \ln(p_K/p_{2,1}), & \cdots & \ln(p_K/p_{2,K_2-1}), & \ln(p_K/p_{2,K_2}), \\
& \vdots & \vdots\vdots\vdots & \vdots & \vdots \\
& \ln(p_K/p_{K_1,1}), & \cdots & \ln(p_K/p_{K_1,K_2-1}) \ ); &
\end{array}
$$

An application of the delta method gives Lemma 5.3 below.

Lemma 5.3 *Let $\mathcal{X}$ and $\mathcal{Y}$ be as in (5.31), let $\mathbf{p}_{X,Y} = \{p_{i,j}\}$ be a joint probability distribution on $\mathcal{X} \times \mathcal{Y}$, and let $\mathbf{v}$ and $\hat{\mathbf{v}}$ be as in (5.33). Suppose $p_{i,j} > 0$ for every pair (i,j) where $i = 1, \ldots, K_1$ and $j = 1, \ldots, K_2$. Then*

$$
\sqrt{n}\left[\begin{pmatrix} \hat{H}(X) \\ \hat{H}(Y) \\ \hat{H}(X,Y) \end{pmatrix} - \begin{pmatrix} H(X) \\ H(Y) \\ H(X,Y) \end{pmatrix}\right] \xrightarrow{L} MVN(0, \Sigma_H) \tag{5.46}
$$

where

$$
\Sigma_H = A\Sigma A^\tau,
$$

Σ is as in (5.35), and A is as in (5.45) according to (5.42)–(5.44).

More generally for the case of $K \le K_1K_2$, that is, $p_{i,j}$ may not be positive for some pairs of (i,j), a result similar to Lemma 5.3 can be derived. For any arbitrary but fixed re-enumeration of the K positive probabilities in $\{p_{i,j}\}$, denoted as

$$\{p_k; k = 1, \dots, K\}, \tag{5.47}$$

consider the following two partitions

$$\{S_1, \dots, S_{K_1}\} \quad \text{and} \quad \{T_1, \dots, T_{K_2}\}$$

of the index set $\{1, 2, \dots, K\}$ such that

1) $\{p_k; k \in S_s\}$ is the collection of all positive probabilities in $\{p_{i,j}; i = s\}$ for each s, $s = 1, \dots, K_1$; and
2) $\{p_k; k \in T_t\}$ is the collection of all positive probabilities in $\{p_{i,j}; j = t\}$ for each t, $t = 1, \dots, K_2$.

By the construction of the partitions,

$$\sum_{k \in S_i} p_k = p_{i,\cdot} \quad \text{and} \quad \sum_{k \in T_j} p_k = p_{\cdot,j}.$$

Without loss of generality, it may be assumed that $K \in S_{K_1} \cap T_{K_2}$. If not, then $K \in S_{i_0} \cap T_{j_0}$ for some i_0 and j_0, by a rearrangement of the indices (i, j), $K \in S_{K_1} \cap T_{K_2}$ will be true.

Letting

$$\mathbf{v} = (p_1, \dots, p_{K-1})^\tau \quad \text{and} \quad \hat{\mathbf{v}} = (\hat{p}_1, \dots, \hat{p}_{K-1})^\tau, \tag{5.48}$$

an application of multivariate central limit theorem gives

$$\sqrt{n}(\hat{\mathbf{v}} - \mathbf{v}) \xrightarrow{L} MVN(0, \Sigma) \tag{5.49}$$

where Σ is the $(K-1) \times (K-1)$ covariance matrix given by

$$\Sigma := \Sigma(\mathbf{v}) := \begin{pmatrix} p_1(1-p_1) & -p_1p_2 & \cdots & -p_1p_{K-1} \\ -p_2p_1 & p_2(1-p_2) & \cdots & -p_2p_{K-1} \\ \cdots & \cdots & \cdots & \cdots \\ -p_{K-1}p_1 & -p_{K-1}p_2 & \cdots & p_{K-1}(1-p_{K-1}) \end{pmatrix}. \tag{5.50}$$

Noting (5.36)–(5.38), the following facts can be verified:

$$\frac{\partial G_X(\mathbf{v})}{\partial p_k} = \frac{\partial H(X)}{\partial p_k} = \begin{cases} \ln p_{K_1,\cdot} - \ln p_{i,\cdot}, & \text{if } k \in S_i \neq S_{K_1} \\ 0, & \text{if } k \in S_{K_1} \end{cases} \tag{5.51}$$

$$\frac{\partial G_Y(\mathbf{v})}{\partial p_k} = \frac{\partial H(Y)}{\partial p_k} = \begin{cases} \ln p_{\cdot,K_2} - \ln p_{\cdot,j}, & \text{if } k \in T_j \neq T_{K_2} \\ 0, & \text{if } k \in T_{K_2} \end{cases} \tag{5.52}$$

$$\frac{\partial G_{XY}(\mathbf{v})}{\partial p_k} = \frac{\partial H(X,Y)}{\partial p_k} = \ln p_K - \ln p_k, \text{ for } 1 \le k \le K-1. \tag{5.53}$$

To show (5.51) is true, consider first

$$\begin{aligned} G_X(\mathbf{v}) &= -\sum_{s=1}^{K_1}\left(\sum_{l\in S_s} p_l\right)\ln\left(\sum_{l\in S_s} p_l\right) \\ &= -\sum_{s=1}^{K_1-1}\left(\sum_{l\in S_s} p_l\right)\ln\left(\sum_{l\in S_s} p_l\right) - \left(\sum_{l\in S_{K_1}} p_l\right)\ln\left(\sum_{l\in S_{K_1}} p_l\right) \\ &= -\sum_{s=1}^{K_1-1}\left(\sum_{l\in S_s} p_l\right)\ln\left(\sum_{l\in S_s} p_l\right) \\ &\quad - \left[1-\sum_{s=1}^{K_1-1}\left(\sum_{l\in S_s} p_l\right)\right]\ln\left(1-\sum_{s=1}^{K_1-1}\left(\sum_{l\in S_s} p_l\right)\right). \end{aligned}$$

For the case of $k \in S_i \neq S_{K_1}$,

$$\begin{aligned} \frac{\partial G_X(\mathbf{v})}{\partial p_k} &= -\frac{\partial}{\partial p_k}\left[\left(\sum_{l\in S_i} p_l\right)\ln\left(\sum_{l\in S_i} p_l\right)\right] \\ &\quad - \frac{\partial}{\partial p_k}\left\{\left[1-\sum_{s=1}^{K_1-1}\left(\sum_{l\in S_s} p_l\right)\right]\ln\left(1-\sum_{s=1}^{K_1-1}\left(\sum_{l\in S_s} p_l\right)\right)\right\} \\ &= -\ln\left(\sum_{l\in S_i} p_l\right) - 1 - \left\{-\ln\left(1-\sum_{s=1}^{K_1-1}\left(\sum_{l\in S_s} p_l\right)\right) - 1\right\} \\ &= -\ln p_{i,\cdot} + \ln\left(1-\sum_{s=1}^{K_1-1}\left(\sum_{l\in S_s} p_l\right)\right) \\ &= -\ln p_{i,\cdot} + \ln p_{K_1,\cdot} = \ln p_{K_1,\cdot} - \ln p_{i,\cdot}. \end{aligned}$$

For the case of $k \in S_{K_1}$, $\partial G_X(\mathbf{v})/\partial p_k = 0$.

A similar argument would give the proof for (5.52) (see Exercise 11). Equation (5.53) is obvious.

For any arbitrary but fixed re-enumeration of the positive terms of $\{p_{i,j}\}$ in (5.47), an application of the delta method, based on (5.49), gives Lemma 5.4 below.

Lemma 5.4 *Let $\mathcal{X}$ and $\mathcal{Y}$ be as in (5.31), let $\mathbf{p}_{X,Y} = \{p_{i,j}\}$ be a joint probability distribution on $\mathcal{X}\times\mathcal{Y}$, and let $\mathbf{v}$ and $\hat{\mathbf{v}}$ be as in (5.48):*

$$\sqrt{n}\left[\begin{pmatrix}\hat{H}(X)\\ \hat{H}(Y)\\ \hat{H}(X,Y)\end{pmatrix} - \begin{pmatrix}H(X)\\ H(Y)\\ H(X,Y)\end{pmatrix}\right] \xrightarrow{L} MVN(0, \Sigma_H) \tag{5.54}$$

where

$$\Sigma_H = A\Sigma A^{\tau}, \tag{5.55}$$

Σ is as in (5.50) and A is as in (5.45) according to (5.51)–(5.53).

Remark 5.1 *Lemma 5.3 is a special case of Lemma 5.4, and it is separately stated for its instructive quality. While symbolically identical, the matrices A and Σ in Lemma 5.4 may take on different forms from those in Lemma 5.3, depending on K, the number of positive probabilities in $\{p_{i,j}\}$, and the re-enumeration in (5.48).*

Note that the covariance matrix Σ_H in Lemma 5.4 (as well as that in Lemma 5.3) may not necessarily be of full rank. For Σ_H to be of full rank, some additional conditions need to be imposed.

To estimate $MI = MI(X, Y)$, noting

$$MI(X,Y) = (1,1,-1)\begin{pmatrix} H(X) \\ H(Y) \\ H(X,Y) \end{pmatrix}$$

and letting

$$\widehat{MI} = \widehat{MI}\,(X,Y) = (1,1,-1)\begin{pmatrix} \hat{H}(X) \\ \hat{H}(Y) \\ \hat{H}(X,Y) \end{pmatrix}, \tag{5.56}$$

an application of the multivariate delta method gives Theorem 5.5 below.

Theorem 5.5 *Under the conditions of Lemma 5.4 and suppose further that*

$$\sigma^2 = (1, 1-1)\Sigma_H(1,1,-1)^\tau > 0 \tag{5.57}$$

where Σ_H is as in (5.35). Then

$$\sqrt{n}(\widehat{MI} - MI) \xrightarrow{L} N(0, \sigma^2). \tag{5.58}$$

Let

$$\hat{\Sigma} = \Sigma(\hat{\mathbf{v}}) = \begin{pmatrix} \hat{p}_1(1-\hat{p}_1) & -\hat{p}_1\hat{p}_2 & \cdots & -\hat{p}_1\hat{p}_{K-1} \\ -\hat{p}_2\hat{p}_1 & \hat{p}_2(1-\hat{p}_2) & \cdots & -\hat{p}_2\hat{p}_{K-1} \\ \cdots & \cdots & \cdots & \cdots \\ -\hat{p}_{K-1}\hat{p}_1 & -\hat{p}_{K-1}\hat{p}_2 & \cdots & \hat{p}_{K-1}(1-\hat{p}_{K-1}) \end{pmatrix}, \tag{5.59}$$

$$\hat{A} = A(\hat{\mathbf{v}}) = \left.\begin{pmatrix} \frac{\partial}{\partial p_1}G_X(\mathbf{v}) & \cdots & \frac{\partial}{\partial p_{K-1}}G_X(\mathbf{v}) \\ \frac{\partial}{\partial p_1}G_Y(\mathbf{v}) & \cdots & \frac{\partial}{\partial p_{K-1}}G_Y(\mathbf{v}) \\ \frac{\partial}{\partial p_1}G_{XY}(\mathbf{v}) & \cdots & \frac{\partial}{\partial p_{K-1}}G_{XY}(\mathbf{v}) \end{pmatrix}\right|_{\mathbf{v}=\hat{\mathbf{v}}}, \tag{5.60}$$

$$\hat{\Sigma}_H = \hat{A}\hat{\Sigma}\hat{A}^\tau, \tag{5.61}$$

$$\hat{\sigma}^2 = (1,1,-1)\hat{\Sigma}_H(1,1,-1)^\tau. \tag{5.62}$$

By the continuous mapping theorem, $\hat{\Sigma} = \Sigma(\hat{\mathbf{v}})$ is a consistent estimator of $\Sigma = \Sigma(\mathbf{v})$, $\hat{A} = A(\hat{\mathbf{v}})$ is a consistent estimator of A, $\hat{\Sigma}_H = \hat{A}\Sigma(\hat{\mathbf{v}})\hat{A}^\tau$ is a consistent estimator of $\Sigma_H = \Sigma(\mathbf{v})$, and $\hat{\sigma}^2$ is a consistent estimator of σ^2. The said consistency and Slutsky's theorem give immediately Corollary 5.2 below.

Corollary 5.2 *Under the conditions of Theorem 5.5 ,*

$$\frac{\sqrt{n}(\widehat{MI} - MI)}{\hat{\sigma}} \xrightarrow{L} N(0,1) \tag{5.63}$$

where $\hat{\sigma}$ is as given in (5.62).

Corollary 5.2 provides the means of a statistical tool to make inferences regarding MI. However, its validity depends on the condition in (5.57). To decipher (5.57), consider first the case of $K = K_1K_2$, that is, $p_{i,j} > 0$ for every pair (i,j).

Since Σ is positive definite (see Exercise 6), (5.57) holds if and only if $(1,1,-1)A \neq 0$, or equivalently X and Y are not independent.

The equivalence of $(1,1,-1)\,A \neq 0$ and that X and Y are not independent is established by the fact that $(1,1,-1)A = 0$ if and only if X and Y are independent. To see this fact, suppose $(1,1,-1)\,A = 0$, that is, for every pair (i,j), by (5.45),

$$\ln(p_{K_1,\cdot}/p_{i,\cdot}) + \ln(p_{\cdot,K_2}/p_{\cdot,j}) = \ln(p_K/p_{i,j}),$$

or after a few algebraic steps,

$$p_{i,j} = \frac{p_K}{p_{K_1,\cdot}p_{\cdot,K_2}} p_{i,\cdot}p_{\cdot,j}.$$

Summing both sides over i and j gives $p_K = p_{K_1,\cdot}p_{\cdot,K_2}$, and therefore

$$p_{i,j} = p_{i,\cdot}p_{\cdot,j}$$

for every pair of (i,j), that is, X and Y are independent. The converse is obviously true.

Next consider the case of $K < K_1K_2$. Without loss of generality, it may be assumed that $p_{i,\cdot} > 0$ for every i and $p_{\cdot,j} > 0$ for every j. (Or else, a re-enumeration of the index i or j would ensure the validity of this assumption.) An similar argument to the case of $K = K_1K_2$ would establish that $p_{i,j} = p_{i,\cdot}p_{\cdot,j}$ for every pair (i,j) such that $p_{i,j} \neq 0$ (see Exercise 12). At this point, it is established in general that (5.57) is equivalent to that X and Y are not independent.

Since MI is a linear combination of three entropies, its estimation is closely related to that of entropy. For entropy, the plug-in estimator $\hat{H}$ based on an *iid* sample of size n, though asymptotically efficient, is known to have a large bias, as given in (3.3). Based on (3.3), it is easy to see that the bias for $\widehat{MI}$ is

$$\mathrm{E}(\widehat{MI} - MI) = \frac{K - (K_1 + K_2) + 1}{2n} + \mathcal{O}(n^{-2})$$

where K is an integer between $\max\{K_1, K_2\}$ and K_1K_2, but often is in the order of K_1K_2, which leads to an overestimation for MI. For this reason, seeking a good MI estimator with smaller bias has been the focal point of much of the research in MI estimation.

In the next section, an estimator of MI based on Turing's perspective is described. This estimator qualitatively alleviates the bias issue observed in the plug-in estimator $\widehat{MI}$.

5.2.2 Estimation in Turing's Perspective

In Chapter 3, an entropy estimator based on Turing's perspective is introduced as $\hat{H}_z$ in (3.27). This estimator naturally extends to the following estimator of MI:

$$\begin{aligned}\widehat{MI}_z &= \widehat{MI}_z(X, Y) \\ &= \hat{H}_z(X) + \hat{H}_z(Y) - \hat{H}_z(X, Y) \\ &= \sum_{v=1}^{n-1} \frac{1}{v} \left\{ \frac{n^{v+1}[n-(v+1)]!}{n!} \sum_{i=1}^{K_1} \left[\hat{p}_{i,\cdot} \prod_{k=0}^{v-1} \left(1 - \hat{p}_{i,\cdot} - \frac{k}{n} \right) \right] \right\} \\ &\quad + \sum_{v=1}^{n-1} \frac{1}{v} \left\{ \frac{n^{v+1}[n-(v+1)]!}{n!} \sum_{j=1}^{K_2} \left[\hat{p}_{\cdot,j} \prod_{k=0}^{v-1} \left(1 - \hat{p}_{\cdot,j} - \frac{k}{n} \right) \right] \right\} \\ &\quad - \sum_{v=1}^{n-1} \frac{1}{v} \left\{ \frac{n^{v+1}[n-(v+1)]!}{n!} \sum_{i=1}^{K_1} \sum_{j=1}^{K_2} \left[\hat{p}_{i,j} \prod_{k=0}^{v-1} \left(1 - \hat{p}_{i,j} - \frac{k}{n} \right) \right] \right\}.\end{aligned} \tag{5.64}$$

By (3.28), it is easy to see that the bias of $\widehat{MI}_z$ in (5.64) is

$$\begin{aligned}\mathrm{E}(\widehat{MI}_z - MI) = &- \sum_{v=n}^{\infty} \frac{1}{v} \sum_{i=1}^{K_1} p_{i,\cdot}(1 - p_{i,\cdot})^v \\ &- \sum_{v=n}^{\infty} \frac{1}{v} \sum_{j=1}^{K_2} p_{\cdot,j}(1 - p_{\cdot,j})^v \\ &+ \sum_{v=n}^{\infty} \frac{1}{v} \sum_{i=1}^{K_1} \sum_{j=1}^{K_2} p_{i,j}(1 - p_{i,j})^v\end{aligned}$$

and it is decaying at least exponentially fast in n since each of the three additive terms is decaying at least exponentially fast in n, as argued in Theorem 3.9. In fact, letting

$$\begin{aligned}p_\vee &= \max_{i,j}\{p_{i,j} > 0\}, \\ p_\wedge &= \min_{i,j}\{p_{i,j} > 0\}, \\ p_{\vee,\cdot} &= \max_{i}\{p_{i,\cdot} > 0\},\end{aligned}$$

$$p_{\wedge,\cdot} = \min_i \{p_{i,\cdot} > 0\},$$
$$p_{\cdot,\vee} = \max_j \{p_{\cdot,j} > 0\}, \quad \text{and}$$
$$p_{\cdot,\wedge} = \min_j \{p_{\cdot,j} > 0\},$$

it is easily verified that

$$\begin{aligned} &-\sum_{v=n}^{\infty}\frac{1}{v}(1-p_{\wedge,\cdot})^v - \sum_{v=n}^{\infty}\frac{1}{v}(1-p_{\cdot,\wedge})^v + \sum_{v=n}^{\infty}\frac{1}{v}(1-p_{\vee})^v \\ &\qquad \le \mathrm{E}(\widehat{MI}_z - MI) \le \\ &-\sum_{v=n}^{\infty}\frac{1}{v}(1-p_{\vee,\cdot})^v - \sum_{v=n}^{\infty}\frac{1}{v}(1-p_{\cdot,\vee})^v + \sum_{v=n}^{\infty}\frac{1}{v}(1-p_{\wedge})^v, \end{aligned} \tag{5.65}$$

and that all the terms of both sides of (5.65) converge to zero exponentially fast since all of $p_\vee, p_\wedge, p_{\vee,\cdot}, p_{\wedge,\cdot}, p_{\cdot,\vee}$, and $p_{\cdot,\wedge}$ are fixed positive constants in $(0,1)$. The inequalities of (5.65) are in fact one of the main advantages of the estimator $\widehat{MI}_z$ in (5.64).

Theorem 5.6 *Under the conditions of Theorem 5.5 and suppose further that X and Y are not independent. Then*

$$\sqrt{n}(\widehat{MI}_z - MI) \xrightarrow{L} N(0, \sigma^2), \tag{5.66}$$

where σ is as in (5.57).

The asymptotic normality of Theorem 5.6 is derived from Theorem 5.5 by showing that $\widehat{MI}$ and $\widehat{MI}_z$ are sufficiently close as $n \to \infty$. Toward that end, consider first the problem of $\hat{H}_z$ in (3.27) estimating Shannon's entropy $H = -\sum_{k=1}^{K'} p_k \ln p_k$ defined with $\{p_k; k = 1, \dots, K'\}$ on alphabet $\mathscr{L} = \{\ell_k; k = 1, \cdot, K'\}$ where K' is finite. Let $\{\hat{p}_k\}$ be the observed relative frequencies of letters in an *iid* sample of size n and $\hat{H} = -\sum_{k=1}^{K'} \hat{p}_k \ln \hat{p}_k$ be the plug-in estimator of H.

Lemma 5.5 *Suppose $\{p_k\}$ is a nonuniform distribution on $\mathscr{L} = \{\ell_k\}$. Then*

$$\sqrt{n}(\hat{H}_z - \hat{H}) \xrightarrow{p} 0.$$

To prove Lemma 5.5, Lemma 3.2 is needed, which is restated as Lemma 5.6 below for easy reference. Let, for any $p \in (0,1)$,

$$g_n(p) = \sum_{v=1}^{n-1}\left\{\frac{1}{v}\left\{\frac{n^{v+1}[n-(v+1)]!}{n!}\right\}\left[\prod_{j=0}^{v-1}\left(1-p-\frac{j}{n}\right)\right]1_{[v\le n(1-p)+1]}\right\},$$

$h_n(p) = p g_n(p)$, and $h(p) = -p \ln p$.

Lemma 5.6 *Let $\hat{p} = X/n$ where X is a binomial random variable with parameters n and p. Then*

1) $\sqrt{n}[h_n(p) - h(p)] \to 0$ *uniformly in* $p \in (c, 1)$ *for any* c, $0 < c < 1$;
2) $\sqrt{n}|h_n(p) - h(p)| < A(n) = \mathcal{O}(n^{3/2})$ *uniformly in* $p \in [1/n, c]$ *for any* c, $0 < c < p$; *and*
3) $\mathrm{P}(\hat{p} \le c) < B(n) = \mathcal{O}(n^{-1/2} \exp\{-nC\})$ *where* $C = (p-c)^2/[p(1-p)]$ *for any* $c \in (0, p)$.

Proof of Lemma 5.5: Without loss of generality, consider the sample proportions of the first letter of the alphabet $\hat{p}_1$. It is of interest to show that $\Delta_n = \sqrt{n}[h_n(\hat{p}_1) - h(\hat{p}_1)] \xrightarrow{p} 0$, that is, for any $\varepsilon > 0$, as $n \to \infty$, $\mathrm{P}(\Delta_n > \varepsilon) \to 0$. Toward that end, observe that for any fixed $c \in (0, p_1)$

$$\begin{aligned}\mathrm{P}(\Delta_n > \varepsilon) &= \mathrm{P}(\sqrt{n}[h_n(\hat{p}_1) - h(\hat{p}_1)] > \varepsilon | \hat{p}_1 \in (c,1))\mathrm{P}(\hat{p}_1 \in (c,1)) \\ &\quad + \mathrm{P}(\sqrt{n}[h_n(\hat{p}_1) - h(\hat{p}_1)] > \varepsilon | \hat{p}_1 \in [1/n, c])\mathrm{P}(\hat{p}_1 \in [1/n, c]) \\ &< \mathrm{P}(\sqrt{n}[h_n(\hat{p}_1) - h(\hat{p}_1)] > \varepsilon | \hat{p}_1 \in (c,1)) + \mathrm{P}(\hat{p}_1 \in [1/n, c]).\end{aligned}$$

By Part 1 of Lemma 5.6, $\mathrm{P}(\sqrt{n}[h_n(\hat{p}_1) - h(\hat{p}_1)] > \varepsilon | \hat{p}_1 \in (c,1)) \xrightarrow{p} 0$; and by Part 3 of Lemma 5.6, $\mathrm{P}(\hat{p}_1 \in [1/n, c]) \xrightarrow{p} 0$. Hence $\Delta_n \xrightarrow{p} 0$. The result of Lemma 5.5 is immediate from the following expression in a finite sum, $\sqrt{n}(\hat{H}_z - \hat{H}) = \sum_{k=1}^{K} \sqrt{n}(h_n(\hat{p}_k) - h(\hat{p}_k))$. □

Lemma 5.5 immediately gives the following corollary.

Corollary 5.3 *Under the conditions of Lemma 5.4,*

$$\sqrt{n}\left[\begin{pmatrix}\hat{H}_z(X) \\ \hat{H}_z(Y) \\ \hat{H}_z(X,Y)\end{pmatrix} - \begin{pmatrix}H(X) \\ H(Y) \\ H(X,Y)\end{pmatrix}\right] \xrightarrow{L} MVN(0, \Sigma_H) \tag{5.67}$$

where

$$\Sigma_H = A\Sigma A^{\tau}, \tag{5.68}$$

Σ is as in (5.50) and A is as in (5.45) according to (5.51)–(5.53).

Proof of Theorem 5.6: In light of Theorem 5.5 and Slutsky's theorem, it suffices to show that $\sqrt{n}(\widehat{MI}_z - \widehat{MI}) \xrightarrow{p} 0$. However, this is trivial since

$$\begin{aligned}\sqrt{n}(\widehat{MI}_z - \widehat{MI}) &= \sqrt{n}(\hat{H}_z(X) - \hat{H}(X)) \\ &\quad + \sqrt{n}(\hat{H}_z(Y) - \hat{H}(Y)) \\ &\quad - \sqrt{n}(\hat{H}_z(X,Y) - \hat{H}(X,Y)).\end{aligned}$$

Applying Lemma 5.5 to each of the three additive terms above, respectively, gives the result of Theorem 5.6. □

From Theorem 5.6 and Slutsky's theorem, Corollary 5.4 follows.

Corollary 5.4 *Under the conditions of Theorem 5.6,*

$$\frac{\sqrt{n}(\widehat{MI}_z - MI)}{\hat{\sigma}} \xrightarrow{L} N(0,1), \tag{5.69}$$

where $\hat{\sigma}$ is as given in (5.62).

Corollary 5.4, like Corollary 5.2, provides a large sample statistical tool for inference about MI, but with a qualitatively lower bias.

5.2.3 Estimation of Standardized Mutual Information

The SMI κ in Definition 5.4 is a practically useful measure that rescales MI to within interval [0, 1], particularly with the necessary and sufficient conditions of Theorem 5.4. In this section, several estimators of κ are described. However, κ is not the only one that has the desirable properties. Since

$$\begin{aligned} 0 \le MI(X,Y) &\le \min\{H(X), H(Y)\} \\ &\le \sqrt{H(X)H(Y)} \\ &\le (H(X)+H(Y))/2 \\ &\le \max\{H(X), H(Y)\} \\ &\le H(X,Y), \end{aligned}$$

(see Exercise 7), each of the following, assuming $H(X,Y) < \infty$,

$$\begin{aligned} \kappa_1 &= \frac{MI(X,Y)}{\min\{H(X),H(Y)\}}, \\ \kappa_2 &= \frac{MI(X,Y)}{\sqrt{H(X)H(Y)}}, \\ \kappa_3 &= \frac{MI(X,Y)}{(H(X)+H(Y))/2}, \\ \kappa_4 &= \frac{MI(X,Y)}{\max\{H(X),H(Y)\}}, \end{aligned}$$

may be considered a SMI and satisfies

$$0 \le \kappa_i \le 1,$$

where $i \in \{1,2,3,4\}$. However the necessary and sufficient condition of Theorem 5.4 holds only for κ_2, κ_3 and κ_4, but not for κ_1. See Exercise 8. Further discussion on these and other variants of SMI may be found in Kvalseth

(1987), Strehl and Ghosh (2002), Yao (2003), and Vinh, Epps, and Bailey (2010). A detailed discussion of the estimation of various standardized mutual information may be found in Zhang and Stewart (2016).

In this section, the plug-in estimators of κ_1, κ_2, and κ_3, as well as those based on Turing's perspective, are described, and their respective asymptotic distributional properties are given. Toward that end, consider the following three functions of triplet (x_1, x_2, x_3), with domain, $x_1 > 0$, $x_2 > 0$, and $x_3 > 0$,

$$\begin{aligned}
\kappa(x_1, x_2, x_3) &= \frac{x_1 + x_2}{x_3} - 1,\\
\kappa_2(x_1, x_2, x_3) &= \frac{x_1 + x_2 - x_3}{\sqrt{x_1 x_2}},\\
\kappa_3(x_1, x_2, x_3) &= 2\left(1 - \frac{x_3}{x_1 + x_2}\right).
\end{aligned}$$

The gradients of these functions are, respectively,

$$\begin{aligned}
g_\kappa(x_1, x_2, x_3) &= \left(\frac{1}{x_3}, \frac{1}{x_3}, -\frac{x_1 + x_2}{x_3^2}\right)^\tau,\\
g_{\kappa_2}(x_1, x_2, x_3) &= \left(\frac{1}{(x_1 x_2)^{1/2}} - \frac{x_2(x_1 + x_2 - x_3)}{2(x_1 x_2)^{3/2}},\right.\\
&\qquad \left.\frac{1}{(x_1 x_2)^{1/2}} - \frac{x_1(x_1 + x_2 - x_3)}{2(x_1 x_2)^{3/2}}, -\frac{1}{(x_1 x_2)^{1/2}}\right)^\tau,\\
g_{\kappa_3}(x_1, x_2, x_3) &= \left(\frac{2x_3}{(x_1 + x_2)^2}, \frac{2x_3}{(x_1 + x_2)^2}, -\frac{2}{x_1 + x_2},\right)^\tau.
\end{aligned}$$

See Exercise 13.

Let the following be the plug-in estimators of κ, κ_2, and κ_3.

$$\hat{\kappa} = \frac{\hat{H}(X) + \hat{H}(Y)}{\hat{H}(X, Y)} - 1 = \frac{\hat{H}(X) + \hat{H}(Y) - \hat{H}(X, Y)}{\hat{H}(X, Y)}, \tag{5.70}$$

$$\hat{\kappa}_2 = \frac{\hat{H}(X) + \hat{H}(Y) - \hat{H}(X, Y)}{\sqrt{\hat{H}(X)\hat{H}(Y)}}, \tag{5.71}$$

$$\hat{\kappa}_3 = 2\left(1 - \frac{\hat{H}(X, Y)}{\hat{H}(X) + \hat{H}(Y)}\right) = \frac{\hat{H}(X) + \hat{H}(Y) - \hat{H}(X, Y)}{(\hat{H}(X) + \hat{H}(Y))/2}. \tag{5.72}$$

Theorem 5.7 *Under the conditions of Lemma 5.4,*

1) $\sqrt{n}(\hat{\kappa} - \kappa) \xrightarrow{L} N(0, \sigma^2_{\hat{\kappa}})$,
2) $\sqrt{n}(\hat{\kappa}_2 - \kappa_2) \xrightarrow{L} N(0, \sigma^2_{\hat{\kappa}_2})$,
3) $\sqrt{n}(\hat{\kappa}_3 - \kappa_3) \xrightarrow{L} N(0, \sigma^2_{\hat{\kappa}_3})$,

where

$$\sigma^2_{\hat\kappa} = g^\tau_\kappa(H(X), H(Y), H(X,Y))\ \Sigma_H\ g_\kappa(H(X), H(Y), H(X,Y)),$$
$$\sigma^2_{\hat\kappa_2} = g^\tau_{\kappa_2}(H(X), H(Y), H(X,Y))\ \Sigma_H\ g_{\kappa_2}(H(X), H(Y), H(X,Y)),$$
$$\sigma^2_{\hat\kappa_3} = g^\tau_{\kappa_3}(H(X), H(Y), H(X,Y))\ \Sigma_H\ g_{\kappa_3}(H(X), H(Y), H(X,Y)),$$

and Σ_H *is as in (5.55).*

Proof: An application of the delta method to (5.54) completes the proof. □

Corollary 5.5 *Under the conditions of Theorem 5.7,*

1) $\dfrac{\sqrt{n}(\hat\kappa - \kappa)}{\hat\sigma_{\hat\kappa}} \xrightarrow{L} N(0,1),$

2) $\dfrac{\sqrt{n}(\hat\kappa_2 - \kappa_2)}{\hat\sigma_{\hat\kappa_2}} \xrightarrow{L} N(0,1),$

3) $\dfrac{\sqrt{n}(\hat\kappa_3 - \kappa_3)}{\hat\sigma_{\hat\kappa_3}} \xrightarrow{L} N(0,1),$

where

$$\hat\sigma^2_{\hat\kappa} = g^\tau_\kappa(\hat H(X), \hat H(Y), \hat H(X,Y))\ \hat\Sigma_H\ g_\kappa(\hat H(X), \hat H(Y), \hat H(X,Y)),$$
$$\hat\sigma^2_{\hat\kappa_2} = g^\tau_{\kappa_2}(\hat H(X), \hat H(Y), \hat H(X,Y))\ \hat\Sigma_H\ g_{\kappa_2}(\hat H(X), \hat H(Y), \hat H(X,Y)),$$
$$\hat\sigma^2_{\hat\kappa_3} = g_{\kappa_3}(\hat H(X), \hat H(Y), \hat H(X,Y))\ \hat\Sigma_H\ g^\tau_{\kappa_3}(\hat H(X), \hat H(Y), \hat H(X,Y)),$$

and $\hat\Sigma_H$ *is as in (5.61).*

Proof: The consistency of $\hat\sigma^2_{\hat\kappa}$, $\hat\sigma^2_{\hat\kappa_2}$, and $\hat\sigma^2_{\hat\kappa_3}$ and the Slutsky's theorem completes the proof. □

Let $\hat H_z(X)$, $\hat H_z(Y)$, and $\hat H_z(X,Y)$ be as in (5.64). Let

$$\hat\kappa_z = \frac{\hat H_z(X) + \hat H_z(Y)}{\hat H_z(X,Y)} - 1 = \frac{\hat H_z(X) + \hat H_z(Y) - \hat H_z(X,Y)}{\hat H_z(X,Y)}, \tag{5.73}$$

$$\hat\kappa_{2z} = \frac{\hat H_z(X) + \hat H_z(Y) - \hat H_z(X,Y)}{\sqrt{\hat H_z(X)\hat H_z(Y)}}, \tag{5.74}$$

$$\hat\kappa_{3z} = 2\left(1 - \frac{\hat H_z(X,Y)}{\hat H_z(X) + \hat H_z(Y)}\right) = \frac{\hat H_z(X) + \hat H_z(Y) - \hat H_z(X,Y)}{(\hat H_z(X) + \hat H_z(Y))/2}. \tag{5.75}$$

Theorem 5.8 *Under the conditions of Corollary 5.3,*

1) $\sqrt{n}(\hat\kappa_z - \kappa) \xrightarrow{L} N(0, \sigma^2_{\hat\kappa_z}),$

2) $\sqrt{n}(\hat{\kappa}_{2z} - \kappa_2) \xrightarrow{L} N(0, \sigma^2_{\hat{\kappa}_{2z}})$,

3) $\sqrt{n}(\hat{\kappa}_{3z} - \kappa_3) \xrightarrow{L} N(0, \sigma^2_{\hat{\kappa}_{3z}})$,

where

$$\begin{aligned}
\sigma^2_{\hat{\kappa}_z} &= g^\tau_\kappa(H(X), H(Y), H(X,Y))\ \Sigma_H\ g_\kappa(H(X), H(Y), H(X,Y)),\\
\sigma^2_{\hat{\kappa}_{2z}} &= g^\tau_{\kappa_2}(H(X), H(Y), H(X,Y))\ \Sigma_H\ g_{\kappa_2}(H(X), H(Y), H(X,Y)),\\
\sigma^2_{\hat{\kappa}_{3z}} &= g^\tau_{\kappa_3}(H(X), H(Y), H(X,Y))\ \Sigma_H\ g_{\kappa_3}(H(X), H(Y), H(X,Y)),
\end{aligned}$$

and Σ_H is as in (5.55).

Proof: An application of the delta method to (5.67) completes the proof. □

Corollary 5.6 *Under the conditions of Theorem 5.8,*

1) $\dfrac{\sqrt{n}(\hat{\kappa}_z - \kappa)}{\hat{\sigma}_{\hat{\kappa}_z}} \xrightarrow{L} N(0, 1)$,

2) $\dfrac{\sqrt{n}(\hat{\kappa}_{z2} - \kappa_2)}{\hat{\sigma}_{\hat{\kappa}_{2z}}} \xrightarrow{L} N(0, 1)$,

3) $\dfrac{\sqrt{n}(\hat{\kappa}_{z3} - \kappa_3)}{\hat{\sigma}_{\hat{\kappa}_{3z}}} \xrightarrow{L} N(0, 1)$,

where

$$\begin{aligned}
\hat{\sigma}^2_{\hat{\kappa}_z} &= g^\tau_\kappa(\hat{H}_z(X), \hat{H}_z(Y), \hat{H}_z(X,Y))\ \hat{\Sigma}_H\ g_\kappa(\hat{H}_z(X), \hat{H}_z(Y), \hat{H}_z(X,Y)),\\
\hat{\sigma}^2_{\hat{\kappa}_{2z}} &= g^\tau_{\kappa_2}(\hat{H}_z(X), \hat{H}_z(Y), \hat{H}_z(X,Y))\ \hat{\Sigma}_H\ g_{\kappa_2}(\hat{H}_z(X), \hat{H}_z(Y), \hat{H}_z(X,Y)),\\
\hat{\sigma}^2_{\hat{\kappa}_{3z}} &= g_{\kappa_3}(\hat{H}_z(X), \hat{H}_z(Y), \hat{H}_z(X,Y))\ \hat{\Sigma}_H\ g^\tau_{\kappa_3}(\hat{H}_z(X), \hat{H}_z(Y), \hat{H}_z(X,Y)),
\end{aligned}$$

and $\hat{\Sigma}_H$ is as in (5.61).

All the plug-in estimators of MI, κ, κ_2, and κ_3 in this section are asymptotically efficient since they are all maximum likelihood estimators. All the estimators based on Turing's perspective in this section are also asymptotically efficient because each of them has the identical asymptotic variance as that of its plug-in counterpart.

5.2.4 An Illustrative Example

A random sample of size $n = 1000$ adults in the United States were selected. The favorite color (among $K_1 = 11$ choices) and the ethnicity (among $K_2 = 4$ choices) for each individual in the selected group were obtained. The survey resulted in the data set in Table 5.1. Letting X be the color preference and Y be the ethnicity of a randomly selected individual, i takes values in

Table 5.1 Frequency data of ethnicity and color preference.

	Black	Asian	White	Hispanic
Black	15	5	46	19
Blue	48	19	285	59
Brown	5	0	7	2
Gray	4	1	7	3
Green	15	4	99	19
Orange	0	3	13	5
Pink	1	3	26	3
Purple	20	4	85	30
Red	15	1	66	25
White	0	1	0	0
Yellow	3	4	26	4

$\{1, \dots, 11 = K_1\}$ corresponding to colors Black to Yellow in the given vertical order in Table 5.1, j takes values in $\{1, \dots, 4 = K_2\}$ corresponding to ethnic groups Black to Hispanic in the given horizontal order of the same table, and $K = K_1K_2 = 11 \times 4 = 44$ assuming $p_{i,j} > 0$ for every pair (i, j). The entries of Table 5.1 are observed frequencies, $f_{i,j}$, for example, $f_{1,1} = 15$.

In practice, n may not necessarily be large enough to ensure that $\hat{p}_{i,j} > 0$ for every pair of (i, j), for which $p_{i,j} > 0$. In fact, the data set in Table 5.1 contains five such cells, $(6, 1)$, $(10, 1)$, $(3, 2)$, $(10, 3)$, and $(10, 4)$. When this is the case, estimating the positive definite covariance matrix $\Sigma = \Sigma(\mathbf{v})$ of (5.50) by means of the plug-in $\hat{\Sigma} = \hat{\Sigma}(\mathbf{v})$ fails to produce a positive definite matrix, which in turn may fail to produce a useful estimated Σ_H of (5.55) by means of the plug-in $\hat{\Sigma}_H = \hat{A}\hat{\Sigma}\hat{A}^\tau$ of (5.61), where $\hat{A}$ is as in (5.60).

To account for such a situation, many techniques may be employed. For example, one may choose to plug an augmented $\hat{\mathbf{v}}$ into $\Sigma(\mathbf{v})$, resulting in $\hat{\Sigma} = \Sigma(\hat{\mathbf{v}}^*)$, where

$$\hat{\mathbf{v}}^* = (\hat{p}^*_{1,1}, \dots, \hat{p}^*_{1,K_2}, \hat{p}^*_{2,1}, \dots, \hat{p}^*_{2,K_2}, \dots, \hat{p}^*_{K_1,1}, \dots, \hat{p}^*_{K_1,K_2-1})^\tau$$

where, for any pair of (i, j),

$$\hat{p}^*_{i,j} = \hat{p}_{i,j} + (1/n)1[\hat{p}_{i,j} = 0]. \tag{5.76}$$

Clearly such an augmentation does not impact any of the asymptotic properties established above but does resolve the computational issue.

In the same configuration as that of Table 5.1, the observed joint and marginal distributions are given in Table 5.2.

Table 5.2 Relative frequency data of ethnicity and color preference.

$\hat{p}_{i,j}$	$j = 1$	$j = 2$	$j = 3$	$j = 4$	$\hat{p}_{i,\cdot}$
$i = 1$	0.015	0.005	0.046	0.019	0.085
$i = 2$	0.048	0.019	0.285	.0059	0.411
$i = 3$	0.005	0.000	0.007	0.002	0.014
$i = 4$	0.004	0.001	0.007	0.003	0.015
$i = 5$	0.015	0.004	0.099	0.019	0.137
$i = 6$	0.000	0.003	0.013	0.005	0.021
$i = 7$	0.001	0.003	0.026	0.003	0.033
$i = 8$	0.020	0.004	0.085	0.030	0.139
$i = 9$	0.015	0.001	0.066	0.025	0.107
$i = 10$	0.000	0.001	0.000	0.000	0.001
$i = 11$	0.003	0.004	0.026	0.004	0.037
$\hat{p}_{\cdot,j}$	0.126	0.045	0.660	0.169	1.000

By (5.39)–(5.41) and (5.56),

$$\hat{H}(X) = 1.8061, \tag{5.77}$$

$$\hat{H}(Y) = 0.9753, \tag{5.78}$$

$$\hat{H}(X, Y) = 2.7534, \quad \text{and} \tag{5.79}$$

$$\widehat{MI} = 0.0280. \tag{5.80}$$

Further by (5.70)–(5.72),

$$\hat{\kappa} = 0.0102, \hat{\kappa}_2 = 0.0211, \quad \text{and} \quad \hat{\kappa}_3 = 0.0201. \tag{5.81}$$

Letting $p_{i,j}$ be enumerated as $\{p_k; k = 1, 2, \ldots, K\}$ by counting the probabilities from left to right in each row, one has the $K = K_1K_2 = 44$ dimensional vector

$$\hat{\mathbf{p}} = (0.015, 0.005, 0.046, 0.019, 0.048, \ldots, 0.026, 0.004)^\tau,$$

and the $K - 1 = 43$ dimensional vector

$$\hat{\mathbf{v}} = (0.015, 0.005, 0.046, 0.019, 0.048, \ldots, 0.026)^\tau.$$

Substituting the zeros in $\hat{\mathbf{v}}$ by $1/n = 0.001$ according to (5.76), at $\hat{v}_{10} = 0$, $\hat{v}_{21} = 0$, $\hat{v}_{37} = 0$, $\hat{v}_{39} = 0$, and $\hat{v}_{40} = 0$, one has $\hat{\mathbf{v}}^*$ given in Table 5.3.

Table 5.3 $\hat{\mathbf{v}}^*$ for data in Table 5.2.

$\hat{\mathbf{v}}^* = ($	0.015,	0.005,	0.046,	0.019,
	0.048,	0.019,	0.285,	0.059,
	0.005,	0.001,	0.007,	0.002,
	0.004,	0.001,	0.007,	0.003,
	0.015,	0.004,	0.099,	0.019,
	0.001,	0.003,	0.013,	0.005,
	0.001,	0.003,	0.026,	0.003,
	0.020,	0.004,	0.085,	0.030,
	0.015,	0.001,	0.066,	0.025,
	0.001,	0.001,	0.001,	0.001,
	0.003,	0.004,	0.026	$)^\tau$

Letting $\hat{A} = A(\hat{\mathbf{v}}^*)$ by means of (5.45), the three rows of $\hat{A}$ are

$$
\begin{array}{lrrrr}
\hat{a}_1^\tau = (& -0.8317, & -0.8317, & -0.8317, & -0.8317, \\
 & -2.4077, & -2.4077, & -2.4077, & -2.4077, \\
 & 0.9719, & 0.9719, & 0.9719, & 0.9719, \\
 & 0.9029, & 0.9029, & 0.9029, & 0.9029, \\
 & -1.3091, & -1.3091, & -1.3091, & -1.3091, \\
 & 0.5664, & 0.5664, & 0.5664, & 0.5664, \\
 & 0.1144, & 0.1144, & 0.1144, & 0.1144, \\
 & -1.3236, & -1.3236, & -1.3236, & -1.3236, \\
 & -1.0619, & -1.0619, & -1.0619, & -1.0619, \\
 & 3.6109, & 3.6109, & 3.6109, & 3.6109, \\
 & 0.0000, & 0.0000, & 0.0000 &); \\
\hat{a}_2^\tau = (& 0.2936 & 1.3232 & -1.3623 & 0.0000, \\
 & 0.2936 & 1.3232 & -1.3623 & 0.0000, \\
 & 0.2936 & 1.3232 & -1.3623 & 0.0000, \\
 & 0.2936 & 1.3232 & -1.3623 & 0.0000, \\
 & 0.2936 & 1.3232 & -1.3623 & 0.0000, \\
 & 0.2936 & 1.3232 & -1.3623 & 0.0000, \\
 & 0.2936 & 1.3232 & -1.3623 & 0.0000, \\
 & 0.2936 & 1.3232 & -1.3623 & 0.0000, \\
 & 0.2936 & 1.3232 & -1.3623 & 0.0000, \\
 & 0.2936 & 1.3232 & -1.3623 & 0.0000, \\
 & 0.2936 & 1.3232 & -1.3623 &);
\end{array}
$$

and

$$\hat{a}_3^\tau = (\begin{array}{rrrr} -1.3216, & -0.2231, & -2.4423, & -1.5581, \\ -2.4849, & -1.5581, & -4.2662, & -2.6912, \\ -0.2231, & 1.3863, & -0.5596, & 0.6931, \\ 0.0000, & 1.3863, & -0.5596, & 0.2877, \\ -1.3218, & 0.0000, & -3.2088, & -1.5581, \\ 1.3863, & 0.2877, & -1.1787, & -0.2231, \\ 1.3863, & 0.2877, & -1.8718, & 0.2877, \\ -1.6094, & 0.0000, & -3.0564, & -2.0149, \\ -1.3218, & 1.3863, & -2.8034, & -1.8326, \\ 1.3863, & 1.3863, & 1.3863, & 1.3863, \\ 0.2877, & 0.0000, & -1.8718 & \end{array}).$$

By (5.61) and (5.62),

$$\hat{\Sigma}_H = \hat{A}\hat{\Sigma}\hat{A}^\tau = \begin{pmatrix} 0.9407 & 0.0661 & 0.9364 \\ 0.0661 & 0.6751 & 0.7041 \\ 0.9364 & 0.7041 & 1.5927 \end{pmatrix},$$

$$\hat{\sigma}^2 = (1, 1, -1)\hat{\Sigma}_H(1, 1, -1)^\tau = 0.0597.$$

Based on the plug-in estimator of MI, an approximate $1 - \alpha = 90\%$ confidence interval for MI is

$$\widehat{MI} \pm z_{\alpha/2}\frac{\hat{\sigma}}{\sqrt{n}} = 0.0280 \pm 1.645\frac{\sqrt{0.0597}}{\sqrt{1000}} = (0.0153, 0.0407).$$

Noting

$$g_\kappa(\hat{H}(X), \hat{H}(Y), \hat{H}(X, Y)) = (0.3632, 0.3632, -0.3669)^\tau,$$
$$g_{\kappa_2}(\hat{H}(X), \hat{H}(Y), \hat{H}(X, Y)) = (0.7477, 0.7477, -0.7535)^\tau,$$
$$g_{\kappa_3}(\hat{H}(X), \hat{H}(Y), \hat{H}(X, Y)) = (0.7119, 0.7119, -0.7191)^\tau,$$

and based on Corollary 5.5,

$$\hat{\sigma}^2_{\hat{\kappa}}(\hat{H}(X), \hat{H}(Y), \hat{H}(X, Y)) = 0.0078,$$
$$\hat{\sigma}^2_{\hat{\kappa}_2}(\hat{H}(X), \hat{H}(Y), \hat{H}(X, Y)) = 0.0330,$$
$$\hat{\sigma}^2_{\hat{\kappa}_3}(\hat{H}(X), \hat{H}(Y), \hat{H}(X, Y)) = 0.0298.$$

Based on the plug-in estimators of κ, κ_2, and κ_3, the following are their respective approximate $1 - \alpha = 90\%$ confidence intervals:

$$\hat{\kappa} \pm z_{\alpha/2}\frac{\hat{\sigma}_{\hat{\kappa}}}{\sqrt{n}} = 0.0102 \pm 1.645\frac{\sqrt{0.0081}}{\sqrt{1000}} = (0.0056, 0.0148),$$
$$\hat{\kappa}_2 \pm z_{\alpha/2}\frac{\hat{\sigma}_{\hat{\kappa}_2}}{\sqrt{n}} = 0.0211 \pm 1.645\frac{\sqrt{0.0341}}{\sqrt{1000}} = (0.0116, 0.0306),$$

$$\hat{\kappa}_3 \pm z_{\alpha/2}\frac{\hat{\sigma}_{\hat{\kappa}_3}}{\sqrt{n}} = 0.0201 \pm 1.645\frac{\sqrt{0.0298}}{\sqrt{1000}} = (0.0111, 0.0291).$$

On the other hand, the corresponding estimators based on Turing's perspective are

$$\hat{H}_z(X) = 1.8112, \tag{5.82}$$

$$\hat{H}_z(Y) = 0.9768, \tag{5.83}$$

$$\hat{H}_z(X, Y) = 2.7731, \quad \text{and} \tag{5.84}$$

$$\widehat{MI}_z = 0.0148. \tag{5.85}$$

Further by (5.70)–(5.72), but with $\hat{H}_z(X)$, $\hat{H}_z(Y)$, and $\hat{H}_z(X, Y)$, in places of $\hat{H}(X)$, $\hat{H}(Y)$, and $\hat{H}(X, Y)$, respectively,

$$\hat{\kappa}_z = 0.0054, \quad \hat{\kappa}_{2z} = 0.0112, \quad \text{and} \quad \hat{\kappa}_{3z} = 0.0106. \tag{5.86}$$

Noting

$$g_{\kappa}(\hat{H}_z(X), \hat{H}_z(Y), \hat{H}_z(X, Y)) = (0.3606, 0.3606, -0.3625)^{\tau},$$

$$g_{\kappa_2}(\hat{H}_z(X), \hat{H}_z(Y), \hat{H}_z(X, Y)) = (0.7488, 0.7488, -0.7518)^{\tau},$$

$$g_{\kappa_3}(\hat{H}_z(X), \hat{H}_z Y), \hat{H}_z(X, Y)) = (0.7136, 0.7136, -0.7174)^{\tau},$$

and based on Corollary 5.6,

$$\hat{\sigma}^2_{\hat{\kappa}_z}(\hat{H}_z(X), \hat{H}_z(Y), \hat{H}_z(X, Y)) = 0.0077,$$

$$\hat{\sigma}^2_{\hat{\kappa}_{2z}}(\hat{H}_z(X), \hat{H}_z(Y), \hat{H}_z(X, Y)) = 0.0333,$$

$$\hat{\sigma}^2_{\hat{\kappa}_{3z}}(\hat{H}_z(X), \hat{H}_z(Y), \hat{H}_z(X, Y)) = 0.0302.$$

The approximate $1 - \alpha = 90\%$ confidence intervals for MI, κ, κ_2, and κ_3 based on Turing's perspective are

$$\widehat{MI}_z \pm z_{\alpha/2}\frac{\hat{\sigma}}{\sqrt{n}} = 0.0148 \pm 1.645\frac{\sqrt{0.0597}}{\sqrt{1000}} = (0.0021, 0.0275),$$

$$\hat{\kappa}_z \pm z_{\alpha/2}\frac{\hat{\sigma}_{\hat{\kappa}_z}}{\sqrt{n}} = 0.0054 \pm 1.645\frac{\sqrt{0.0077}}{\sqrt{1000}} = (0.0008, 0.0100),$$

$$\hat{\kappa}_{2z} \pm z_{\alpha/2}\frac{\hat{\sigma}_{\hat{\kappa}_{2z}}}{\sqrt{n}} = 0.0112 \pm 1.645\frac{\sqrt{0.0333}}{\sqrt{1000}} = (0.0017, 0.0207),$$

$$\hat{\kappa}_{3z} \pm z_{\alpha/2}\frac{\hat{\sigma}_{\hat{\kappa}_{3z}}}{\sqrt{n}} = 0.0106 \pm 1.645\frac{\sqrt{0.0302}}{\sqrt{1000}} = (0.0016, 0.0196).$$

In addition to the illustration of the example, at least two points are worth noting. First, the sample data suggest a very low level of association between ethnicity and color preference, as measured by MI and its derivatives, and yet the association is supported with fairly strong statistical evidence.

Second, it may be of interesting to note again that the plug-in estimator of entropy, $\hat{H}$, has a much larger negative bias compared to $\hat{H}_z$, which also has a negative bias. Consequently, the entropy estimators in the left block are consistently lower than those in the right block below.

$\hat{H}(X)$	1.8061
$\hat{H}(Y)$	0.9753
$\hat{H}(X,Y)$	2.7534

$\hat{H}_z(X)$	1.8112
$\hat{H}_z(Y)$	0.9768
$\hat{H}_z(X,Y)$	2.7731

On the other hand, due to the dominance of the negative bias of the plug-in estimator $\hat{H}(X,Y)$, $\widehat{MI}$ has a pronounced positive bias as numerically evidenced by the following blocks in comparison.

$\widehat{MI}$	0.0280
$\hat{\kappa}$	0.0102
$\hat{\kappa}_2$	0.0211
$\hat{\kappa}_3$	0.0201

$\widehat{MI}_z$	0.0148
$\hat{\kappa}_z$	0.0054
$\hat{\kappa}_{2z}$	0.0112
$\hat{\kappa}_{3z}$	0.0106

For more examples regarding the over-estimation of mutual information using $\widehat{MI}$ and $\widehat{MI}_z$, readers may wish to refer to Zhang and Zheng (2015).

5.3 Estimation of Kullback–Leibler Divergence

Let $\mathbf{p} = \{p_k\}$ and $\mathbf{q} = \{q_k\}$ be two discrete probability distributions on a same finite alphabet, $\mathscr{X} = \{\ell_k; k = 1, \dots, K\}$, where $K \geq 2$ is a finite integer. Suppose it is of interest to estimate the Kullback–Leibler divergence, also known as relative entropy,

$$\begin{aligned} D &= D(\mathbf{p}\|\mathbf{q}) \\ &= \sum_{k=1}^{K} p_k \ln(p_k/q_k) \\ &= \sum_{k=1}^{K} p_k \ln(p_k) - \sum_{k=1}^{K} p_k \ln(q_k). \end{aligned} \tag{5.87}$$

Here and throughout, the standard conventions that $0 \log (0/q) = 0$ if $q \geq 0$ and $p \ln(p/0) = \infty$ if $p > 0$ are adopted. The Kullback–Leibler divergence is a measure of the difference between two probability distributions, which was first introduced by Kullback and Leibler (1951), and is an important measure of information in information theory. This measure is not a metric since it does not satisfy the triangle inequality and it is not symmetric. A symmetric modification of Kullback–Leibler divergence is considered in Section 5.3.3 below.

Assume that two independent *iid* samples of sizes n and m are to be available, respectively, according to two unknown distributions $\mathbf{p}$ and $\mathbf{q}$.

Let

$$\begin{array}{lllll} \{ & X_1, & X_2, & \cdots, & X_K & \} \\ \{ & Y_1, & Y_2, & \cdots, & Y_K & \} \end{array}$$

be the sequences of observed frequencies of letters $\{\ell_1, \ell_2, \ldots, \ell_K\}$ in the two samples, respectively, and let

$$\begin{array}{l} \hat{\mathbf{p}} = \{\hat{p}_k\} = \{ \quad X_1/n, \quad X_2/n, \quad \cdots, \quad X_K/n \quad \} \\ \hat{\mathbf{q}} = \{\hat{q}_k\} = \{ \quad Y_1/m, \quad Y_2/m, \quad \cdots, \quad Y_K/m \quad \} \end{array}$$

be the sequences of corresponding relative frequencies.

For simplicity, the following is further imposed.

Condition 5.1 *The probability distributions,* $\mathbf{p}$ *and* $\mathbf{q}$, *and the observed sample distributions,* $\hat{\mathbf{p}}$ *and* $\hat{\mathbf{q}}$, *satisfy*

A1. $p_k > 0$ *and* $q_k > 0$ *for each* $k = 1, \ldots, K$, *and*
A2. there exists a $\lambda \in (0, \infty)$ *such that* $n/m \to \lambda$, *as* $n \to \infty$.

Note that it is not assumed that K is known, only that it is some finite integer greater than or equal to two.

Perhaps the most intuitive estimator of $D(\mathbf{p}||\mathbf{q})$ is the "plug-in" estimator given by

$$\hat{D} = \hat{D}(\mathbf{p}||\mathbf{q}) = \sum_{k=1}^{K} \hat{p}_k \ln(\hat{p}_k) - \sum_{k=1}^{K} \hat{p}_k \ln(\hat{q}_k). \tag{5.88}$$

In Section 5.3.1, it is shown that this estimator is consistent and asymptotically normal, but with an infinite bias. Then a modification of this estimator is introduced and is shown that it is still consistent and asymptotically normal, but now with a finite bias. It is also shown that this bias decays no faster than $\mathcal{O}(1/n)$. In Section 5.3.2, the following estimator of $D(\mathbf{p}||\mathbf{q})$,

$$\begin{aligned} \hat{D}^{\sharp} &= \hat{D}^{\sharp}(\mathbf{p}||\mathbf{q}) \\ &= \sum_{k=1}^{K} \hat{p}_k \left[\sum_{v=1}^{m-Y_k} \frac{1}{v} \prod_{j=1}^{v} \left(1 - \frac{Y_k}{m-j+1}\right) - \sum_{v=1}^{n-X_k} \frac{1}{v} \prod_{j=1}^{v} \left(1 - \frac{X_k - 1}{n-j}\right) \right], \end{aligned} \tag{5.89}$$

based on Turing's perspective and proposed by Zhang and Grabchak (2014), is discussed and is shown to have asymptotic consistency and normality.

This estimator is also shown to have a bias decaying exponentially fast in sample sizes n and m. Finally, in Section 5.3.3, the estimation of the symmetrized Kullback–Leibler divergence is discussed.

Before proceeding let

$$A = \sum_{k=1}^{K} p_k \ln(p_k) \quad \text{and} \quad B = -\sum_{k=1}^{K} p_k \ln(q_k), \tag{5.90}$$

and note that $D(\mathbf{p}||\mathbf{q}) = A + B$. The first term, A, is the negative of the entropy of the distribution of $\mathbf{p}$. In the subsequent text of this section, an estimator of $D(\mathbf{p}||\mathbf{q})$ is occasionally expressed as the sum of an estimator of A and an estimator of B.

5.3.1 The Plug-In Estimator

In this section, it is shown that $\hat{D}$, the plug-in estimator given in (5.88), is a consistent and asymptotically normal estimator of D, but with an infinite bias. Define the $(2K-2)$-dimensional vectors

$$\mathbf{v} = (p_1, \dots, p_{K-1}, q_1, \dots, q_{K-1})^\tau \text{ and } \hat{\mathbf{v}} = (\hat{p}_1, \dots, \hat{p}_{K-1}, \hat{q}_1, \dots, \hat{q}_{K-1})^\tau,$$

and note that $\hat{\mathbf{v}} \xrightarrow{p} \mathbf{v}$ as $n \to \infty$. Moreover, by the multivariate normal approximation to the multinomial distribution

$$\sqrt{n}(\hat{\mathbf{v}} - \mathbf{v}) \xrightarrow{L} MVN(0, \Sigma(\mathbf{v})), \tag{5.91}$$

where $\Sigma(\mathbf{v})$ is the $(2K-2) \times (2K-2)$ covariance matrix given by

$$\Sigma(\mathbf{v}) = \begin{pmatrix} \Sigma_1(\mathbf{v}) & 0 \\ 0 & \Sigma_2(\mathbf{v}) \end{pmatrix}. \tag{5.92}$$

Here $\Sigma_1(\mathbf{v})$ and $\Sigma_2(\mathbf{v})$ are $(K-1) \times (K-1)$ matrices given by

$$\Sigma_1(\mathbf{v}) = \begin{pmatrix} p_1(1-p_1) & -p_1p_2 & \cdots & -p_1p_{K-1} \\ -p_2p_1 & p_2(1-p_2) & \cdots & -p_2p_{K-1} \\ \cdots & \cdots & \cdots & \cdots \\ -p_{K-1}p_1 & -p_{K-1}p_2 & \cdots & p_{K-1}(1-p_{K-1}) \end{pmatrix}$$

and

$$\Sigma_2(\mathbf{v}) = \lambda \begin{pmatrix} q_1(1-q_1) & -q_1q_2 & \cdots & -q_1q_{K-1} \\ -q_2q_1 & q_2(1-q_2) & \cdots & -q_2q_{K-1} \\ \cdots & \cdots & \cdots & \cdots \\ -q_{K-1}q_1 & -q_{K-1}q_2 & \cdots & q_{K-1}(1-q_{K-1}) \end{pmatrix}.$$

Let

$$G(\mathbf{v}) = \sum_{k=1}^{K-1} p_k(\ln p_k - \ln q_k)$$
$$+ \left(1 - \sum_{k=1}^{K-1} p_k\right)\left[\ln\left(1 - \sum_{k=1}^{K-1} p_k\right) - \ln\left(1 - \sum_{k=1}^{K-1} q_k\right)\right] \tag{5.93}$$

and

$$g(\mathbf{v}) := \nabla G(\mathbf{v})$$
$$= \left(\frac{\partial}{\partial p_1}G(\mathbf{v}), \ldots, \frac{\partial}{\partial p_{K-1}}G(\mathbf{v}), \frac{\partial}{\partial q_1}G(\mathbf{v}), \ldots, \frac{\partial}{\partial q_{K-1}}G(\mathbf{v})\right)^{\tau}. \tag{5.94}$$

For each j, $j = 1, \ldots, K - 1$,

$$\frac{\partial}{\partial p_j}G(\mathbf{v}) = \ln p_j - \ln q_j - \left[\ln\left(1 - \sum_{k=1}^{K-1} p_k\right) - \ln\left(1 - \sum_{k=1}^{K-1} q_k\right)\right]$$
$$= \ln p_j - \ln q_j - (\ln p_K - \ln q_K)$$
$$= \ln\frac{p_j}{q_j} - \ln\frac{p_K}{q_K} - \frac{p_j}{q_j} + \frac{1 - \sum_{k=1}^{K-1} p_k}{1 - \sum_{k=1}^{K-1} q_k}$$
$$= -\frac{p_j}{q_j} + \frac{p_K}{q_K}.$$

The delta method immediately gives the following result.

Theorem 5.9 *Provided that* $g^{\tau}(\mathbf{v})\Sigma(\mathbf{v})g(\mathbf{v}) > 0$,

$$\frac{\sqrt{n}(\hat{D} - D)}{\sqrt{g^{\tau}(\mathbf{v})\Sigma(\mathbf{v})g(\mathbf{v})}} \xrightarrow{L} N(0, 1). \tag{5.95}$$

By the continuous mapping theorem, $\Sigma(\hat{\mathbf{v}})$ is a consistent estimator of $\Sigma(\mathbf{v})$ and $g(\hat{\mathbf{v}})$ is a consistent estimator of $g(\mathbf{v})$. Corollary 5.7 follows from Slutsky's theorem.

Corollary 5.7 *Provided that* $g^{\tau}(\mathbf{v})\Sigma(\mathbf{v})g(\mathbf{v}) > 0$,

$$\frac{\sqrt{n}(\hat{D} - D)}{\hat{\sigma}_{\hat{D}}} \xrightarrow{L} N(0, 1), \tag{5.96}$$

where $\hat{\sigma}^2_{\hat{D}} = g^{\tau}(\hat{\mathbf{v}})\Sigma(\hat{\mathbf{v}})g(\hat{\mathbf{v}})$.

Although $\hat{D}$ is a consistent and asymptotically normal estimator of D, it has an infinite bias. To see this, consider $\mathrm{P}(\{\hat{p}_k > 0\} \cap \{\hat{q}_k = 0\}) > 0$ for every k, $k = 1, \ldots, K$. For each of such k, $\mathrm{E}(\hat{p}_k \ln(\hat{q}_k)) = \infty$. The problem is clearly caused by the fact that $\hat{q}_k$ may be zero even though $q_k > 0$. To deal with this minor issue, the following augmentation is added:

$$\hat{q}_k^* = \hat{q}_k + \frac{1[Y_k = 0]}{m}, \tag{5.97}$$

where $k = 1, \dots, K$.

5.3.2 Properties of the Augmented Plug-In Estimator

Define an augmented plug-in estimator of D by

$$\hat{D}^* = \sum_{k=1}^{K} \hat{p}_k \ln(\hat{p}_k) - \sum_{k=1}^{K} \hat{p}_k \ln(\hat{q}_k^*) =: \hat{A} + \hat{B}^*. \tag{5.98}$$

It is shown that $\hat{D}^*$ is a consistent and asymptotically normal estimator of D, with a bias that decays no faster than $\mathcal{O}(1/n)$.

Theorem 5.10 *Provided that $g^\tau(\mathbf{v})\Sigma(\mathbf{v})g(\mathbf{v}) > 0$,*

$$\frac{\sqrt{n}(\hat{D}^* - D)}{\sqrt{g^\tau(\mathbf{v})\Sigma(\mathbf{v})g(\mathbf{v})}} \xrightarrow{L} N(0, 1). \tag{5.99}$$

Proof: In light of Theorem 5.9, it suffices to show that $\sqrt{n}(\hat{D}^* - \hat{D}) \xrightarrow{p} 0$. The fact that, for any $\varepsilon > 0$,

$$\begin{aligned} \mathrm{P}(|\sqrt{n}(\hat{D}^* - \hat{D})| > \varepsilon) &\le \mathrm{P}\left(\bigcup_{k=1}^{K}[Y_k = 0]\right) \le \sum_{k=1}^{K} \mathrm{P}(Y_k = 0) \\ &= \sum_{k=1}^{K} (1 - q_k)^m \to 0, \end{aligned}$$

gives the result. □

Arguments similar to these for Corollary 5.7 give the following corollary.

Corollary 5.8 *Provided that $g^\tau(\mathbf{v})\Sigma(\mathbf{v})g(\mathbf{v}) > 0$,*

$$\frac{\sqrt{n}(\hat{D}^* - D)}{\hat{\sigma}_{\hat{D}^*}} \xrightarrow{L} N(0, 1), \tag{5.100}$$

where $\hat{\sigma}^2_{\hat{D}^*} = g^\tau(\hat{\mathbf{v}}^*)\Sigma(\hat{\mathbf{v}}^*)g(\hat{\mathbf{v}}^*)$ and $\hat{\mathbf{v}}^* = (\hat{p}_1, \dots, \hat{p}_{K-1}, \hat{q}_1^*, \dots, \hat{q}_{K-1}^*)^\tau$.

Example 5.10 *A commonly used special case is when $K = 2$. In this case, $\mathbf{p}$ and $\mathbf{q}$ are both Bernoulli distributions with probabilities of success being p and q (note that $q \neq 1 - p$), respectively. In this case,*

$$\mathbf{v} = (p, q)^\tau,$$

$$\Sigma(\mathbf{v}) = \begin{pmatrix} p(1-p) & 0 \\ 0 & \lambda q(1-q) \end{pmatrix},$$

$$g(\mathbf{v}) = \left(\ln \left(\frac{p(1-q)}{q(1-p)} \right), \frac{q-p}{q(1-q)} \right)^{\tau},$$

and $\sigma^2 := g^{\tau}(\mathbf{v})\Sigma(\mathbf{v})g(\mathbf{v})$ *reduces to*

$$\sigma^2 = p(1-p)\left[\ln \left(\frac{p(1-q)}{q(1-p)} \right) \right]^2 + \frac{\lambda(p-q)^2}{q(1-q)}. \tag{5.101}$$

By Theorems 5.9 and 5.10, it follows that

$$\sqrt{n}(\hat{D} - D) \xrightarrow{L} N(0, \sigma^2) \quad \textit{and} \quad \sqrt{n}(\hat{D}^* - D) \xrightarrow{L} N(0, \sigma^2).$$

Next, consider the bias of $\hat{D}^*$. In Miller (1955) (see also Paninski (2003) for a more formal treatment), it is shown that

$$\mathrm{E}(\hat{A}) - A \geq 0 \quad \text{and} \quad \mathrm{E}(\hat{A}) - A = \mathcal{O}(1/n). \tag{5.102}$$

It will be shown below that, for sufficiently large n,

$$\mathrm{E}(\hat{B}^*) - B \geq 0 \quad \text{and} \quad \mathrm{E}(\hat{B}^*) - B \geq \mathcal{O}(1/n). \tag{5.103}$$

Once this is established, so is the following theorem.

Theorem 5.11 *For a sufficiently large n,*

$$\mathrm{E}(\hat{D}^*) - D \geq 0 \quad \textit{and} \quad \mathrm{E}(\hat{D}^*) - D \geq \mathcal{O}(1/n). \tag{5.104}$$

Before proving (5.103), and hence the theorem, the following result for general "plug-in" estimators of B is first given.

Lemma 5.7 *Let* $\tilde{q}_k$ *be any estimator of* q_k *such that* $\tilde{q}_k$ *is independent of* $\hat{p}_k$ *and* $0 < \tilde{q}_k \leq 2$ *a.s. for all* $k = 1, \ldots, K$. *If* $c_k = \mathrm{E}(\tilde{q}_k)$ *and*

$$\tilde{B} = -\sum_{k=1}^{K} \hat{p}_k \ln \tilde{q}_k$$

then

$$\mathrm{E}(\tilde{B}) - B = \sum_{k=1}^{K} p_k \left[\ln(q_k/c_k) + \frac{1}{2}\mathrm{Var}(\tilde{q}_k) + R_k \right],$$

where

$$R_k = \sum_{\nu=3}^{\infty} \frac{1}{\nu} \mathrm{E}[(1-\tilde{q}_k)^{\nu} - (1-c_k)^{\nu})].$$

If, in addition, $0 < \tilde{q}_k \leq 1$ *a.s., then* $R_k \geq 0$ *a.s.*

Proof: By the Taylor expansion of the logarithm

$$E(\ln(c_k) - \ln(\tilde{q}_k)) = \sum_{\nu=1}^{\infty} \frac{1}{\nu} E((1-\tilde{q}_k)^\nu - (1-c_k)^\nu) = \frac{1}{2}\mathrm{Var}(\tilde{q}_k) + R_k.$$

Since $\hat{p}_k$ is an unbiased estimator of p_k and it is independent of $\tilde{q}_k$

$$\begin{aligned} E(\tilde{B}) - B &= \sum_{k=1}^{K} p_k E(\ln(q_k) - \ln(c_k) + \ln(c_k) - \ln(\tilde{q}_k)) \\ &= \sum_{k=1}^{K} p_k \left(\ln(q_k/c_k) + \frac{1}{2}\mathrm{Var}(\tilde{q}_k) + R_k \right). \end{aligned}$$

If $0 < \tilde{q}_k \le 1$, then Jensen's inequality implies that

$$E((1-\tilde{q}_k)^\nu) \ge (E(1-\tilde{q}_k))^\nu = (1-c_k)^\nu,$$

and thus $R_k \ge 0$. □

Proof of Theorem 5.11: It suffices to show that both inequalities in (5.103) are true. Note that $Y_k \sim \mathrm{Bin}(m, q_k)$ and $c_k := E(\hat{q}_k^*) = q_k + e_k$, where $e_k = (1-q_k)^m/m$. By Lemma 5.7,

$$\begin{aligned} E(\hat{B}_n^*) - B = \sum_{k=1}^{K} p_k &\left[\ln\left(\frac{q_k}{q_k + e_k}\right) + \frac{1}{2}\mathrm{Var}(\hat{q}_k^*) \right. \\ &\left. + \sum_{\nu=3}^{\infty} \frac{1}{\nu} E((1-\hat{q}_k^*)^\nu - (1-q_k-e_k)^\nu) \right]. \end{aligned}$$

Note that

$$\begin{aligned} \frac{1}{2}\mathrm{Var}(\hat{q}_k^*) &= \frac{\mathrm{Var}(Y_k) + \mathrm{Var}(1[Y_k=0]) + 2\mathrm{Cov}(Y_k, 1[Y_k=0])}{2m^2} \\ &= \frac{q_k(1-q_k)}{2m} + \frac{1}{2m^2}\left[(1-q_k)^m - (1-q_k)^{2m} - 2mq_k(1-q_k)^m\right] \\ &= \frac{q_k(1-q_k)}{2m} + \frac{e_k}{2m} - \frac{e_k^2}{2} - q_k e_k, \end{aligned}$$

and for sufficiently large m, by the Taylor expansion of the logarithm,

$$\ln\left(\frac{q_k}{q_k+e_k}\right) = -\ln(1+e_k/q_k) = -e_k/q_k + r_k,$$

where the remainder term $r_k \ge 0$ by properties of alternating series. This means that, for a sufficiently large m,

$$\begin{aligned} E(\hat{B}^*) - B = \sum_{k=1}^{K} p_k &\left(-e_k(q_k + 1/q_k) + r_k + \frac{q_k(1-q_k)}{2m} + \frac{e_k}{2m} - \frac{e_k^2}{2} \right. \\ &\left. + \sum_{\nu=3}^{\infty} \frac{1}{\nu} E((1-\tilde{q}_k)^\nu - (1-q_k-e_k)^\nu) \right), \end{aligned}$$

where the only negative terms are $-e_k(q_k + 1/q_k)$ and $-e_k^2/2$. But

$$\frac{e_k}{2m} - \frac{e_k^2}{2} \geq 0$$

and for sufficiently large m

$$\begin{aligned}\frac{q_k(1-q_k)}{2m} - e_k(q_k + 1/q_k) &= \frac{q_k(1-q_k)}{4m} \\ &\quad + e_k(q_k + 1/q_k)\left(\frac{q_k(1-q_k)}{4(q_k + 1/q_k)me_k} - 1\right) \\ &\geq \frac{q_k(1-q_k)}{4m},\end{aligned}$$

where the final inequality follows by the fact that $\lim_{m\to\infty} me_k = 0$. Thus, for sufficiently large m

$$\mathrm{E}(\hat{B}_n^*) - B \geq \frac{1}{4m}\sum_{k=1}^{K} p_k q_k(1-q_k) = \mathcal{O}(1/m) = \mathcal{O}(1/n),$$

which completes the proof. □

5.3.3 Estimation in Turing's Perspective

In this section, the estimator, $\hat{D}^\sharp$, of D as given in (5.89) is motivated, and is shown to have asymptotic normality and a bias decaying exponentially fast in sample size. Recall that $D(\mathbf{p}\|\mathbf{q}) = A + B$. The estimator of $D(\mathbf{p}\|\mathbf{q})$ is derived by that of A and that of B separately.

Since A is the negative of the entropy of $\mathbf{p}$, it can be estimated by the negative of an entropy estimator, specifically the negative of the estimator based on Turing's perspective, $-\hat{H}_z$, which has exponentially decaying bias. This leads to the estimator

$$\hat{A}^\sharp = -\sum_{v=1}^{n-1} \frac{1}{v} Z_{1,v}, \tag{5.105}$$

where

$$Z_{1,v} = \frac{n^{1+v}[n-(1+v)]!}{n!}\sum_{k=1}^{K}\left[\hat{p}_k \prod_{j=0}^{v-1}\left(1 - \hat{p}_k - \frac{j}{n}\right)\right]. \tag{5.106}$$

Let $B_1 := \mathrm{E}(\hat{A}^\sharp) - A$. It is demonstrated in Chapter 3 that

$$0 \leq B_1 = \sum_{v=n}^{\infty}\frac{1}{v}\sum_{k=1}^{K} p_k(1-p_k)^v \leq \mathcal{O}(n^{-1}(1-p_\wedge)^n), \tag{5.107}$$

and therefore the bias decays exponentially fast in n.

Remark 5.2 *The product term in* (5.106) *assumes a value of 0 when* $v \geq n - X_k + 1$. *This is because if* $v \geq n - X_k + 1$, *then when* $j = n - X_k$ *one has* $(1 - \hat{p}_k - j/n) = 0$. *Combining this observation with some basic algebra gives*

$$\hat{A}^\sharp = -\sum_{k=1}^{K} \hat{p}_k \sum_{v=1}^{n-X_k} \frac{1}{v} \prod_{j=1}^{v} \left(1 - \frac{X_k - 1}{n - j}\right). \tag{5.108}$$

See Exercise 9.

Next an estimator of B is derived. By the Taylor expansion of the logarithmic function

$$\begin{aligned} B &= \sum_{v=1}^{m} \frac{1}{v} \sum_{k=1}^{K} p_k(1 - q_k)^v + \sum_{v=m+1}^{\infty} \frac{1}{v} \sum_{k=1}^{K} p_k(1 - q_k)^v \\ &=: \eta_m + B_{2,m}, \end{aligned} \tag{5.109}$$

where m is the size of the second sample.

First, for each v, $v = 1, \ldots, m$, an unbiased estimator of $(1 - q_k)^v$ is derived. For each fixed k, $k = 1, \ldots, K$, and each fixed v, $v = 1, \ldots, m$, consider a subsample of size v of the sample of size-m from $\mathbf{q}$. Let $\{Y_1^\circ, \ldots, Y_K^\circ\}$ and $\{\hat{q}_1^\circ, \ldots, \hat{q}_K^\circ\}$ be the counts and the relative frequencies of the subsample. Note that $\mathrm{E}(1[Y_k^\circ = 0]) = (1 - q_k)^v$, thus $1[Y_k^\circ = 0]$ is an unbiased estimator of $(1 - q_k)^v$. Since there are $\binom{m}{v}$ different subsamples of size $v \leq m$ from the sample of size m, the U-statistic

$$U_{k,v} = \binom{m}{v}^{-1} \sum_{*} 1[Y_k^\circ = 0], \tag{5.110}$$

where the sum, $\sum_*$, is taken over all possible subsamples of size $v \leq m$, is also an unbiased estimator of $(1 - q_k)^v$. Note that $\sum_* 1[Y_k^\circ = 0]$ is simply counting the number of subsamples of size v in which ℓ_k, the kth letter of the alphabet, is missing. This count can be re-expressed and summarized in the following cases:

1) if $v \leq m - Y_k$, the count is

$$\binom{m - Y_k}{v} = \frac{m^v}{v!} \prod_{j=0}^{v-1} \left(1 - \hat{q}_k - \frac{j}{m}\right), \tag{5.111}$$

2) if $v \geq m - Y_k + 1$, the count is zero.

Since, in the second case,

$$\prod_{j=0}^{v-1} \left(1 - \hat{q}_k - \frac{j}{m}\right) = 0,$$

a general form for both cases becomes

$$U_{k,v} = \binom{m}{v}^{-1} \frac{m^v}{v!} \prod_{j=0}^{v-1} \left(1 - \hat{q}_k - \frac{j}{m}\right) = \prod_{j=0}^{v-1} \left(1 - \frac{Y_k}{m-j}\right). \tag{5.112}$$

Noting that $\hat{p}_k$ is an unbiased estimator of p_k and it is independent of $U_{k,v}$, for each v, $1 \le v \le m$,

$$\sum_{k=1}^{K} \hat{p}_k U_{k,v}$$

is an unbiased estimator of $\sum_{k=1}^{K} p_k (1-q_k)^v$.

This leads to an estimator of B given by

$$\hat{B}^\sharp = \sum_{v=1}^{m} \frac{1}{v} \sum_{k=1}^{K} \hat{p}_k U_{k,v}. \tag{5.113}$$

By construction $\mathrm{E}(\hat{B}^\sharp) = \eta_m$, and by (5.109) the bias $B_{2,m} = B - \mathrm{E}(\hat{B}^\sharp)$ satisfies

$$0 \le B_{2,m} \le \frac{1}{m+1} \sum_{v=m+1}^{\infty} (1-q_\wedge)^v = \frac{(1-q_\wedge)^{m+1}}{(m+1)q_\wedge}, \tag{5.114}$$

where $q_\wedge = \min\{q_1, \ldots, q_K\}$. Thus, the bias of $\hat{B}^\sharp$ decays exponentially in m, and hence in n.

Combining the fact that the product term in (5.112) assumes a value of 0 for any $v \ge m - Y_k + 1$ and some basic algebra shows that

$$\hat{B}^\sharp = \sum_{k=1}^{K} \hat{p}_k \sum_{v=1}^{m-Y_k} \frac{1}{v} \prod_{j=1}^{v} \left(1 - \frac{Y_k}{m-j+1}\right). \tag{5.115}$$

Let an estimator of $D(\mathbf{p}\|\mathbf{q})$ be given by

$$\hat{D}^\sharp = \hat{A}^\sharp + \hat{B}^\sharp. \tag{5.116}$$

Combining (5.108) and (5.115), it can be shown that this estimator is as in (5.89). By (5.107) and (5.114), the bias of $\hat{D}_n^\sharp$ is given by

$$\begin{aligned} \mathrm{E}\left(\hat{D}_n^\sharp - D\right) &= B_{1,n} - B_{2,m} \\ &= \sum_{k=1}^{K} p_k \left(\sum_{v=n}^{\infty} \frac{1}{v}(1-p_k)^v - \sum_{v=m+1}^{\infty} \frac{1}{v}(1-q_k)^v \right). \end{aligned} \tag{5.117}$$

Since the bias in parts decays exponentially fast, the bias in whole decays at least exponentially as well. In particular when the distributions and the sample sizes are similar, the bias tends to be smaller.

Two normal laws of $\hat{D}^\sharp$ are given in Theorem 5.12 and Corollary 5.9.

Theorem 5.12 *Provided that* $g^\tau(\mathbf{v})\Sigma(\mathbf{v})g(\mathbf{v}) > 0$,

$$\frac{\sqrt{n}(\hat{D}^\sharp - D)}{\sqrt{g^\tau(\mathbf{v})\Sigma(\mathbf{v})g(\mathbf{v})}} \xrightarrow{L} N(0,1).$$

Corollary 5.9 *Provided that* $g^\tau(\mathbf{v})\Sigma(\mathbf{v})g(\mathbf{v}) > 0$,

$$\frac{\sqrt{n}(\hat{D}^\sharp - D)}{\hat{\sigma}_{\hat{D}^\sharp}} \xrightarrow{L} N(0,1),$$

where $\hat{\sigma}^2_{\hat{D}^\sharp} = g^\tau(\hat{\mathbf{v}}^*)\Sigma(\hat{\mathbf{v}}^*)g(\hat{\mathbf{v}}^*)$ *and* $\hat{\mathbf{v}}^*$ *is as in Corollary 5.8.*

Given Theorem 5.12, the proof of Corollary 5.9 is similar to that of Corollary 5.7.

Toward proving Theorem 5.12, Lemmas 5.8 and 5.9 are needed.

Lemma 5.8 *If* $0 < a < b < 1$, *then for any integer* $m \geq 0$,

$$b^m - a^m \leq m b^{m-1}(b-a).$$

Proof: Noting that $f(x) = x^m$ is a convex function on the interval $(0,1)$ and $f'(b) = mb^{m-1}$, the result follows immediately. □

Lemma 5.9 *If* $Y \sim \mathrm{Bin}(n,p)$ *and* $X = Y + 1[Y = 0]$, *then*

$$\limsup_{n\to\infty} \mathrm{E}(n/X) < \infty.$$

Proof: Let

$$h_n(x) = \frac{x+1}{x(n-x)}$$

for $x = 1, 2, \ldots, n-1$. Note that h_n attains its maximum at a value x_n, which is either $\lfloor\sqrt{n+1}\rfloor - 1$ or $\lfloor\sqrt{n+1}\rfloor$ (here $\lfloor\cdot\rfloor$ is the floor of a nonnegative value). Clearly $x_n = \mathcal{O}(\sqrt{n})$ and $h_n(x_n) = \mathcal{O}(n^{-1})$. Therefore,

$$\begin{aligned}
\mathrm{E}(n/X) &= n(1-p)^n + n\sum_{x=1}^{n} \frac{1}{x}\binom{n}{x} p^x(1-p)^{n-x} \\
&= n(1-p)^n + p^n \\
&\quad + n\sum_{x=1}^{n-1} \frac{n!}{(x+1)!(n-x-1)!}\,\frac{x+1}{x(n-x)} p^x(1-p)^{n-x} \\
&\leq n(1-p)^n + p^n \\
&\quad + nh_n(x_n)\frac{1-p}{p}\sum_{x=1}^{n-1} \frac{n!}{(x+1)![n-(x+1)]!} p^{x+1}(1-p)^{n-(x+1)}
\end{aligned}$$

$$= n(1-p)^n + p^n + nh_n(x_n)\frac{1-p}{p}\sum_{x=2}^{n}\frac{n!}{x!(n-x)!}p^x(1-p)^{n-x}$$
$$= n(1-p)^n + p^n + nh_n(x_n)\frac{1-p}{p}[1-(1-p)^n - np(1-p)^{n-1}].$$

From here, the result follows. □

A proof of Theorem 5.12 is given in the appendix of this chapter.

5.3.4 Symmetrized Kullback–Leibler Divergence

The symmetrized Kullback–Leibler divergence is defined to be

$$S = S(\mathbf{p},\mathbf{q}) = \frac{1}{2}(D(\mathbf{p}||\mathbf{q}) + D(\mathbf{q}||\mathbf{p})) \tag{5.118}$$
$$= \frac{1}{2}\left(\sum_{k=1}^{K} p_k \ln p_k - \sum_{k=1}^{K} p_k \ln q_k\right) + \frac{1}{2}\left(\sum_{k=1}^{K} q_k \ln q_k - \sum_{k=1}^{K} q_k \ln p_k\right).$$

The plug-in estimator of $S(\mathbf{p},\mathbf{q})$ is defined by

$$\hat{S}_n = \hat{S}_n(\mathbf{p},\mathbf{q}) = \frac{1}{2}\left(\hat{D}_n(\mathbf{p}||\mathbf{q}) + \hat{D}_m(\mathbf{q}||\mathbf{p})\right) \tag{5.119}$$
$$= \frac{1}{2}\left(\sum_{k=1}^{K} \hat{p}_k \ln \hat{p}_k - \sum_{k=1}^{K} \hat{p}_k \ln \hat{q}_k\right) + \frac{1}{2}\left(\sum_{k=1}^{K} \hat{q}_k \ln \hat{q}_k - \sum_{k=1}^{K} \hat{q}_k \ln \hat{p}_k\right),$$

and the augmented plug-in estimator of $S(\mathbf{p},\mathbf{q})$ is defined by

$$\hat{S}^*_n = \hat{S}^*_n(\mathbf{p},\mathbf{q}) = \frac{1}{2}\left(\hat{D}^*_n(\mathbf{p}||\mathbf{q}) + \hat{D}^*_m(\mathbf{q}||\mathbf{p})\right) \tag{5.120}$$
$$= \frac{1}{2}\left(\sum_{k=1}^{K} \hat{p}_k \ln \hat{p}_k - \sum_{k=1}^{K} \hat{p}_k \ln \hat{q}^*_k\right) + \frac{1}{2}\left(\sum_{k=1}^{K} \hat{q}_k \ln \hat{q}_k - \sum_{k=1}^{K} \hat{q}_k \ln \hat{p}^*_k\right)$$

where q^*_k is as in (5.97) and

$$\hat{p}^*_k = \hat{p}_k + 1[x_k = 0]/n$$

for $k = 1, \ldots, K$.

Let $G_s(\mathbf{v})$ be a function such that

$$2G_s(\mathbf{v}) = \sum_{k=1}^{K-1} p_k(\ln p_k - \ln q_k)$$
$$+ \left(1 - \sum_{k=1}^{K-1} p_k\right)\left[\ln\left(1 - \sum_{k=1}^{K-1} p_k\right) - \ln\left(1 - \sum_{k=1}^{K-1} q_k\right)\right]$$
$$+ \sum_{k=1}^{K-1} q_k(\ln q_k - \ln p_k)$$

$$+\left(1-\sum_{k=1}^{K-1} q_k\right)\left[\ln\left(1-\sum_{k=1}^{K-1} q_k\right)-\ln\left(1-\sum_{k=1}^{K-1} p_k\right)\right]$$

and let $g_s(\mathbf{v}) = \nabla G_s(\mathbf{v})$.

For each j, $j = 1, \ldots, K-1$,

$$\frac{\partial}{\partial p_j} G_s(\mathbf{v}) = \frac{1}{2}\left(\ln\frac{p_j}{q_j} - \ln\frac{p_K}{q_K}\right) - \frac{1}{2}\left(\frac{q_j}{p_j} - \frac{q_K}{p_K}\right) \text{ and}$$

$$\frac{\partial}{\partial q_j} G_s(\mathbf{v}) = \frac{1}{2}\left(\ln\frac{q_j}{p_j} - \ln\frac{q_K}{p_K}\right) - \frac{1}{2}\left(\frac{p_j}{q_j} - \frac{p_K}{q_K}\right).$$

By arguments similar to the proofs of Theorems 5.9 and 5.10 and Corollaries 5.7 and 5.8, one gets Theorem 5.13 and Corollary 5.10.

Theorem 5.13 *Provided that* $g_s^\tau(v)\Sigma(v)g_s(v) > 0$,

$$\frac{\sqrt{n}(\hat{S} - S)}{\sqrt{g_s^\tau(\mathbf{v})\Sigma(\mathbf{v})g_s(\mathbf{v})}} \xrightarrow{L} N(0, 1)$$

and

$$\frac{\sqrt{n}(\hat{S}^* - S)}{\sqrt{g_s^\tau(\mathbf{v})\Sigma(\mathbf{v})g_s(\mathbf{v})}} \xrightarrow{L} N(0, 1).$$

Corollary 5.10 *Provided that* $g_s^\tau(\mathbf{v})\Sigma(\mathbf{v})g_s(\mathbf{v}) > 0$,

$$\frac{\sqrt{n}(\hat{S} - S)}{\hat{\sigma}_{\hat{S}}} \xrightarrow{L} N(0, 1),$$

where $\hat{\sigma}^2_{\hat{S}} = g_s^\tau(\hat{\mathbf{v}})\Sigma(\hat{\mathbf{v}})g_s(\hat{\mathbf{v}})$, *and*

$$\frac{\sqrt{n}(\hat{S}^* - S)}{\hat{\sigma}_{\hat{S}^*}} \xrightarrow{L} N(0, 1),$$

where $\hat{\sigma}^2_{\hat{S}^*} = g_s^\tau(\hat{\mathbf{u}}^*)\Sigma(\hat{\mathbf{u}}^*)g_s(\hat{\mathbf{u}}^*)$ *and* $\hat{\mathbf{u}}^* = (\hat{p}_1^*, \ldots, \hat{p}_{K-1}^*, \hat{q}_1^*, \ldots, \hat{q}_{K-1}^*)^\tau$.

Example 5.11 *Revisiting Example 5.10 where both* **p** *and* **q** *have Bernoulli distributions, with probabilities of success being p and q, one has*

$$\sqrt{n}(\hat{S} - S) \xrightarrow{L} N(0, \sigma_s^2) \quad \textit{and} \quad \sqrt{n}(\hat{S}^* - S) \xrightarrow{L} N(0, \sigma_s^2),$$

where

$$\sigma_s^2 = \frac{p(1-p)}{4}\left[\ln\left(\frac{p(1-q)}{q(1-p)}\right) + \frac{p-q}{p(1-p)}\right]^2 + \frac{\lambda q(1-q)}{4}\left[\ln\left(\frac{q(1-p)}{p(1-q)}\right) + \frac{q-p}{q(1-q)}\right]^2.$$

Remark 5.3 *Since $\hat{D}(\mathbf{p}||\mathbf{q})$ has an infinite bias as an estimator of $D(\mathbf{p}||\mathbf{q})$ and $\hat{D}(\mathbf{q}||\mathbf{p})$ has an infinite bias as an estimator of $D(\mathbf{q}||\mathbf{p})$, it follows that $\hat{S}$ has an infinite bias as an estimator of $S(\mathbf{p}, \mathbf{q})$. Similarly, from Theorem 5.11, it follows that $\hat{S}^*$ has a bias that decays no faster than $\mathcal{O}(1/n)$.*

By modifying the estimator given in (5.89), an estimator of $S(\mathbf{p}, \mathbf{q})$, whose bias decays exponentially fast, may be comprised as follows.

Let

$$\begin{aligned}\hat{S}^\sharp = \hat{S}^\sharp(\mathbf{p}, \mathbf{q}) &= \frac{1}{2}\left(\hat{D}^\sharp(\mathbf{p}||\mathbf{q}) + \hat{D}^\sharp(\mathbf{q}||\mathbf{p})\right) \qquad (5.121)\\ &= \frac{1}{2}\sum_{k=1}^{K}\hat{p}_n\left[\sum_{v=1}^{m-Y_k}\frac{1}{v}\prod_{j=1}^{v}\left(1-\frac{Y_k}{m-j+1}\right)-\sum_{v=1}^{n-X_k}\frac{1}{v}\prod_{j=1}^{v}\left(1-\frac{X_k-1}{n-j}\right)\right]\\ &\quad+\frac{1}{2}\sum_{k=1}^{K}\hat{q}_n\left[\sum_{v=1}^{n-X_k}\frac{1}{v}\prod_{j=1}^{v}\left(1-\frac{X_k}{n-j+1}\right)-\sum_{v=1}^{m-Y_k}\frac{1}{v}\prod_{j=1}^{v}\left(1-\frac{Y_k-1}{m-j}\right)\right].\end{aligned}$$

By (5.117), the bias of this estimator is

$$\begin{aligned}\mathrm{E}(\hat{S}^\sharp - S) &= \sum_{k=1}^{K}p_k\left(\sum_{v=n}^{\infty}\frac{1}{v}(1-p_k)^v - \sum_{v=m+1}^{\infty}\frac{1}{v}(1-q_k)^v\right)\\ &\quad+\sum_{k=1}^{K}q_k\left(\sum_{v=m}^{\infty}\frac{1}{v}(1-q_k)^v - \sum_{v=n+1}^{\infty}\frac{1}{v}(1-p_k)^v\right),\end{aligned}$$

which decays at least exponentially fast as $n, m \to \infty$ since each part does.

Theorem 5.14 *Provided that $g_s^\tau(\mathbf{v})\Sigma(\mathbf{v})g_s(\mathbf{v}) > 0$,*

$$\frac{\sqrt{n}(\hat{S}^\sharp - S)}{\sqrt{g_s^\tau(\mathbf{v})\Sigma(\mathbf{v})g_s(\mathbf{v})}} \xrightarrow{L} N(0, 1).$$

Proof: By Theorem 5.13 and Slutsky's theorem, it suffices to show that

$$\sqrt{n}(\hat{S}^\sharp - \hat{S}^*) \xrightarrow{p} 0.$$

This however follows from the facts that

$$\sqrt{n}(\hat{D}^\sharp(\mathbf{p}||\mathbf{q}) - \hat{D}^*(\mathbf{p}||\mathbf{q})) \xrightarrow{p} 0$$

and

$$\sqrt{n}(\hat{D}^\sharp(\mathbf{q}||\mathbf{p}) - \hat{D}^*(\mathbf{q}||\mathbf{p})) \xrightarrow{p} 0,$$

which follow from the proof of Theorem 5.12. □

By arguments similar to the proof of Corollary 5.7, the following corollary is established.

Corollary 5.11 *Provided that* $g_s^\tau(\mathbf{v})\Sigma(\mathbf{v})g_s(\mathbf{v}) > 0$,

$$\frac{\sqrt{n}(\hat{S}^\sharp - S)}{\hat{\sigma}_{\hat{S}^\sharp}} \xrightarrow{L} N(0,1),$$

where $\hat{\sigma}^2_{\hat{S}^\sharp} = g_s^\tau(\hat{\mathbf{u}}^*)\Sigma(\hat{\mathbf{u}}^*)g_s(\hat{\mathbf{u}}^*)$ *and* $\hat{\mathbf{u}}^*$ *is as in Corollary 5.10.*

5.4 Tests of Hypotheses

In the preceding sections of this chapter, estimators of various information indices are discussed and their respective asymptotic normalities are presented under various conditions. Of MI-related indices MI, κ, κ_2, and κ_3, the main estimators discussed are

$$\widehat{MI}, \quad \hat{\kappa}, \quad \hat{\kappa}_2, \quad \text{and} \quad \hat{\kappa}_3,$$

respectively, given in (5.56), (5.70), (5.71), and (5.72), and

$$\widehat{MI}_z, \quad \hat{\kappa}_z, \quad \hat{\kappa}_{2z}, \quad \text{and} \quad \hat{\kappa}_{3z},$$

respectively, given in (5.64), (5.73), (5.74), and (5.75). Of Kullback–Leibler divergence indices, $D = D(\mathbf{p}||\mathbf{q})$ and $S = S(\mathbf{p}, \mathbf{q})$, the main estimators discussed are

$$\hat{D} \quad \text{and} \quad \hat{S},$$

respectively, given in (5.88) and (5.119), and

$$\hat{D}^\sharp \quad \text{and} \quad \hat{S}^\sharp.$$

respectively, given in (5.89) and (5.121).

Each of these estimators could lead to a large sample test of hypothesis if its asymptotic normality holds. The key conditions supporting the asymptotic normalities are

$$(1, 1-1)\Sigma_H(1, 1, -1)^\tau > 0 \tag{5.122}$$

in Theorem 5.5 for MI, κ, κ_2, and κ_3, and

$$g^\tau(\mathbf{v})\Sigma(\mathbf{v})g(\mathbf{v}) > 0 \tag{5.123}$$

in Theorem 5.9 for D and S. It may be verified that if X and Y are independent random elements on joint alphabet $\mathscr{X} \times \mathscr{Y}$, that is, $MI = 0$, then (5.122) is not satisfied, and that if $\mathbf{p} = \mathbf{q}$ on a same finite alphabet $\mathscr{X}$, that is, $D = D(\mathbf{p}||\mathbf{q}) = 0$, then (5.123) is not satisfied (see Exercises 14 and 15). This implies that the asymptotic normalities of this chapter do not directly render valid large sample tests for hypotheses such as $H_0 : MI = 0$ or $H_0 : D = 0$. However, they do support tests for $H_0 : MI = c_0$ $(H_0 : \kappa = c_0$, $H_0 : \kappa_2 = c_0$ or $H_0 : \kappa_3 = c_0)$ or $H_0 : D = c_0$ $(H_0 : S = c_0)$ where $c_0 > 0$ is some appropriately scaled fixed positive value.

For completeness as well as conciseness, a general form of such a test is given. Let $(\theta, \hat{\theta})$ be any one of the following pairs:

$$(MI, \widehat{MI}), (MI, \widehat{MI}_z), (\kappa, \hat{\kappa}), (\kappa, \hat{\kappa}_z), (\kappa_2, \hat{\kappa}_2), (\kappa_2, \hat{\kappa}_{2z}), (\kappa_3, \hat{\kappa}_3), (\kappa_3, \hat{\kappa}_{3z}),$$
$$(D, \hat{D}), (D, \hat{D}^*), (D, \hat{D}^\sharp), (S, \hat{S}), (S, \hat{S}^*), (S, \hat{S}^\sharp).$$

Under $H_0 : \theta = \theta_0 > 0$, the following statistic is approximately a standard normal random variable:

$$Z = \frac{\sqrt{n}(\hat{\theta} - \theta_0)}{\hat{\sigma}_{\hat{\theta}}}$$

where $\hat{\sigma}^2_{\hat{\theta}}$ is as in Table 5.4. It is to be specifically noted that this large sample normal test is valid only for $\theta_0 > 0$. When $\theta_0 = 0$, the underlying asymptotic normality does not hold for any of the listed $\hat{\theta}$.

In practice, it is often of interest to test the hypothesis of $H_0 : MI = 0$ versus $H_a : MI > 0$, or just as often to test the hypothesis $H_0 : D(\mathbf{p}||\mathbf{q}) = 0$ versus $H_a : D(\mathbf{p}||\mathbf{q}) > 0$. For these hypotheses, provided that the sizes of the underlying alphabets, K_1 and K_2, are finite, there exist well-known goodness-of-fit tests in the standard statistical literature.

For example, by Theorem 5.1, $D(\mathbf{p}||\mathbf{q}) = 0$ if and only if $\mathbf{p} = \mathbf{q}$. Therefore, to test $H_0 : D(\mathbf{p}||\mathbf{q}) = 0$ versus $H_a : D(\mathbf{p}||\mathbf{q}) > 0$, one may use Pearson's goodness-of-fit test. In the case of a one-sample goodness-of-fit test

Table 5.4 Estimated information indices with estimated variances.

θ	$\hat{\theta}$	$\hat{\sigma}^2_{\hat{\theta}}$	As in
MI	$\widehat{MI}$	$\hat{\sigma}^2$	*Equation* (5.62)
MI	$\widehat{MI}_z$	$\hat{\sigma}^2$	*Equation* (5.62)
κ	$\hat{\kappa}$	$\hat{\sigma}^2_{\hat{\kappa}}$	Corollary 5.5
κ	$\hat{\kappa}_z$	$\hat{\sigma}^2_{\hat{\kappa}_z}$	Corollary 5.6
κ_2	$\hat{\kappa}_2$	$\hat{\sigma}^2_{\hat{\kappa}_2}$	Corollary 5.5
κ_2	$\hat{\kappa}_{2z}$	$\hat{\sigma}^2_{\hat{\kappa}_{2z}}$	Corollary 5.6
κ_3	$\hat{\kappa}_3$	$\hat{\sigma}^2_{\hat{\kappa}_3}$	Corollary 5.5
κ_3	$\hat{\kappa}_{3z}$	$\hat{\sigma}^2_{\hat{\kappa}_{3z}}$	Corollary 5.6
D	$\hat{D}_n$	$\hat{\sigma}^2_{\hat{D}_n}$	Corollary 5.7
D	$\hat{D}^*_n$	$\hat{\sigma}^2_{\hat{D}^*_n}$	Corollary 5.8
D	$\hat{D}^\sharp_n$	$\hat{\sigma}^2_{\hat{D}^\sharp_n}$	Corollary 5.9
S	$\hat{S}_n$	$\hat{\sigma}^2_{\hat{S}_n}$	Corollary 5.10
S	$\hat{S}^*_n$	$\hat{\sigma}^2_{\hat{S}^*_n}$	Corollary 5.10
S	$\hat{S}^\sharp_n$	$\hat{\sigma}^2_{\hat{S}^\sharp_n}$	Corollary 5.11

for $H_0 : \mathbf{p} = \mathbf{q}$ where $\mathbf{q} = \{q_1, \dots, q_K\}$ is a pre-fixed known probability distribution,

$$Q_1 = \sum_{k=1}^{K} \frac{(F_k - nq_k)^2}{nq_k}, \tag{5.124}$$

where F_k is the observed frequency of the kth letter in the sample of size n, behaves approximately as a chi-squared random variable with degrees of freedom $m = K - 1$.

In the case of a two-sample goodness-of-fit test for $H_0 : \mathbf{p} = \mathbf{q}$ where neither $\mathbf{p}$ or $\mathbf{q}$ is known *a priori*,

$$Q_2 = \sum_{i=1}^{2} \sum_{k=1}^{K} \frac{[F_{i,k} - n_i F_{\cdot,k}/(n_1 + n_2)]^2}{n_i F_{\cdot,k}/(n_1 + n_2)}, \tag{5.125}$$

where K is the common effective cardinality of the underlying finite alphabet, i indicates the ith sample, k indicates the kth letter of the alphabet, $F_{i,k}$ is the observed frequency of the kth letter in the ith sample, $F_{\cdot,k} = F_{1,k} + F_{2,k}$, and n_i is size of the ith sample. The statistic Q_2 is known as a generalization of the Pearson goodness-of-fit statistic, Q_1. Under H_0, Q_2 is asymptotically a chi-squared random variable with degrees of freedom $df = K - 1$. One would reject H_0 if Q_2 takes a large value, say greater than $\chi^2_\alpha(K-1)$, the $(1-\alpha) \times 100$th quantile of chi-squared distribution with degrees of freedom $K - 1$. A discussion of Q_1 and Q_2 is found in many textbooks, in particular, Mood, Graybill, and Boes (1974).

Similar to Theorem 5.2, $MI = 0$ if and only if X and Y are independent, that is, $p_{i,j} = p_{i,\cdot} p_{\cdot,j}$ for every pair (i, j). Therefore, to test $H_0 : MI = 0$ versus $H_a : MI > 0$, one may use the Pearson chi-squared statistic for independence in two-way contingency tables,

$$Q_3 = \sum_{i=1}^{K_1} \sum_{j=1}^{K_2} \frac{(F_{i,j} - n\hat{p}_{i,\cdot}\hat{p}_{\cdot,j})^2}{n\hat{p}_{i,\cdot}\hat{p}_{\cdot,j}} \tag{5.126}$$

where $F_{i,j}$ is the observed frequency of $(x_i, y_j) \in \mathscr{X} \times \mathscr{Y}$ in an *iid* sample of size n; $\hat{p}_{i,j} = F_{i,j}/n$; $\hat{p}_{i,\cdot} = \sum_{j=1}^{K_2} \hat{p}_{i,j}$; $\hat{p}_{\cdot,j} = \sum_{i=1}^{K_1} \hat{p}_{i,j}$; Under $H_0 : MI = 0$, Q_3 is asymptotically a chi-square random variable with degrees of freedom $(K_1 - 1)(K_2 - 1)$. One would reject H_0 if Q_3 takes a large value, say greater than $\chi^2_\alpha((K_1 - 1)(K_2 - 1))$, the $(1-\alpha) \times 100$th quantile of chi-square distribution with degrees of freedom $(K_1 - 1)(K_2 - 1)$. This well-known test is also credited to Pearson (1900, 1922) and sometimes to Fisher and Tippett (1922).

However it is well known that Pearson's goodness-of-fit tests perform poorly when there are many low-frequency cells in the sample data. A rule of thumb widely adopted in statistical practice for the adequacy of Pearson's goodness-of-fit test is that the expected (or similarly, the observed) cell frequency is greater or equal to 5. This rule is largely credited to

Cochran (1952). Because of such a characteristic, tests based on Q_1, Q_2, and Q_3 often become less reliable in cases of small sample sizes, relative to the effective cardinality of the underlying alphabet, or a lack of sample coverage. It would be of great interest to develop better-performing testing procedures for $H_0 : MI = 0$ and $H_0 : D = 0$ with sparse data.

5.5 Appendix

5.5.1 Proof of Theorem 5.12

Proof: Since

$$\sqrt{n}(\hat{D}^\sharp - D) = \sqrt{n}(\hat{D}^\sharp - \hat{D}^*) + \sqrt{n}(\hat{D}^* - D),$$

by Theorem 5.10 and Slutsky's theorem, it suffices to show that

$$\sqrt{n}(\hat{D}^\sharp - \hat{D}^*) \xrightarrow{p} 0.$$

Note that

$$\sqrt{n}(\hat{D}^\sharp - \hat{D}^*) = \sqrt{n}(\hat{A}^\sharp - \hat{A}) + \sqrt{n}(\hat{B}^\sharp - \hat{B}^*) =: \mathcal{A}_1 + \mathcal{A}_2,$$

where $\hat{A}$ and $\hat{B}^*$ are as in (5.98), $\hat{A}^\sharp$ and $\hat{B}^\sharp$ are as in (5.108) and (5.115) respectively. It is shown that $\mathcal{A}_1 \xrightarrow{p} 0$ and $\mathcal{A}_2 \xrightarrow{p} 0$. For a better presentation, the proof is divided into two parts: Part 1 is the proof of $\mathcal{A}_1 \xrightarrow{p} 0$, and Part 2 is the proof of $\mathcal{A}_2 \xrightarrow{p} 0$.

Part 1: First consider $\mathcal{A}_1$. By the Taylor expansion of the logarithm,

$$\begin{aligned}\mathcal{A}_1 &= -\sqrt{n}\left(-\hat{A}^\sharp - \sum_{k=1}^{K}\sum_{v=1}^{n-X_k}\frac{1}{v}\hat{p}_k(1-\hat{p}_k)^v\right)\\ &\quad + \sqrt{n}\sum_{k=1}^{K}\sum_{v=n-X_k+1}^{\infty}\frac{1}{v}\hat{p}_k(1-\hat{p}_k)^v\\ &=: -\mathcal{A}_{1,1} + \mathcal{A}_{1,2}.\end{aligned}$$

Since

$$\begin{aligned}0 \le \mathcal{A}_{1,2} &\le \sqrt{n}\sum_{k=1}^{K}\frac{\hat{p}_k}{n-X_k+1}\sum_{v=n-X_k+1}^{\infty}(1-\hat{p}_k)^v\\ &= \sqrt{n}\sum_{k=1}^{K}\frac{1}{n-X_k+1}(1-\hat{p}_k)^{n-X_k+1}\\ &= \frac{1}{\sqrt{n}}\sum_{k=1}^{K}\frac{1}{1-\hat{p}_k+1/n}(1-\hat{p}_k)^{n-X_k+1}1[\hat{p}_k<1]\end{aligned}$$

$$\leq \frac{1}{\sqrt{n}} \sum_{k=1}^{K} (1-\hat{p}_k)^{n-X_k} 1[\hat{p}_k < 1] \leq \frac{K}{\sqrt{n}}, \tag{5.127}$$

$\mathcal{A}_{1,2} \to 0$ a.s. and therefore $\mathcal{A}_{1,2} \xrightarrow{p} 0$.

Before considering $\mathcal{A}_{1,1}$, note that for $j = 0, 1, \dots, n-1$

$$\left(1 - \frac{X_k - 1}{n - j}\right) \geq \left(1 - \frac{X_k}{n}\right) = (1 - \hat{p}_k) \tag{5.128}$$

if and only if

$$0 \leq j \leq \frac{n}{X_k + 1[X_k = 0]} =: \frac{1}{\hat{p}_k^*}.$$

Let $J_k = \lfloor 1/\hat{p}_k^* \rfloor$. One can write

$$\begin{aligned}
\mathcal{A}_{1,1} = {} & \sqrt{n} \sum_{k=1}^{K} \hat{p}_k \sum_{v=1}^{n-X_k} \frac{1}{v} \left[\prod_{j=1}^{v} \left(1 - \frac{X_k - 1}{n - j}\right) \right. \\
& \left. - \prod_{j=1}^{J_k \wedge v} \left(1 - \frac{X_k - 1}{n - j}\right) (1 - \hat{p}_k)^{0 \vee (v - J_k)} \right] \\
& + \sqrt{n} \sum_{k=1}^{K} \hat{p}_k \sum_{v=1}^{n-X_k} \frac{1}{v} \left[\prod_{j=1}^{J_k \wedge v} \left(1 - \frac{X_k - 1}{n - j}\right) (1 - \hat{p}_k)^{0 \vee (v - J_k)} \right. \\
& \left. - (1 - \hat{p}_k)^{v} \right] \\
=: {} & \mathcal{A}_{1,1,1} + \mathcal{A}_{1,1,2}.
\end{aligned}$$

By (5.128), $\mathcal{A}_{1,1,1} \leq 0$ and $\mathcal{A}_{1,1,2} \geq 0$. It is to show that $\mathrm{E}(\mathcal{A}_{1,1,1}) \to 0$ and $\mathrm{E}(\mathcal{A}_{1,1,2}) \to 0$ respectively. Toward that end,

$$\begin{aligned}
\mathcal{A}_{1,1,2} \leq {} & \sqrt{n} \sum_{k=1}^{K} \hat{p}_k \sum_{v=1}^{n-X_k} \frac{1}{v} \left[\prod_{j=1}^{J_k \wedge v} \left(1 - \frac{X_k - 1}{n - 1}\right) (1 - \hat{p}_k)^{0 \vee (v - J_k)} \right. \\
& \left. - (1 - \hat{p}_k)^{v} \right] \\
\leq {} & \sqrt{n} \sum_{k=1}^{K} \hat{p}_k \sum_{v=1}^{n-X_k} \frac{1}{v} \left[\left(1 - \frac{X_k - 1}{n - 1}\right)^{J_k \wedge v} - (1 - \hat{p}_k)^{J_k \wedge v} \right] \\
\leq {} & \sqrt{n} \sum_{k=1}^{K} \hat{p}_k \sum_{v=1}^{n-X_k} \frac{1}{v} \left[(J_k \wedge v) \left(1 - \frac{X_k - 1}{n - 1}\right)^{(J_k \wedge v) - 1} \frac{n - X_k}{n(n-1)} \right] \\
\leq {} & \frac{\sqrt{n}}{n-1} \sum_{k=1}^{K} \sum_{v=1}^{n} \frac{1}{v} (J_k \wedge v)
\end{aligned}$$

$$
\begin{aligned}
&= \frac{\sqrt{n}}{n-1}\sum_{k=1}^{K}\left(J_k + J_k\sum_{v=J_k+1}^{n}\frac{1}{v}\right)\\
&\le \frac{\sqrt{n}}{n-1}\sum_{k=1}^{K}J_k\left(1+\sum_{v=2}^{n}\frac{1}{v}\right)\\
&\le \frac{\sqrt{n}}{n-1}\sum_{k=1}^{K}J_k(1+\ln n),
\end{aligned}
$$

where the third line follows by Lemma 5.8 and the fact that

$$
1-\frac{X_k-1}{n-1} > 1-\hat{p}_k.
$$

Thus, by (5.128) and Lemma 5.9,

$$
0 \le \mathrm{E}(\mathcal{A}_{1,1,2}) \le \mathcal{O}\left(\frac{\sqrt{n}\ln(n)}{n}\right) \to 0.
$$

Furthermore,

$$
\begin{aligned}
\mathcal{A}_{1,1,1} &= \sqrt{n}[A-\hat{A}^{\sharp}]\\
&\quad - \sqrt{n}\sum_{k=1}^{K}\hat{p}_k\sum_{v=1}^{n-X_k}\frac{1}{v}\left[\prod_{j=1}^{J_k\wedge v}\left(1-\frac{X_k-1}{n-j}\right)(1-\hat{p}_k)^{0\vee(v-J_k)}\right.\\
&\quad \left. -(1-\hat{p}_k)^{v}\right]\\
&\quad - \sqrt{n}\left[\sum_{k=1}^{K}\hat{p}_k\sum_{v=1}^{n-X_k}\frac{1}{v}(1-\hat{p}_k)^{v}+A\right]\\
&:= \mathcal{A}_{1,1,1,1}-\mathcal{A}_{1,1,2}-\mathcal{A}_{1,1,1,3}.
\end{aligned}
$$

It has already been shown that $\mathrm{E}(\mathcal{A}_{1,1,2}) \to 0$, and from (5.107) it follows that $\mathrm{E}(\mathcal{A}_{1,1,1,1}) = \sqrt{n}B_1 \to 0$. Still further

$$
\begin{aligned}
\mathcal{A}_{1,1,1,3} &= \sqrt{n}\left[\sum_{k=1}^{K}\hat{p}_k\sum_{v=1}^{\infty}\frac{1}{v}(1-\hat{p}_k)^{v}+A\right]\\
&\quad - \sqrt{n}\sum_{k=1}^{K}\hat{p}_k\sum_{v=n-X_k+1}^{\infty}\frac{1}{v}(1-\hat{p}_k)^{v}\\
&= \sqrt{n}(A-\hat{A}_n) - \sqrt{n}\sum_{k=1}^{K}\hat{p}_k\sum_{v=n-X_k+1}^{\infty}\frac{1}{v}(1-\hat{p}_k)^{v}\\
&:= \mathcal{A}_{1,1,1,3,1}-\mathcal{A}_{1,2},
\end{aligned}
$$

where the second line follows by the Taylor expansion of the logarithm. Since $\mathrm{E}(\mathcal{A}_{1,1,1,3,1}) \to 0$ by (5.102) and $\mathrm{E}(\mathcal{A}_{1,2}) \to 0$ by (5.127). Thus, $\mathrm{E}(\mathcal{A}_{1,1,1,3}) \to 0$. This means that $\mathcal{A}_{1,1,2} \xrightarrow{p} 0$. From here, it follows that $\mathcal{A}_1 \xrightarrow{p} 0$.

Part 2: Consider now $\mathcal{A}_2$. Note that

$$\mathcal{A}_2 = \sqrt{\frac{n}{m}} \sum_{k=1}^{K} \hat{p}_k \sqrt{m} \left[\sum_{v=1}^{m-Y_k} \frac{1}{v} U_{k,v} + \ln \hat{q}_k^* \right].$$

Since $\hat{p}_k \xrightarrow{p} p_k$ for every $k = 1, \dots, K$ and $n/m \to \lambda$, by Slutsky's theorem it suffices to show that for

$$\sqrt{m} \left[\sum_{v=1}^{m-Y_k} \frac{1}{v} U_{k,v} + \ln(\hat{q}_k^*) \right] \xrightarrow{p} 0$$

for each $k = 1, 2, \dots, K$. By the Taylor expansion of the logarithm

$$\begin{aligned}
\sqrt{m} \left[\sum_{v=1}^{m-Y_k} \frac{1}{v} U_{k,v} + \ln(\hat{q}_k^*) \right] &= \sqrt{m} \left[\sum_{v=1}^{m-Y_k} \frac{1}{v} U_{k,v} - \sum_{v=1}^{\infty} \frac{1}{v} (1 - \hat{q}_k^*)^v \right] \\
&= \sqrt{m} \left\{ \sum_{v=1}^{m-Y_k} \frac{1}{v} [U_{k,v} - (1 - \hat{q}_k^*)^v] \right\} 1[\hat{q}_k > 0] \\
&\quad + \sqrt{m} \left\{ \sum_{v=1}^{m-Y_k} \frac{1}{v} [U_{k,v} - (1 - \hat{q}_k^*)^v] \right\} 1[\hat{q}_k = 0] \\
&\quad - \sqrt{m} \sum_{v=m-Y_k+1}^{\infty} \frac{1}{v} (1 - \hat{q}_k^*)^v \\
&=: \mathcal{A}_{2,1} + \mathcal{A}_{2,2} - \mathcal{A}_{2,3}.
\end{aligned}$$

Since $\hat{q}_k \xrightarrow{p} q_k$ and $q_k^* \xrightarrow{p} q_k$, Slutsky's theorem implies that

$$\begin{aligned}
0 \le \mathcal{A}_{2,3} &\le \frac{\sqrt{m}}{m - Y_k + 1} \sum_{v=m-Y_k+1}^{\infty} (1 - \hat{q}_k^*)^v \\
&= \frac{\sqrt{m}}{m - Y_k + 1} \frac{1}{\hat{q}_k^*} (1 - \hat{q}_k^*)^{m-Y_k+1} \\
&= \frac{1}{\sqrt{m}\hat{q}_k^*} \frac{1}{1 - \hat{q}_k + 1/m} (1 - \hat{q}_k^*)^{m-Y_k+1} \\
&\le \frac{1}{\sqrt{m}\hat{q}_k^*} \frac{1}{1 - \hat{q}_k + 1/m} \xrightarrow{p} 0
\end{aligned}$$

as $m \to \infty$, and hence $\mathcal{A}_{2,3} \xrightarrow{p} 0$.

Next, it is to show that $\mathcal{A}_{2,2} \xrightarrow{p} 0$. When $\hat{q}_k = 0$, one has $U_{k,v} = 1$, $\hat{q}_k^* = 1/m$, and

$$\begin{aligned} 0 \le \mathcal{A}_{2,2} &\le \sqrt{m}\left\{\sum_{v=1}^{m}\frac{1}{v}\left[1-\left(1-\frac{1}{m}\right)^{v}\right]\right\}1[\hat{q}_k = 0] \\ &\le m^{3/2}1[\hat{q}_k = 0]. \end{aligned}$$

Thus, $0 \le \mathrm{E}(\mathcal{A}_{2,2}) \le m^{3/2}(1-q_k)^m \to 0$, and therefore $\mathcal{A}_{2,2} \xrightarrow{p} 0$. Next, it is to show that $\mathcal{A}_{2,1} \xrightarrow{p} 0$. Since that in this case $\hat{q}_k = \hat{q}_k^*$, one has

$$\begin{aligned} \mathcal{A}_{2,1} &= \sqrt{m}\sum_{v=1}^{m-Y_k}\frac{1}{v}\left\{\left[\prod_{j=0}^{v-1}\left(1-\frac{Y_k}{m-j}\right)\right]-(1-\hat{q}_k^*)^v\right\}1[\hat{q}_k > 0] \\ &\le \sqrt{m}\sum_{v=1}^{m-Y_k}\frac{1}{v}[(1-\hat{q}_k)^v-(1-\hat{q}_k^*)^v]1[\hat{q}_k > 0] = 0. \end{aligned}$$

To complete the proof, it is to show that $\mathrm{E}(\mathcal{A}_{2,1}) \to 0$. Toward that end,

$$\begin{aligned} \mathcal{A}_{2,1} + \mathcal{A}_{2,2} &= \sqrt{m}\left[\sum_{v=1}^{m-Y_k}\frac{1}{v}U_{k,v} - \sum_{v=1}^{m}\frac{1}{v}(1-q_k)^v\right] \\ &\quad - \sqrt{m}\left[\sum_{v=1}^{\infty}\frac{1}{v}(1-\hat{q}_k^*)^v - \sum_{v=1}^{\infty}\frac{1}{v}(1-q_k)^v\right] \\ &\quad + \sqrt{m}\sum_{v=m-Y_k+1}^{\infty}\frac{1}{v}(1-\hat{q}_k^*)^v - \sqrt{m}\sum_{v=m+1}^{\infty}\frac{1}{v}(1-q_k)^v \\ &=: \mathcal{A}_{2,1,1} - \mathcal{A}_{2,1,2} + \mathcal{A}_{2,1,3} - \mathcal{A}_{2,1,4}. \end{aligned}$$

$\mathrm{E}(\mathcal{A}_{2,1,1}) = 0$ because

$$\sum_{v=1}^{m-Y_k}\frac{1}{v}U_{k,v}$$

is an unbiased estimator of

$$\sum_{v=1}^{m}\frac{1}{v}(1-q_k)^v$$

by construction.

By the Taylor expansion of the logarithm $\mathcal{A}_{2,1,2} = \sqrt{m}[\ln(q_k) - \ln(\hat{q}_k^*)]$ and $\mathrm{E}(\mathcal{A}_{2,1,2}) \to 0$ by an argument similar to the proof of Proposition 5.11. To show that $\mathrm{E}(\mathcal{A}_{2,1,3}) \to 0$, note that

$$0 \le \mathcal{A}_{2,1,3} \le \frac{\sqrt{m}}{m-Y_k+1}\frac{1}{\hat{q}_k^*}(1-\hat{q}_k^*)^{m-Y_k+1}$$

$$\begin{aligned}
&\leq \frac{1}{\sqrt{m}(1-\hat{q}_k+1/m)}\frac{1}{\hat{q}_k^*}(1-\hat{q}_k)^{m-Y_k+1}\\
&= \frac{1}{\sqrt{m}\hat{q}_k^*}\frac{1-\hat{q}_k}{(1-\hat{q}_k+1/m)}(1-\hat{q}_k)^{m-Y_k}\\
&\leq \frac{1}{\sqrt{m}\hat{q}_k^*}.
\end{aligned}$$

Thus, by Lemma 5.9,

$$0 \leq \mathrm{E}(\mathcal{A}_{2,1,3}) \leq \frac{1}{m^{3/2}}\mathrm{E}\left(\frac{m}{\hat{q}_k^*}\right) \to 0$$

as $m \to \infty$. Finally, $\mathrm{E}(\mathcal{A}_{2,1,4}) \to 0$ because $\mathcal{A}_{2,1,4}$ is deterministic and

$$0 \leq \mathcal{A}_{2,1,4} \leq \sqrt{m}\sum_{v=m}^{\infty}\frac{1}{v}(1-q_k)^v \leq \frac{1}{\sqrt{m}}\frac{(1-q_k)^m}{q_k} \to 0.$$

Hence, $\mathrm{E}(\mathcal{A}_{2,1}+\mathcal{A}_{2,2}) \to 0$, but since $\mathrm{E}(\mathcal{A}_{2,2}) \to 0$ is already established, it follows that $\mathrm{E}(\mathcal{A}_{2,1}) \to 0$, which implies $\mathcal{A}_2 \overset{p}{\to} 0$. The result of the theorem follows. □

5.6 Exercises

1 Prove Lemma 5.2, Jensen's inequality.

2 Verify both parts in (5.16) of Example 5.2.

3 Verify (5.17)–(5.20) in Example 5.3.

4 Verify (5.21)–(5.24) in Example 5.4.

5 Verify (5.27) and (5.28) in Example 5.7.

6 Show that Σ in (5.50) is positive definite.

7 Prove each of the following six inequalities.

$$\begin{aligned}
0 \leq MI(X,Y) &\leq \min\{H(X),H(Y)\}\\
&\leq \sqrt{H(X)H(Y)}\\
&\leq (H(X)+H(Y))/2\\
&\leq \max\{H(X),H(Y)\}\\
&\leq H(X,Y).
\end{aligned}$$

8 Let

$$\kappa_1 = \frac{MI(X,Y)}{\min\{H(X),H(Y)\}},$$
$$\kappa_2 = \frac{MI(X,Y)}{\sqrt{H(X)H(Y)}},$$
$$\kappa_3 = \frac{MI(X,Y)}{(H(X)+H(Y))/2}, \quad \text{and}$$
$$\kappa_4 = \frac{MI(X,Y)}{\max\{H(X),H(Y)\}},$$

and assume $H(X,Y) < \infty$. Show that Theorem 5.4 holds for
a) κ_2,
b) κ_3,
c) κ_4,
d) but not for κ_1.

9 Show that $Z_{1,v}$ of (5.106) is identical to $\hat{A}_n^\sharp$ of (5.108).

10 In mathematics, a distance function on a given set $\mathscr{P}$ is a function

$$d : \mathscr{P} \times \mathscr{P} \to R,$$

where R denotes the set of real numbers, which satisfies the following conditions: for any three objects, $\mathbf{p}_X \in \mathscr{P}$, $\mathbf{p}_Y \in \mathscr{P}$, and $\mathbf{p}_Z \in \mathscr{P}$,
i) $d(\mathbf{p}_X, \mathbf{p}_Y) \geq 0$;
ii) $d(\mathbf{p}_X, \mathbf{p}_Y) = 0$ if and only if $\mathbf{p}_X = \mathbf{p}_Y$;
iii) $d(\mathbf{p}_X, \mathbf{p}_Y) = d(\mathbf{p}_Y, \mathbf{p}_X)$; and
iv) $d(\mathbf{p}_X, \mathbf{p}_Z) \leq d(\mathbf{p}_X, \mathbf{p}_Y) + d(\mathbf{p}_Y, \mathbf{p}_Z)$.
Show that
a) the Kullback–Leibler divergence $D(\mathbf{p}\|\mathbf{q})$ of (5.87) is not a distance function on $\mathscr{P}$, where $\mathscr{P}$ is the collection of all probability distributions on a countable alphabet $\mathscr{X}$; but
b) the symmetrized Kullback–Leibler divergence $S(\mathbf{p},\mathbf{q})$ of (5.118) is a distance function on $\mathscr{P}$.

11 Suppose a joint probability distribution $\mathbf{p}_{X,Y} = \{p_{i,j}\}$ on

$$\mathscr{X} \times \mathscr{Y} = \{x_i; i = 1, \dots, K_1\} \times \{y_j; j = 1, \dots, K_2\}$$

is such that $p_{i,j} > 0$ for every pair (i,j). Let $K = K_1K_2$. Re-enumerating $\{p_{i,j}\}$ to be indexed by a single index k, that is,

$$\begin{aligned} v &= (p_1, \dots, p_{K-1})^\tau \\ &= (p_{1,1}, \dots, p_{1,K_2}, p_{2,1}, \dots, p_{2,K_2}, \dots, p_{K_1,1}, \dots, p_{K_1,K_2-1})^\tau, \end{aligned}$$

show that each of (5.42)–(5.44) is true.

12 Let $\{p_{i,j}; i = 1, \dots, K_1, j = 1, \dots, K_2\}$ be a joint probability distribution on $\mathscr{X} \times \mathscr{Y}$ where $\mathscr{X}$ and $\mathscr{Y}$ are as in (5.31). Suppose $p_{i,\cdot} > 0$ for each i, $p_{\cdot,j} > 0$ for each j, $K_1 \geq 2$ and $K_2 \geq 2$. Show that (5.57) of Theorem 5.5 holds if and only if $p_{i,j} = p_{i,\cdot} p_{\cdot,j}$, or every pair of (i,j) such that $p_{i,j} > 0$.

13 Consider the following three functions of triplet (x_1, x_2, x_3), with the same domain, $x_1 > 0$, $x_2 > 0$, and $x_3 > 0$,

$$\kappa(x_1, x_2, x_3) = \frac{x_1 + x_2}{x_3} - 1,$$
$$\kappa_2(x_1, x_2, x_3) = \frac{x_1 + x_2 - x_3}{\sqrt{x_1 x_2}},$$
$$\kappa_3(x_1, x_2, x_3) = 2\left(1 - \frac{x_3}{x_1 + x_2}\right).$$

Show that the gradients of these functions are, respectively,

$$g_\kappa(x_1, x_2, x_3) = \left(\frac{1}{x_3}, \frac{1}{x_3}, -\frac{x_1 + x_2}{x_3^2}\right)^\tau,$$
$$g_{\kappa_2}(x_1, x_2, x_3) = \left(\frac{1}{(x_1x_2)^{1/2}} - \frac{x_2(x_1 + x_2 - x_3)}{2(x_1x_2)^{3/2}},\right.$$
$$\left.\frac{1}{(x_1x_2)^{1/2}} - \frac{x_1(x_1 + x_2 - x_3)}{2(x_1x_2)^{3/2}}, -\frac{1}{\sqrt{x_1x_2}}\right)^\tau,$$
$$g_{\kappa_3}(x_1, x_2, x_3) = \left(\frac{2x_3}{(x_1 + x_2)^2}, \frac{2x_3}{(x_1 + x_2)^2}, -\frac{2}{x_1 + x_2},\right)^\tau.$$

14 In Theorem 5.5, show that if X and Y are independent random elements on joint alphabet $\mathscr{X} \times \mathscr{Y}$, that is, $MI = 0$, then

$$(1, 1 - 1)\Sigma_H(1, 1, -1)^\tau = 0.$$

15 In Theorem 5.9, show that if $\mathbf{p} = \mathbf{q}$ on a same finite alphabet $\mathscr{X}$, that is, $D = D(\mathbf{p}\|\mathbf{q}) = 0$, then $g^\tau(\mathbf{v})\Sigma(\mathbf{v})g(\mathbf{v}) = 0$.

16 Use the data in Table 5.1 and Pearson's goodness-of-fit statistic, Q_3 of (5.126), to test the hypothesis that the two random elements, Ethnicity and Favorite Color, are independent, that is, $H_0 : MI = 0$, at $\alpha = 0.05$.

17 Suppose $n = 600$ tosses of a die yield the following results.

X	1	2	3	4	5	6
f	91	101	105	89	124	90

Use the data and Pearson's goodness-of-fit statistic, Q_1 of (5.124), to test the hypothesis that the dice is a balanced die, that is, $H_0 : D(\mathbf{p}\|\mathbf{q}) = 0$, where $\mathbf{q}$ is the hypothesized uniform distribution for the six sides ($q_k = 1/6$ for $k = 1, \ldots, 6$) and $\mathbf{p}$ is the true probability distribution of the die tossed, at $\alpha = 0.05$.

18 Two dice, A and B, are independently tossed $n_A = 600$ and $n_B = 300$ times, respectively, with the given results.

X	1	2	3	4	5	6
f_A	91	101	105	89	124	90
f_B	54	53	52	41	64	36

Let the underlying probability distributions of the two dice be $\mathbf{p}$ for A and $\mathbf{q}$ for B, respectively.

a) Test the hypothesis that the two distributions are identical, that is, $H_0 : \mathbf{p} = \mathbf{q}$, at $\alpha = 0.05$.
b) Test the hypothesis that both distributions are uniform, that is, $H_0 : p_k = q_k = 1/6$ for $k = 1, \ldots, 6$, at $\alpha = 0.05$.

19 Let X_m be a sequence of chi-squared random variables with degrees of freedom m. Show that $\sqrt{m}(X_m - m) \xrightarrow{L} N(0, 2)$ as $m \to \infty$.

20 Suppose the joint probability distributions of (X, Y) on $\{0, 1\} \times \{0, 1\}$ is

	$X = 0$	$X = 1$
$Y = 0$	$0.50 - 0.01\,m$	$0.01\,m$
$Y = 1$	$0.01\,m$	$0.50 - 0.01\,m$

for some integer $m \geq 1$. Show that

$$\kappa = \frac{\ln 2 - [(m/50)\ln(50/m - 1) - \ln(1 - m/50)]}{\ln 2 + [(m/50)\ln(50/m - 1) - \ln(1 - m/50)]}.$$

6

Domains of Attraction on Countable Alphabets

6.1 Introduction

A domain of attraction is a family of probability distributions whose members share a set of common properties, more specifically a set of common properties pertaining to the tail of a probability distribution. In the probability and statistics literature, domains of attraction are usually discussed in the context of extreme value theory. The simplest setup involves a sample of *iid* random variables, $X_1, \dots, X_n$, under a probability distribution with a differentiable cumulative distribution function (*cdf*), $F(x)$. Let $X_{(n)} = \max\{X_1, \dots, X_n\}$. The asymptotic behavior of $X_{(n)}$ may be characterized by that of a properly normalized $X_{(n)}$, namely $Y_n = (X_{(n)} - b_n)/a_n$ where $\{a_n > 0\}$ and $\{b_n\}$ are normalizing sequences, in the sense that

$$P(Y_n \le y) = P(X_{(n)} \le a_n y + b_n) = [F(a_n y + b_n)]^n \to G(y)$$

where $G(y)$ is a nondegenerated distribution function. Many distributions on the real line may be categorized into three families or domains. The members in each of the three domains have a $G(y)$ belonging to a same parametric family. These three families are often identified as Fréchet for thick-tailed distributions (e.g., with density function $f(x) = x^{-2}1[x \ge 1]$), Gumbel for thin-tailed distributions (e.g., with density function $f(x) = e^{-x}1[x \ge 0]$), and Weibull for distributions with finite support (e.g., with density function $f(x) = 1[0 \le x \le 1]$). Extreme value theory is one of the most important topics in probability and statistics, partially because of its far-reaching implications in applications. Early works in this area include (Fréchet, 1927; Fisher and Tippett, 1928, and von Mises, 1936). However, Gnedenko (1943) and Gnedenko (1948) are often thought to be the first rigorous mathematical treatment of the topic. For a comprehensive introduction to extreme value theory, readers may wish to refer to de Haan and Ferreira (2006).

Consider a countable alphabet $\mathscr{X} = \{\ell_k; k \ge 1\}$ and its associated probability distribution $\mathbf{p} = \{p_k; k \ge 1\} \in \mathscr{P}$ where $\mathscr{P}$ is the collection of all probability distributions on $\mathscr{X}$. Let $X_1, \dots, X_n$ be an *iid* random sample from $\mathscr{X}$ under $\mathbf{p}$. Let $\{Y_k; k \ge 1\}$ and $\{\hat{p}_k = Y_k/n; k \ge 1\}$ be the observed frequencies and relative

Statistical Implications of Turing's Formula, First Edition. Zhiyi Zhang.

frequencies of the letters in the *iid* sample of size n. This chapter presents a discussion of domains of attraction in $\mathscr{P}$.

Unlike random variables on the real line, the letters in $\mathscr{X}$ are not necessarily on a numerical scale or even ordinal. The nature of the alphabet presents an immediate issue inhibiting an intuitive parallel to the notions well conceived with random variables on the real line, say for example, the maximum order statistic $X_{(n)}$, which may be considered as an extreme value in an *iid* sample of size n, or the notion of a "tail" of a distribution, $P(X_{(n)} > x) = 1 - F^n(x)$ for large x. It is much less clear what an extreme value means on an alphabet. Can an imitation game still be played? If so, in what sense?

For a random variable, the word "extreme" suggests an extremely large value on the real line. However, an extremely large value on the real line necessarily corresponds to an extremely small probability density. In that sense, an extreme value could be taken as a value in a sample space that carries an extremely small probability. If this notion is adopted, then it becomes reasonable to perceive an extreme value on an alphabet to be a letter with extremely small probability. In that sense, a subset of $\mathscr{X}$ with low probability letters may be referred to as a "tail," a subset of $\mathscr{X}$ with very low probability letters may be referred to as a "distant tail," and a distribution on a finite alphabet has "no tail."

By the same argument, one may choose to associate the notion of extremity in an *iid* sample with the rarity of the letter in the sample. The values that are observed exactly once in the sample are rarer than any other values in the sample; and there could be and usually are many more than one such observed value in a sample. There exist however rarer values than those even with frequency one, and these would be the letters with frequency zero, that is, the letters in $\mathscr{X}$ that are not represented in the sample. Though not in the sample, the letters with zero observed frequencies are, nevertheless, associated with and completely specified by the sample.

To bring the above-described notion of extremity into a probabilistic argument, consider a two-step experiment as follows:

Step 1: draw an *iid* sample of a large size n from $\mathscr{X}$, and then
Step 2: draw another independent *iid* sample of size m from $\mathscr{X}$.

Let the samples be denoted as

$$\{X_1, \dots, X_n\} \qquad \text{and} \qquad \{X_{n+1}, \dots, X_{n+m}\}.$$

When $m = 1$, this experiment is the same as that described in Robbins' Claim in Chapter 1. Let E_1 be the event that X_{n+1} assumes a value from $\mathscr{X}$ that is not assumed by any of the X_i in the first sample, i.e.,

$$E_1 = \cap_{i=1}^{n} \{X_{n+1} \neq X_i\}.$$

Event E_1 may be thought of as a new discovery, and in this regard, an occurrence of a rare event, and therefore is pertaining to the tail of the distribution. The argument (1.60) in Chapter 1 gives $\mathrm{P}(E_1) = \zeta_{1,n} =: \zeta_n$. Clearly $\zeta_n \to 0$ as $n \to \infty$ for any probability distribution $\{p_k\}$ on $\mathscr{X}$ (see Exercise 1).

When $m = n$, a quantity of interest is the number of letters found in the second sample $\{X_{n+1}, \dots, X_{n+n}\}$ that are not found in the first sample $\{X_1, \dots, X_n\}$, that is,

$$t_n = \sum_{j=1}^{n} 1[E_j] \tag{6.1}$$

where $E_j = \cap_{i=1}^{n} \{X_{n+j} \neq X_i\}$ for $j = 1, \dots, n$. The expected value of t_n in (6.1) is

$$\tau_n = \mathrm{E}\,(t_n) = n\zeta_n. \tag{6.2}$$

τ_n is referred to as the tail index and plays a central role in this chapter. The relevance of τ_n to the tail of the underlying distribution manifests in the relevance of ζ_n to the tail.

Furthermore, it is to be noted that, for any given integer $k_0 \geq 1$, the first k_0 terms in the re-expression of τ_n, as in

$$\tau_n = \sum_{1 \leq k \leq k_0} np_k(1-p_k)^n + \sum_{k>k_0} np_k(1-p_k)^n,$$

converges to zero exponentially fast as $n \to \infty$. Therefore, the asymptotic behavior of τ_n has essentially nothing to do with how the probabilities are distributed over any fixed and finite subset of $\mathscr{X}$. Also to be noted is that τ_n is invariant under any permutation on the index set $\{k; k \geq 1\}$.

In contrast to the fact that $\zeta_n \to 0$ for any probability distribution $\{p_k\}$, the multiplicatively inflated version $\tau_n = n\zeta_n$ has different asymptotic behavior under various distributions. $\mathscr{P}$, the total collection of all probability distributions, on $\mathscr{X}$ splinters into different domains of attraction characterized by the asymptotic features of τ_n.

Finally, a notion of thinner/thicker tail between two distributions, $\mathbf{p} = \{p_k; k \geq 1\}$ and $\mathbf{q} = \{q_k; k \geq 1\}$, is to be mentioned. Although there is no natural ordering among the letters in $\mathscr{X}$, there is one on the index set $\{k; k \geq 1\}$. There therefore exists a natural notion of a distribution $\mathbf{p}$ having a thinner tail than that of another distribution $\mathbf{q}$, in the sense of $p_k \leq q_k$ for every $k \geq k_0$ for some integer $k_0 \geq 1$, when $\mathbf{p}$ and $\mathbf{q}$ share a same alphabet $\mathscr{X}$ and are enumerated by a same index set $\{k; k \geq 1\}$. Whenever this is the case in the subsequent text, $\mathbf{p}$ is said to have a thinner tail than $\mathbf{q}$ *in the usual sense.*

Definition 6.1 *A distribution* $\mathbf{p} = \{p_k\}$ *on* $\mathscr{X}$ *is said to belong to*

1) *Domain 0 if* $\lim_{n\to\infty} \tau_n = 0$,
2) *Domain 1 if* $\limsup_{n\to\infty} \tau_n = c_{\mathbf{p}}$ *for some constant* $c_{\mathbf{p}} > 0$,
3) *Domain 2 if* $\lim_{n\to\infty} \tau_n = \infty$, *and*
4) *Domain T, or Domain Transient, if it does not belong to Domains 0, 1, or 2.*

The four domains so defined above form a partition of $\mathscr{P}$. The primary results presented in this chapter include the following:

1) Domain 0 includes only probability distributions with finite support, that is, there are only finitely many letters in $\mathscr{X}$ carrying positive probabilities.
2) Domain 1 includes distributions with thin tails such as $p_k \propto a^{-\lambda k}$, $p_k \propto k^r a^{-\lambda k}$, and $p_k \propto a^{-\lambda k^2}$, where $a > 1$, $\lambda > 0$, and $r \in (-\infty, \infty)$ are constants.
3) Domain 2 includes distributions with thick tails satisfying $p_{k+1}/p_k \to 1$ as $k \to \infty$, for example, $p_k \propto k^{-\lambda}$, $p_k \propto (k \ln^\lambda k)^{-1}$ where $\lambda > 1$, and $p_k \propto e^{-k^\delta}$ where $\delta \in (0, 1)$.
4) A relative regularity condition between two distributions (one dominates the other) is defined. Under this condition, all distributions on a countably infinite alphabet, which are dominated by a Domain 1 distribution, must also belong to Domain 1; and all distributions on a countably infinite alphabet, which dominate a Domain 2 distribution, must also belong to Domain 2.
5) Domain T is not empty.

Other relevant results presented in this chapter include the following:

1) In Domain 0, $\tau_n \to 0$ exponentially fast for every distribution.
2) The tail index τ_n of a distribution with tail $p_k =\propto e^{-\lambda k}$ where $\lambda > 0$ in Domain 1 perpetually oscillates between two positive constants and does not have a limit as $n \to \infty$.
3) There is a uniform positive lower bound for $\lim\ \sup_{n\to\infty} \tau_n$ for all distributions with positive probabilities on infinitely many letters of $\mathscr{X}$.

Remark 6.1 *To honor the great minds of mathematics whose works mark the trail leading to the results of this chapter, Zhang (2017) named the distributions in Domain 0 as members of the* Gini–Simpson *family, after Corrado Gini and Edward Hugh Simpson; those in Domain 1 as members of the* Molchanov *family, after Stanislav Alekseevich Molchanov; and those in Domain 2 as the* Turing–Good *family, after Alan Mathison Turing and Irving John Good. In the subsequent text of this chapter, the domains are so identified according to Zhang (2017).*

6.2 Domains of Attraction

Let K be the effective cardinality, or simply the cardinality when there is no ambiguity, of $\mathscr{X}$, that is, $K = \sum_{k\geq 1} 1[p_k > 0]$. Let $\mathbb{N}$ be the set of all positive integers.

Lemma 6.1 *If $K = \infty$, then there exist a constant $c > 0$ and a subsequence $\{n_k; k \geq 1\}$ in $\mathbb{N}$, satisfying $n_k \to \infty$ as $k \to \infty$, such that $\tau_{n_k} > c$ for all sufficiently large k.*

Proof: Assume without loss of generality that $p_k > 0$ for all $k \ge 1$. Since ζ_n is invariant with respect to any permutation on the index set $\{k; k \ge 1\}$, it can be assumed without loss of generality that $\{p_k\}$ is nonincreasing in k. For every k, let $n_k = \lfloor 1/p_k \rfloor$. With n_k so defined,

$$\frac{1}{n_k + 1} < p_k \le \frac{1}{n_k}$$

for every k and $\lim_{k\to\infty} n_k = \infty$ though $\{n_k\}$ may not necessarily be strictly increasing. By construction, the following are true about the n_k, $k \ge 1$.

1) $\{n_k; k \ge 1\}$ is an infinite subset of $\mathbb{N}$.
2) Every p_k is covered by the interval $(1/(n_k + 1), 1/n_k]$.
3) Every interval $(1/(n_k + 1), 1/n_k]$ covers at least one p_k and at most finitely many p_ks.

Let $f_n(x) = nx(1 - x)^n$ for $x \in [0, 1]$. $f_n(x)$ attains its maximum at $x = (n + 1)^{-1}$ with value

$$f_n\left(\frac{1}{n+1}\right) = \frac{n}{n+1}\left(1 - \frac{1}{n+1}\right)^n = \left(1 - \frac{1}{n+1}\right)^{n+1} \to e^{-1}.$$

Also

$$f_n\left(\frac{1}{n}\right) = \left(1 - \frac{1}{n}\right)^n \to e^{-1}.$$

Furthermore, since $f_n'(x) < 0$ for all x satisfying $1/(n + 1) < x < 1$, the following is true

$$f_n\left(\frac{1}{n}\right) < f_n(x) < f_n\left(\frac{1}{n+1}\right)$$

for all x satisfying $1/(n + 1) < x < 1/n$.

Since $f_n(1/n) \to e^{-1}$ and $f_n(1/(n + 1)) \to e^{-1}$, for any arbitrarily small but fixed $\varepsilon > 0$, there exists a positive N_ε such that for any $n > N_\varepsilon$,

$$f_n\left(\frac{1}{n+1}\right) > f_n\left(\frac{1}{n}\right) > e^{-1} - \varepsilon.$$

Since $\lim_{k\to\infty} n_k = \infty$ and $\{n_k\}$ is nondecreasing, there exists an integer $K_\varepsilon > 0$ such that $n_k > N_\varepsilon$ for all $k > K_\varepsilon$. Consider the subsequence $\{\tau_{n_k}; k \ge 1\}$. For any $k > K_\varepsilon$,

$$\tau_{n_k} = \sum_{i=1}^{\infty} n_k p_i (1 - p_i)^{n_k} > f_{n_k}(p_k).$$

Since $p_k \in (1/(n_k + 1), 1/n_k]$ and $f_{n_k}(x)$ is decreasing on the same interval,

$$f_{n_k}(p_k) > f_{n_k}\left(\frac{1}{n_k}\right) \ge e^{-1} - \varepsilon,$$

and hence

$$\tau_{n_k} > f_{n_k}(p_k) \ge c = e^{-1} - \varepsilon$$

for all $k > K_\varepsilon$. □

Theorem 6.1 *$K < \infty$ if and only if*

$$\lim_{n\to\infty} \tau_n = 0. \tag{6.3}$$

Proof: Assuming that $\mathbf{p} = \{p_k; 1 \le k \le K\}$ where K is finite and $p_k > 0$ for all k, $1 \le k \le K$, and denoting $p_\wedge = \min\{p_k; 1 \le k \le K\} > 0$, the necessity of (6.3) follows the fact that as $n \to \infty$

$$\tau_n = n\sum_k^K p_k(1-p_k)^n \le n\sum_k^K p_k(1-p_\wedge)^n = n(1-p_\wedge)^n \to 0.$$

The sufficiency of (6.3) follows the fact that if $K = \infty$, then Lemma 6.1 would provide a contradiction to (6.3). □

In fact, the proof of Theorem 6.1 also establishes the following corollary.

Corollary 6.1 *$K < \infty$ if and only if $\tau_n \le \mathcal{O}(nq_0^n)$ where q_0 is a constant in* $(0, 1)$.

Theorem 6.1 and Corollary 6.1 firmly characterize the members of the Gini–Simpson family as the distributions on finite alphabets. All distributions outside of the Gini–Simpson family must have positive probabilities on infinitely many letters of $\mathscr{X}$. The entire class of such distributions is denoted as $\mathscr{P}_+$. In fact, in the subsequent text when there is no ambiguity, $\mathscr{P}_+$ will denote the entire class of distributions with a positive probability on every ℓ_k in $\mathscr{X}$. For all distributions in $\mathscr{P}_+$, a natural group would be those for which $\lim_{n\to\infty}\tau_n = \infty$ and so the Turing–Good family is defined.

The next lemma includes two trivial but useful facts.

Lemma 6.2

1) *For any real number $x \in [0, 1)$,*

$$1 - x \ge \exp\left(-\frac{x}{1-x}\right).$$

2) *For any real number $x \in (0, 1/2)$,*

$$\frac{1}{1-x} < 1 + 2x.$$

Proof: For Part 1, the function $y = (1+t)^{-1}e^t$ is strictly increasing over $[0, \infty)$ and has value 1 at $t = 0$. Therefore, $(1+t)^{-1}e^t \ge 1$ for $t \in [0, \infty)$. The desired inequality follows the change of variable $x = t/(1+t)$. For Part 2, the proof is trivial. □

Lemma 6.3 *For any given probability distribution* $\mathbf{p} = \{p_k; k \geq 1\}$ *and two constants* $c > 0$ *and* $\delta \in (0, 1)$, *as* $n \to \infty$,

$$n^{1-\delta} \sum_{k\geq 1} p_k(1-p_k)^n \to c > 0$$

if and only if

$$n^{1-\delta} \sum_{k\geq 1} p_k e^{-np_k} \to c > 0.$$

A proof of Lemma 6.3 is given in the appendix of this chapter.

Theorem 6.2 *For any given probability distribution* $\mathbf{p} = \{p_k; k \geq 1\}$, *if there exists constants* $\lambda > 1$, $c > 0$ *and integer* $k_0 \geq 1$ *such that for all* $k \geq k_0$

$$p_k \geq ck^{-\lambda}, \tag{6.4}$$

then $\lim_{n\to\infty} \tau_n = \infty$.

A proof of Theorem 6.2 is given in the appendix of this chapter.

Theorem 6.2 puts distributions with power decaying tails, for example, $p_k = c_\lambda k^{-\lambda}$, and those with much more slowly decaying tails, for example, $p_k = c_\lambda (k \ln^{\lambda} k)^{-1}$, where $\lambda > 1$ and $c_\lambda > 0$ is a constant, which may depend on λ, in the Turing–Good family.

An interesting question at this point is whether there exist distributions with thinner tails, thinner than those with power decaying tails, in the Turing–Good family. An affirmative answer is given by Theorem 6.3 below with a sufficient condition on $\mathbf{p} = \{p_k; k \geq 1\}$.

Theorem 6.3 *If* $p_{k+1}/p_k \to 1$ *as* $k \to \infty$, *then* $\tau_n \to \infty$ *as* $n \to \infty$.

The proof of Theorem 6.3 requires the following lemma.

Lemma 6.4 *For every sufficiently large* n, *let* $k_n = \max\{k; p_k \geq 1/n\}$. *If there exist a constant* $c \in (0, 1)$ *and a positive integer* K_c *such that for all* k, $k \geq K_c$, $p_{k+1}/p_k \geq c$, *then* $1 \leq np_{k_n} < c^{-1}$.

Proof: $1 \leq np_{k_n}$ holds by definition. It suffices to show only that $np_{kn} \leq c^{-1}$. Toward that end, let n_0 be sufficiently large so that $p_{k_{n_0}+1}/p_{k_{n_0}} \geq c$, and therefore $p_{k_n+1}/p_{k_n} \geq c$ for every $n \geq n_0$. But since $cp_{k_n} \leq p_{k_n+1} < 1/n$ by definition, it follows that $np_{k_n} < c^{-1}$. □

Proof of Theorem 6.3: For any given arbitrarily small $\varepsilon > 0$, there exists a K_ε such that for all $k \geq K_\varepsilon$, $p_{k+1}/p_k > 1 - \varepsilon$. There exists a sufficiently large

n_ε such that for any $n > n_\varepsilon$, $k_n > K_\varepsilon$ where $k_n = \max\{k; p_k \geq 1/n\}$. For every $n > n_\varepsilon$,

1) a) $p_{k_n+1} > (1-\varepsilon)p_{k_n}$;
   b) $p_{k_n+2} > (1-\varepsilon)p_{k_n+1} > (1-\varepsilon)^2 p_{k_n}, \cdots$; and hence
   c) $p_{k_n+j} > (1-\varepsilon)^j p_{k_n}$, for $j = 1, 2, \cdots$;
   and
2) $p_{k_n+j} < 1/n$, for $j = 1, 2, \cdots$.

Therefore, for every $n > n_\varepsilon$,

$$\begin{aligned}
\tau_n &> n \sum_{k \geq k_n} p_k(1-p_k)^n \\
&= n \sum_{j \geq 0} p_{k_n+j}(1-p_{k_n+j})^n \\
&> n \sum_{j \geq 0} (1-\varepsilon)^j p_{k_n}(1-1/n)^n \\
&= np_{k_n}(1-1/n)^n \sum_{j \geq 0} (1-\varepsilon)^j \\
&= [np_{k_n}] \; [(1-1/n)^n] \; \left[\frac{1}{\varepsilon}\right].
\end{aligned}$$

In the last expression above, the first factor is bounded below by 1, the second factor is bounded below by $e^{-1}/2$ for large n, and the third factor can be taken over all bounds since ε is arbitrarily small. Hence, $\tau_n \to \infty$. □

Example 6.1 *Let $p_k = ck^{-\lambda}$ for some constants $c > 0$ and $\lambda > 1$. Since $p_{k+1}/p_k = [k/(k+1)]^\lambda \to 1$, Theorem 6.3 applies and $\tau_n \to \infty$.*

Example 6.2 *Let $p_k = ck^{-1}(\ln k)^{-\lambda}$ for some constants $c > 0$ and $\lambda > 1$. Since $p_{k+1}/p_k = [k/(k+1)][\ln k/\ln(k+1)]^\lambda \to 1$, Theorem 6.3 applies and $\tau_n \to \infty$.*

Example 6.3 *Let $p_k = ce^{-k^\delta}$ for some constants $c > 0$ and $\delta \in (0,1)$. Since, applying L'Hôpital's rule whenever necessary,*

$$p_{k+1}/p_k = \exp(k^\delta - (k+1)^\delta) = \exp\left(\frac{1-\left(1+\frac{1}{k}\right)^\delta}{k^{-\delta}}\right) \to e^0 = 1,$$

Theorem 6.3 applies and $\tau_n \to \infty$.

Example 6.4 *Let $p_k = ce^{-k/\ln k}$ for some constants $c > 0$. Since, applying L'Hôpital's rule whenever necessary,*

$$\begin{aligned}
p_{k+1}/p_k &= \exp\left(\frac{k}{\ln k} - \frac{k+1}{\ln(k+1)}\right)\\
&= \exp\left(\frac{k}{\ln k} - \frac{k}{\ln(k+1)} - \frac{1}{\ln(k+1)}\right)\\
&\sim \exp\left(\frac{k}{\ln k} - \frac{k}{\ln(k+1)}\right)\\
&= \exp\left(\frac{\ln\left(1+\frac{1}{k}\right)}{\left(\frac{\ln k \ln(k+1)}{k}\right)}\right)\\
&\sim \exp\left(\frac{\left(1+\frac{1}{k}\right)^{-1}}{\ln k \ln(k+1) - \ln(k+1) - \frac{k}{k+1}\ln k}\right) \to e^0 = 1,
\end{aligned}$$

Theorem 6.3 applies and $\tau_n \to \infty$.

Example 6.5 *Let $p_k = ce^{-k}$ for some constant $c > 0$. Since $p_{k+1}/p_k = e^{-1} < 1$, Theorem 6.3 does not apply. In fact, it can be shown that τ_n is bounded above by a constant (see Lemma 6.6).*

Examples 6.3 and 6.4 are particularly interesting and somewhat surprising. In the extreme value theory for continuous random variables, the maximum of a sample under the density function $f(x) = ce^{-x^{\delta}}$ where $\delta \in (0, 1)$ or $f(x) = ce^{-x/\ln x}$ converges weakly to a member of the Gumbel family to which extreme values under other thin-tailed distributions converge. In the current setting, where the domains are defined by the limiting behavior of τ_n, $p_k = ce^{-k^{\delta}}$ and $p_k = ce^{-k/\ln k}$ belong to the domain shared by thick-tailed discrete distributions such as $p_k = ck^{-\lambda}$ where $c > 0$ and $\lambda > 1$.

In view of Lemma 6.3 and Theorems 6.1 and 6.2, Domain 1, or the Molchanov family, has a more intuitive definition as given in the following lemma.

Lemma 6.5 *A distribution* $\mathbf{p}$ *on* $\mathscr{X}$ *belongs to the Molchanov family if and only if*

1) *the effective cardinality of $\mathscr{X}$ is countably infinite, that is, $K = \infty$, and*
2) *$\tau_n \le u_{\mathbf{p}}$ for all n, where $u_{\mathbf{p}} > 0$ is a constant that may depend on* $\mathbf{p}$.

The proof is left as an exercise.

Lemma 6.6 *For any $\mathbf{p} = \{p_k\} \in \mathscr{P}_+$, if there exists an integer $k_0 \ge 1$ such that $p_k = c_0 e^{-k}$, where $c_0 > 0$ is a constant, for all $k \ge k_0$, then*

1) $\tau_n \le u$ *for some upper bound* $u > 0$*; and*
2) $\lim_{n\to\infty} \tau_n$ *does not exist.*

A proof of Lemma 6.6 is given in the appendix of this chapter.

A similar proof to that of Lemma 6.6 immediately gives Theorem 6.4 with a slightly more general statement.

Theorem 6.4 *For any given probability distribution* $\mathbf{p} = \{p_k; k \ge 1\}$*, if there exists constants* $a > 1$ *and integer* $k_0 \ge 1$ *such that for all* $k \ge k_0$

$$p_k = ca^{-k}, \tag{6.5}$$

then

1) $\tau_n \le u_a$ *for some upper bound* $u_a > 0$*, which may depend on* a*; and*
2) $\lim_{n\to\infty} \tau_n$ *does not exist.*

Theorem 6.4 puts distributions with tails of geometric progression, for example, $p_k = c_\lambda e^{-\lambda k}$ where $\lambda > 0$ and $c_\lambda > 0$ are constants or $p_k = 2^{-k}$ in the Molchanov family, a family of distributions with perpetually oscillating tail indices.

Next, a notion of relative dominance of one probability distribution over another is defined on a countable alphabet within $\mathcal{P}_+$. Let $\#A$ denote the cardinality of a set A.

Definition 6.2 *Let* $\mathbf{q}^* \in \mathcal{P}_+$ *and* $\mathbf{p} \in \mathcal{P}_+$ *be two distributions on* $\mathcal{X}$*, and let* $\mathbf{q} = \{q_k\}$ *be a nonincreasingly ordered version of* $\mathbf{q}^*$*.* $\mathbf{q}^*$ *is said to dominate* $\mathbf{p}$ *if*

$$\#\{i; p_i \in (q_{k+1}, q_k], i \ge 1\} \le M < \infty$$

for every $k \ge 1$*, where* M *is a finite positive integer.*

It is easy to see that the notion of dominance by Definition 6.2 is a tail property, and that it is transitive, that is, if $\mathbf{p}_1$ dominates $\mathbf{p}_2$ and $\mathbf{p}_2$ dominates $\mathbf{p}_3$, then $\mathbf{p}_1$ dominates $\mathbf{p}_3$. It says in essence that if $\mathbf{p}$ is dominated by $\mathbf{q}$, then the p_is do not get overly congregated locally into some intervals defined by the q_ks.

The following examples illustrate the notion of dominance by Definition 6.2.

Example 6.6 *Let* $p_k = c_1 e^{-k^2}$ *and* $q_k = c_2 e^{-k}$ *for all* $k \ge k_0$ *for some integer* $k_0 \ge 1$ *and other two constants* $c_1 > 0$ *and* $c_2 > 0$*. For every sufficiently large* k*, suppose*

$$p_j = c_1 e^{-j^2} \le q_k = c_2 e^{-k},$$

then

$$-j^2 \le \ln(c_2/c_1) - k$$

and

$$j+1 \geq [k+\ln(c_1/c_2)\ ]^{1/2}+1.$$

It follows that

$$\begin{aligned}
p_{j+1} &= c_1 e^{-(j+1)^2} \\
&\leq c_1 e^{-\left(\sqrt{k+\ln(c_1/c_2)}+1\right)^2} \\
&= c_1 e^{-\,(k+\ln(c_1/c_2)+1)\,-2\sqrt{k+\ln(c_1/c_2)}} \\
&= c_2 e^{-\,(k+1)\,-2\sqrt{k+\ln(c_1/c_2)}} \\
&= c_2 e^{-\,(k+1)}\ e^{-2\sqrt{k+\ln(c_1/c_2)}} \\
&\leq c_2 e^{-\,(k+1)}\ = q_{k+1}.
\end{aligned}$$

This means that if $p_j \in (q_{k+1}, q_k]$, then necessarily $p_{j+1} \notin (q_{k+1}, q_k]$, which implies that each interval $(q_{k+1}, q_k]$ can contain only one p_j at most for a sufficiently large k, that is, $k \geq k_{00} := \max\{k_0, \ln(c_2/c_1)\}$. Since there are only finite p_js covered by $\cup_{1\leq k<k_{00}}(q_k, q_{k+1}]$, $\mathbf{q} = \{q_k\}$ dominates $\mathbf{p} = \{p_i\}$.

Example 6.7 *Let $p_k = c_1 a^{-k}$ and $q_k = c_2 b^{-k}$ for all $k \geq k_0$ for some integer $k_0 \geq 1$ and other two constants $a > b > 1$. For every sufficiently large k, suppose $p_j = c_1 a^{-j} \leq q_k = c_2 b^{-k}$, then $-j\ln a \leq \ln(c_2/c_1)\ - k\ln b$ and $j+1 \geq k(\ln b/\ln a) + 1 + \ln(c_1/c_2)\ /\ln a$. It follows that*

$$\begin{aligned}
p_{j+1} &= c_1 a^{-\ \left(k\frac{\ln b}{\ln a}+1+\frac{\ln(c_1/c_2)}{\ln a}\right)} \\
&= c_1 a^{-\ \left(k\log_a b+1+\frac{\ln(c_1/c_2)}{\ln a}\right)} \\
&= c_1 b^{-k} a^{-1} a^{-\frac{\ln(c_1/c_2)}{\ln a}} \\
&\leq c_1 b^{-(k+1)} a^{-\log_a (c_1/c_2)} \\
&= c_2 b^{-(k+1)} \\
&= q_{k+1}.
\end{aligned}$$

By a similar argument as that in Example 6.6, $\mathbf{q} = \{q_k\}$ dominates $\mathbf{p} = \{p_i\}$.

Example 6.8 *Let $p_k = c_1 k^{-r} e^{-\lambda k}$ for some integer $k_0 \geq 1$ and constants $\lambda > 0$ and $r > 0$, and $q_k = c_2 e^{-\lambda k}$ for all $k \geq k_0$. Suppose for a $k \geq k_0$ there is a j such that $p_j = c_1 j^{-r} e^{-\lambda j} \in (q_{k+1} = c_2 e^{-\lambda(k+1)}, q_k = c_2 e^{-\lambda k}]$, then*

$$\begin{aligned}
p_{j+1} &= c_1 (j+1)^{-r} e^{-\lambda(j+1)} \\
&= c_1 (j+1)^{-r} e^{-\lambda j} e^{-\lambda} \\
&\leq c_1 j^{-r} e^{-\lambda j} e^{-\lambda} \\
&\leq c_2 e^{-\lambda k} e^{-\lambda} = q_{k+1},
\end{aligned}$$

which implies that there is at most one p_j in $(q_{k+1}, q_k]$ for every sufficiently large k. Therefore, $\mathbf{q} = \{q_k\}$ dominates $\mathbf{p} = \{p_i\}$.

Example 6.9 *Let $p_k = c_1 k^r e^{-\lambda k}$ for some integer $k_0 \geq 1$ and constants $\lambda > 0$ and $r > 0$, and $q_k = c_2 e^{-(\lambda/2)k}$ for all $k \geq k_0$. Suppose for any sufficiently large j, $j \geq j_0 := [e^{\lambda/(2r)} - 1]^{-1}$,*

$$p_j = c_1 j^r e^{-\lambda j} \in (q_{k+1} = c_2 e^{-(\lambda/2)(k+1)}, q_k = c_2 e^{-(\lambda/2)k}]$$

for some sufficiently large $k \geq k_0$, then

$$\begin{aligned}
p_{j+1} &= c_1 (j+1)^r e^{-\lambda(j+1)} \\
&= c_1 (j+1)^r e^{-\lambda j} e^{-\lambda} \\
&= c_1 j^r e^{-\lambda j} e^{-\lambda} \frac{(j+1)^r}{j^r} \\
&\leq c_2 e^{-\frac{\lambda}{2}k} e^{-\lambda} \left(\frac{j+1}{j}\right)^r \\
&= c_2 e^{-\frac{\lambda}{2}(k+1)} e^{-\frac{\lambda}{2}} \left(\frac{j+1}{j}\right)^r \\
&\leq q_{k+1} e^{-\frac{\lambda}{2}} \left(\frac{j_0+1}{j_0}\right)^r = q_{k+1}
\end{aligned}$$

which implies that there is at most one p_j in $(q_{k+1}, q_k]$ for every sufficiently large k. Therefore, $\mathbf{q} = \{q_k\}$ dominates $\mathbf{p} = \{p_i\}$.

Example 6.10 *Let $p_k = q_k$ for all $k \geq 1$. $\mathbf{q} = \{q_k\}$ and $\mathbf{p} = \{p_k\}$ dominate each other.*

While in each of Examples 6.6 through 6.9, the dominating distribution $\mathbf{q}$ has a thicker tail than $\mathbf{p}$ in the usual sense, the dominance of Definition 6.2 in general is not implied by such a thinner/thicker tail relationship. This is so because a distribution $\mathbf{p} \in \mathscr{P}_+$, satisfying $p_k \leq q_k$ for all sufficiently large k, could exist yet congregate irregularly to have an unbounded $\sup_{k\geq 1} \#\{p_i; p_i \in (q_{k+1}, q_k], i \geq 1\}$. One such example is given in Section 6.3. In this regard, the dominance of Definition 6.2 is more appropriately considered as a regularity condition. However, it may be interesting to note that the said regularity is a relative one in the sense that the behavior of $\mathbf{p}$ is regulated by a reference distribution $\mathbf{q}$. This relative regularity gives an umbrella structure in the Molchanov family as well as in the Turing–Good family, as demonstrated by Theorems 6.5 and 6.6 below.

Theorem 6.5 *If two distributions $\mathbf{p}$ and $\mathbf{q}$ in $\mathscr{P}_+$ on a same countably infinite alphabet $\mathscr{X}$ are such that $\mathbf{q}$ is in the Molchanov family and $\mathbf{q}$ dominates $\mathbf{p}$, then $\mathbf{p}$ belongs to the Molchanov family.*

Proof: Without loss of generality, it may be assumed that $\mathbf{q}$ is nonincreasingly ordered. For every sufficiently large n, there exists a k_n such that $\frac{1}{n+1} \in (q_{k_n+1}, q_{k_n}]$. Noting that the function $np(1-p)^n$ increases in p over $(0, 1/(n+1)]$, attains its maximum value of $[1-1/(n+1)]^{n+1} < e^{-1}$ at $p = 1/(n+1)$, and decreases over $[1/(n+1), 1]$, consider

$$\begin{aligned}
\tau_n(\mathbf{p}) &= \sum_{k\geq 1} np_k(1-p_k)^n \\
&= \sum_{k:p_k\leq q_{k_n+1}} np_k(1-p_k)^n + \sum_{k:q_{k_n+1}<p_k\leq q_{k_n}} np_k(1-p_k)^n \\
&\quad + \sum_{k:p_k>q_{k_n}} np_k(1-p_k)^n \\
&\leq M \sum_{k\geq k_n+1} nq_k(1-q_k)^n + \sum_{k;q_{k_n+1}<p_k\leq q_{k_n}} e^{-1} + M \sum_{1\leq k\leq k_n} nq_k(1-q_k)^n \\
&= M \sum_{k\geq 1} nq_k(1-q_k)^n + \sum_{k:q_{k_n+1}<p_k\leq q_{k_n}} e^{-1} \\
&\leq M\tau_n(\mathbf{q}) + Me^{-1} < \infty,
\end{aligned}$$

where M is as in Definition 6.2 and it exists because the assumed condition that $\mathbf{q}$ dominates $\mathbf{p}$. The desired result immediately follows. □

Corollary 6.2 *Any distribution* $\mathbf{p}$ *on a countably infinite alphabet* $\mathcal{X}$ *satisfying* $p_k = ae^{-\lambda k}$, $p_k = be^{-\lambda k^2}$, *or* $p_k = ck^r e^{-\lambda k}$ *for all* $k \geq k_0$, *where* $k_0 \geq 1$, $\lambda > 0$, $r \in (-\infty, +\infty)$, $a > 0$, $b > 0$, *and* $c > 0$ *are constants, is in the Molchanov family.*

Proof: The result is immediate following Theorem 6.5 and Examples 6.6 through 6.9. □

Theorem 6.6 *If two distributions* $\mathbf{p}$ *and* $\mathbf{q}$ *in* $\mathscr{P}_+$ *on a same countably infinite alphabet* $\mathcal{X}$ *are such that* $\mathbf{p}$ *is in the Turing–Good family and* $\mathbf{q}$ *dominates* $\mathbf{p}$, *then* $\mathbf{q}$ *belongs to the Turing–Good family.*

Proof: In the proof of Theorem 6.5, it is established that if $\mathbf{q}$ dominates $\mathbf{p}$, then

$$\tau_n(\mathbf{p}) \leq M\tau_n(\mathbf{q}) + Me^{-1}$$

for some positive constant M. The fact $\tau_n(\mathbf{p}) \to \infty$ implies $\tau_n(\mathbf{q}) \to \infty$, as $n \to \infty$. □

Figures 6.1–6.3 show graphic representations of τ_n for several distributions in various domains. Figure 6.1 plots τ_n for $p_k = 0.01$ for $k = 1, \dots, 100$ and $q_k = 0.02$ for $k = 1, \dots, 50$. The tail indices are plotted on a same vertical scale from 0 to 40, ranging from $n = 1$ to $n = 1000$. When n increases indefinitely, the rapid decay in both indices, as suggested by Theorem 6.1 and Lemma 6.1, is visible.

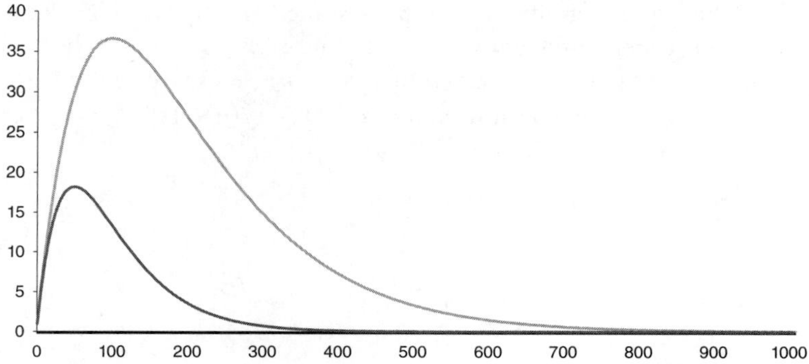

Figure 6.1 τ_n of $p_k = 0.01$ (upper) for $k = 1, \dots, 100$ and $q_k = 0.02$ (lower) for $k = 1, \dots, 50$.

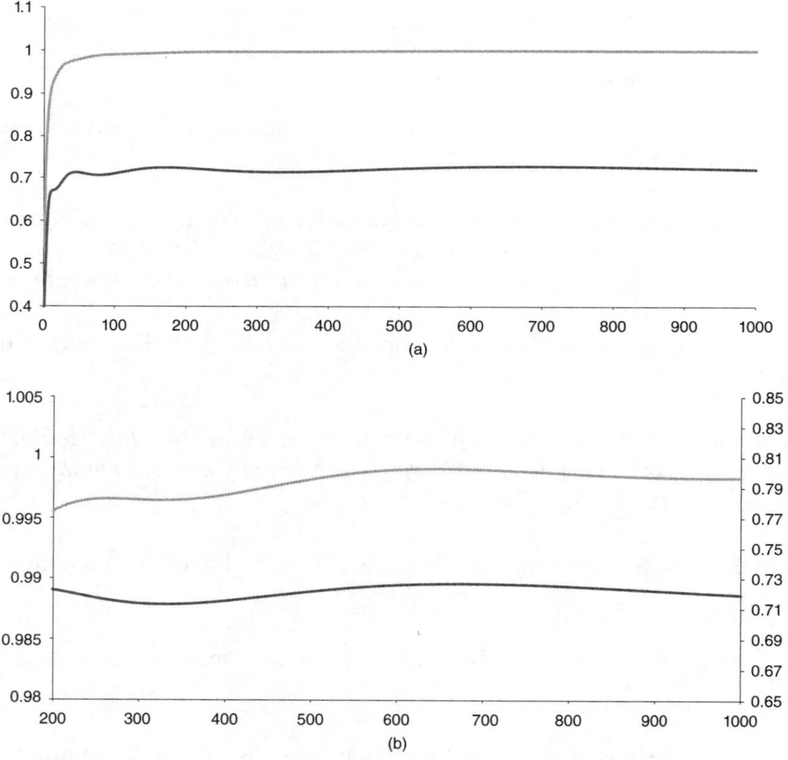

Figure 6.2 τ_n of $p_k = (e-1)e^{-k}$ (upper) and $p_k = 3 \times 4^{-k}$ (lower), $k \geq 1$. (a) τ_n from $n = 1$ to $n = 1000$, one scale. (b) τ_n from $n = 200$ to $n = 1000$, two scales.

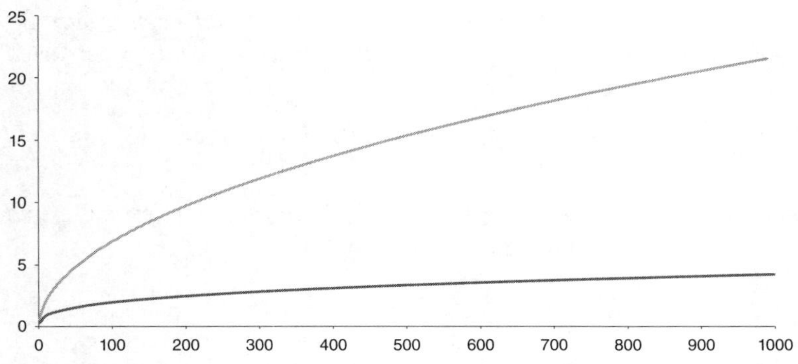

Figure 6.3 τ_n of $p_k = c_1 k^{-2}$ (upper) and $q_k = c_2 k^{-3}$ (lower), $k \geq 1$.

Figure 6.2 gives τ_n for $p_k = (e-1)e^{-k}$ and $q_k = 3 \times 4^{-k}$, $k \geq 1$. In Figure 6.2a, the pair of tail indices are plotted on the same vertical scale for τ_n from 0.4 to 1.1, ranging from $n = 1$ to $n = 1000$. The values for τ_n seem stable for large values of n in Figure 6.2a. However, when the tail indices are plotted on different vertical scales, on the left from 0.98 to 1.005 for p_k and on the right from 0.65 to 0.85 for q_k, ranging from $n = 200$ to $n = 1000$, the oscillating patterns suggested by Theorem 6.4 become visible.

Figure 6.3 gives τ_n for $p_k = c_1 k^{-2}$ and $q_k = c_2 k^{-3}$, $k \geq 1$. The pair of tail indices are plotted on the same vertical scale from 0 to 25, ranging from $n = 1$ to $n = 1000$. The divergent patterns suggested by Theorem 6.2 are visible.

6.3 Examples and Remarks

Three constructed examples are given in this section, and each illustrates a point of interest. The first constructed example shows that the notion of thinner tail, in the usual sense of $p_k \leq q_k$ for $k \geq k_0$ where $k_0 \geq 1$ is some fixed integer and $\mathbf{p} = \{p_k\}$ and $\mathbf{q} = \{q_k\}$ are two distributions, does not imply dominance of $\mathbf{q}$ over $\mathbf{p}$.

Example 6.11 *Consider any strictly decreasing distribution* $\mathbf{q} = \{q_k; k \geq 1\} \in \mathscr{P}_+$ *and the following grouping of the index set* $\{k; k \geq 1\}$.

$$
\begin{aligned}
G_1 &= \{1\}, \\
G_2 &= \{2, 3\}, \\
&\vdots, \\
G_m &= \left\{ \frac{m(m-1)}{2} + 1, \ldots, \frac{m(m-1)}{2} + m \right\}, \\
&\vdots .
\end{aligned}
$$

$\{G_m; m \geq 1\}$ is a partition of the index set $\{k; k \geq 1\}$ and each group G_m contains m consecutive indices. A new distribution $\mathbf{p} = \{p_k\}$ is constructed according to the following steps:

1) For each $m \geq 2$, let $p_k = q_{m(m-1)/2+m}$ for all $k \in G_m$.
2) $p_1 = 1 - \sum_{k \geq 2} p_k$.

In the first step, $m(m-1)/2 + m = m(m+1)/2$ is the largest index in G_m, and therefore $q_{m(m+1)/2}$ is the smallest q_k with index $k \in G_m$. Since

$$0 \leq \sum_{k \geq 2} p_k = \sum_{m \geq 2} m q_{m(m+1)/2} < \sum_{k \geq 2} q_k \leq 1,$$

p_1 so assigned is a probability. The distribution $\mathbf{p} = \{p_k\}$ satisfies $p_k \leq q_k$ for every $k \geq 2 = k_0$. However, the number of terms of p_i in the interval $(q_{m(m+1)/2+1}, q_{m(m+1)/2}]$ is at least m and it increases indefinitely as $m \to \infty$; and hence $\mathbf{q}$ does not dominate $\mathbf{p}$.

The second constructed example shows that the notion of dominance of $\mathbf{q} = \{q_k\}$ over $\mathbf{p} = \{p_k\}$, as defined in Definition 6.2, does not imply that $\mathbf{p}$ has thinner tail than $\mathbf{q}$, in the usual sense of $p_k \leq q_k$ for $k \geq k_0$ where $k_0 \geq 1$ is some fixed integer.

Example 6.12 *Consider any strictly decreasing distribution $\mathbf{q} = \{q_k; k \geq 1\} \in \mathscr{P}_+$ and the following grouping of the index set $\{k; k \geq 1\}$.*

$$G_1 = \{1, 2\}, G_2 = \{3, 4\}, \ldots, G_m = \{2m-1, 2m\}, \ldots .$$

$\{G_m; m \geq 1\}$ is a partition of the index set $\{k; k \geq 1\}$ and each group G_m contains two consecutive indices, the first one odd and the second one even. The construction of a new distribution $\mathbf{p} = \{p_k\}$ is as follows: for each group G_m with its two indices $k = 2m - 1$ and $k + 1 = 2m$, let $p_k = p_{k+1} = (q_k + q_{k+1})/2$. With the new distribution $\mathbf{p} = \{p_k\}$ so defined, one has $p_{2m} < q_{2m}$ and $p_{2m-1} > q_{2m-1}$ for all $m \geq 1$. Clearly $\mathbf{q}$ dominates $\mathbf{p}$ ($\mathbf{p}$ dominates $\mathbf{q}$ as well), but $\mathbf{p}$ does not have a thinner tail in the usual sense.

At this point, it becomes clear that the notation of dominance of Definition 6.2 and the notation of thinner/thicker tail in the usual sense are two independent notions.

The next constructed example shows that there exists a distribution such that the associated τ_n approaches infinity along one subsequence of $\mathbb{N}$ and is bounded above along another subsequence of $\mathbb{N}$, hence belonging to Domain T. Domain T is not empty.

Example 6.13 *Consider the probability sequence $q_j = 2^{-j}$, for $j = 1, 2, \ldots$, along with a diffusion sequence $d_i = 2^i$, for $i = 1, 2, \ldots$. A probability sequence $\{p_k\}$, for $k = 1, 2, \ldots$, is constructed by the following steps:*

1st:

a) *Take the first value of d_i, $d_1 = 2^1$, and assign the first $2d_1 = 2^2 = 4$ terms of q_j,*

$$q_1 = 2^{-1},\ q_2 = 2^{-2},\ q_3 = 2^{-3},\ q_4 = 2^{-4},$$

to the first four terms of p_k,

$$p_1 = 2^{-1},\ p_2 = 2^{-2},\ p_3 = 2^{-3},\ p_4 = 2^{-4}.$$

b) *Take the next unassigned term in q_j, $q_5 = 2^{-5}$, and diffuse it into $d_1 = 2$ equal terms, 2^{-6} and 2^{-6}.*
   i) *Starting at q_5 in the sequence $\{q_j\}$, look forwardly ($j > 5$) for terms greater or equal to 2^{-6}, if any, continue to assign them to p_k. In this case, there is only one such term $q_6 = 2^{-6}$ and it is assigned to $p_5 = 2^{-6}$.*
   ii) *Take the $d_1 = 2$ diffused terms and assign them to $p_6 = 2^{-6}$ and $p_7 = 2^{-6}$. At this point, the first few terms of the partially assigned sequence $\{p_k\}$ are*

$$\begin{array}{llll} p_1 = 2^{-1}, & p_2 = 2^{-2}, & p_3 = 2^{-3}, & p_4 = 2^{-4}, \\ p_5 = 2^{-6}, & p_6 = 2^{-6}, & p_7 = 2^{-6}. & \end{array}$$

2nd:

a) *Take the next value of d_i, $d_2 = 2^2$, and assign the next $2d_2 = 2^3 = 8$ unused terms of q_j,*

$$q_7 = 2^{-7},\ \ldots,\ q_{14} = 2^{-14},$$

to the next eight terms of p_k,

$$p_8 = 2^{-7},\ \ldots,\ p_{15} = 2^{-14}.$$

b) *Take the next unassigned term in q_j, $q_{15} = 2^{-15}$, and diffuse it into $d_2 = 4$ equal terms of 2^{-17} each.*
   i) *Starting at q_{15} in the sequence of $\{q_j\}$, look forwardly ($j > 15$) for terms greater or equal to 2^{-17}, if any, continue to assign them to p_k. In this case, there are 2 such terms*

$$q_{16} = 2^{-16},\ q_{17} = 2^{-17},$$

and they are assigned to

$$p_{16} = 2^{-16},\ p_{17} = 2^{-17}.$$

   ii) *Take the $d_2 = 2^2 = 4$ diffused terms and assign them to*

$$p_{18} = 2^{-17},\ \ldots,\ p_{21} = 2^{-17}.$$

At this point, the first few terms of the partially assigned sequence $\{p_k\}$ *are*

$$\begin{array}{llll} p_1 = 2^{-1}, & p_2 = 2^{-2}, & p_3 = 2^{-3}, & p_4 = 2^{-4}, \\ p_5 = 2^{-6}, & p_6 = 2^{-6}, & p_7 = 2^{-6}, & p_8 = 2^{-7}, \\ p_9 = 2^{-8}, & \dots, & p_{15} = 2^{-14}, & p_{16} = 2^{-16}, \\ p_{17} = 2^{-17}, & p_{18} = 2^{-17}, & \dots, & p_{21} = 2^{-17}. \end{array}$$

ith:

a) *In general, take the next value of* d_i*, say* $d_i = 2^i$*, and assign the next* $2d_i = 2^{i+1}$ *unused terms of* q_j*, say*

$$q_{j_0} = 2^{-j_0}, \ \dots, \ q_{j_0+2^{i+1}-1} = 2^{-(j_0+2^{i+1}-1)},$$

to the next $2d_i = 2^{i+1}$ *terms of* p_k*, say*

$$p_{k_0} = 2^{-j_0}, \ \dots, \ p_{k_0+2^{i+1}-1} = 2^{-(j_0+2^{i+1}-1)}.$$

b) *Take the next unassigned term in* q_j*,*

$$q_{j_0+2^{i+1}} = 2^{-(j_0+2^{i+1})},$$

and diffuse it into $d_i = 2^i$ *equal terms,* $2^{-(j_0+i+2^{i+1})}$ *each.*

i) *Starting at* $q_{j_0+2^{i+1}}$ *in the sequence of* $\{q_j\}$*, look forwardly* ($j > j_0 + 2^{i+1}$) *for terms greater or equal to* $2^{-(j_0+i+2^{i+1})}$*, if any, continue to assign them to* p_k*. Denote the last assigned* p_k *as* p_{k_0}*.*

ii) *Take the* $d_i = 2^i$ *diffused terms and assign them to*

$$p_{k_0+1} = 2^{-(j_0+i+2^{i+1})}, \ \dots, \ p_{k_0+2^i} = 2^{-(j_0+i+2^{i+1})}.$$

In essence, the sequence $\{p_k\}$ *is generated based on the sequence* $\{q_j\}$ *with infinitely many selected j's at each of which* q_j *is diffused into increasingly many equal probability terms according a diffusion sequence* $\{d_i\}$*. The diffused sequence is then rearranged in a nonincreasing order.*

By construction, it is clear that the sequence $\{p_k; k \geq 1\}$ *satisfies the following properties:*

$\mathcal{A}_1$: $\{p_k\}$ *is a probability sequence in a nonincreasing order.*

$\mathcal{A}_2$: *As k increases,* $\{p_k\}$ *is a string of segments alternating between two different types: (i) a strictly decreasing segment and (ii) a segment (a run) of equal probabilities.*

$\mathcal{A}_3$: *As k increases, the length of the last run increases and approaches infinity.*

$\mathcal{A}_4$: *In each run, there are exactly* $d_i + 1$ *equal terms,* d_i *of which are diffused terms and 1 of which belongs to the original sequence* q_j*.*

$\mathcal{A}_5$: *Between two consecutive runs (with lengths* $d_i + 1$ *and* $d_{i+1} + 1$*, respectively), the strictly decreasing segment in the middle has at least* $2d_{i+1}$ *terms and satisfies*

$$2d_{i+1} = 4d_i = d_i + 3d_i > d_i + d_{i+1}.$$

$\mathcal{A}_6$: *For any k,* $1/p_k$ *is a positive integer.*

Next it is to show that there is a subsequence $\{n_i\} \in \mathbb{N}$ such that τ_{n_i} defined with $\{p_k\}$ approaches infinity. Toward that end, consider the subsequence $\{p_{k_i}; i \geq 1\}$ of $\{p_k\}$ where the index k_i is such that p_{k_i} is the first term in the ith run segment. Let $\{n_i\} = \{1/p_{k_i}\}$ which by $\mathcal{A}_6$ is a subsequence of $\mathbb{N}$. By $\mathcal{A}_3$ and $\mathcal{A}_4$,

$$\begin{aligned}\tau_{n_i} &= n_i \sum_{k\geq 1} p_k(1-p_k)^{n_i} \\ &> n_i(d_i+1)p_{k_i}(1-p_{k_i})^{n_i} \\ &= (d_i+1)\left(1-\frac{1}{n_i}\right)^{n_i} \to \infty.\end{aligned}$$

Consider next the subsequence $\{p_{k_i-(d_i+1)}; i \geq 1\}$ of $\{p_k\}$ where the index k_i is such that p_{k_i} is the first term in the ith run segment, and therefore $p_{k_i-(d_i+1)}$ is the (d_i+1)th term counting backwards from p_{k_i-1}, into the preceding segment of at least $2d_i$ strictly decreasing terms. Let

$$\{m_i\} = \left\{\frac{1}{p_{k_i-(d_i+1)}} - 1\right\}$$

(so $p_{k_i-(d_i+1)} = (m_i+1)^{-1}$) which by $\mathcal{A}_6$ is a subsequence of $\mathbb{N}$.

$$\begin{aligned}\tau_{m_i} &= m_i \sum_{k\geq 1} p_k(1-p_k)^{m_i} \\ &= m_i \sum_{k\leq k_i-(d_i+1)} p_k(1-p_k)^{m_i} + m_i \sum_{k\geq k_i-d_i} p_k(1-p_k)^{m_i} \\ &:= \tau_{m_i,1} + \tau_{m_i,2}.\end{aligned}$$

Before proceeding further, let several detailed facts be noted.

1) *The function $np(1-p)^n$ increases in $[0, 1/(n+1)]$, attains maximum at $p = 1/(n+1)$, and decreases in $[1/(n+1), 1]$.*
2) *Since $p_{k_i-(d_i+1)} = (m_i+1)^{-1}$, by $\mathcal{A}_1$ each summand in $\tau_{m_i,1}$ is bounded above by*

$$m_i p_{k_i-(d_i+1)}(1-p_{k_i-(d_i+1)})^{m_i}$$

   and each summand in $\tau_{m_i,2}$ is bounded above by

$$m_i p_{k_i-d_i}(1-p_{k_i-d_i})^{m_i}.$$

3) *By $\mathcal{A}_4$ and $\mathcal{A}_5$, for each diffused term of $p_{k'}$ with $k' \leq k_i-(d_i+1)$ in a run, there is a different nondiffused term $p_{k''}$ with $k'' \leq k_i-(d_i+1)$ such that $p_{k'} > p_{k''}$ and therefore*

$$m_i\, p_{k'}(1-p_{k'})^{m_i} \leq m_i\, p_{k''}(1-p_{k''})^{m_i};$$

and similarly, for each diffused term of $p_{k'}$ with $k' \geq k_i - d_i$ in a run, there is a different nondiffused term $p_{k''}$ with $k'' \geq k_i - d_i$ such that $p_{k'} < p_{k''}$ and therefore

$$m_i p_{k'}(1 - p_{k'})^{m_i} \leq m_i p_{k''}(1 - p_{k''})^{m_i}.$$

These facts imply that

$$\begin{aligned}\tau_{m_i} &= \tau_{m_i,1} + \tau_{m_i,2}\\ &= m_i \sum_{k \leq k_i-(d_i+1)} p_k(1-p_k)^{m_i} + m_i \sum_{k \geq k_i - d_i} p_k(1-p_k)^{m_i}\\ &\leq 2m_i \sum_{j \geq 1} q_j(1-q_j)^{m_i} < \infty\end{aligned}$$

and the last inequality above is due to Corollary 6.2.

While the domains of attraction on alphabets have probabilistic merit, the statistical implication is also quite significant. $Z_{1,v}$ in (3.26) is an unbiased estimator of $\zeta_{1,v}$, for $v = 1, \dots, n-1$, as is demonstrated in Chapter 2, and therefore

$$\hat{\tau}_v = vZ_{1,v}. \tag{6.6}$$

is an unbiased estimator of $\tau_v = v\zeta_v$. In fact, Zhang and Zhou (2010) established several useful statistical properties of $\hat{\tau}_v$, including the asymptotic normality.

The availability of $\hat{\tau}_v$ gives much added merit to the discussion of the domains of attraction on alphabets as presented in this chapter. Specifically the fact that the asymptotic behavior of τ_n characterizes the tail probability of the underlying $\mathbf{p}$ and the fact that the trajectory of τ_v up to $v = n-1$ is estimable suggest that much could be revealed by a sufficiently large sample.

6.4 Appendix

6.4.1 Proof of Lemma 6.3

Proof: Let $\delta^* = \delta/8$. Consider the partition of the index set

$$\{k; k \geq 1\} = I \cup II$$

where

$$I = \{k : p_k \leq 1/n^{1-\delta^*}\} \qquad \text{and} \qquad II = \{k : p_k > 1/n^{1-\delta^*}\}.$$

Since pe^{-np} has a negative derivative with respect to p on interval $(1/n, 1]$ and hence on $(1/n^{1-\delta^*}, 1]$ for large n, $p_k e^{-np_k}$ attains its maximum at $p_k = 1/n^{1-\delta^*}$ for every $k \in II$. Therefore, noting that there are at most $n^{1-\delta^*}$ indices in II,

$$
\begin{aligned}
0 &\le n^{1-\delta} \sum_{k \in II} p_k (1-p_k)^n \\
&\le n^{1-\delta} \sum_{k \in II} p_k e^{-np_k} \\
&\le n^{1-\delta} \sum_{k \in II} \left(\frac{1}{n^{1-\delta^*}} e^{-\frac{n}{n^{1-\delta^*}}} \right) \le n^{1-\delta} n^{1-\delta^*} \left(\frac{1}{n^{1-\delta^*}} e^{-\frac{n}{n^{1-\delta^*}}} \right) \\
&= n^{1-\delta} e^{-n^{\delta^*}} \to 0.
\end{aligned}
$$

Thus

$$
\lim_{n\to\infty} n^{1-\delta} \sum_{k\ge 1} p_k(1-p_k)^n = \lim_{n\to\infty} n^{1-\delta} \sum_{k\in I} p_k(1-p_k)^n \tag{6.7}
$$

and

$$
\lim_{n\to\infty} n^{1-\delta} \sum_{k\ge 1} p_k e^{-np_k} = \lim_{n\to\infty} n^{1-\delta} \sum_{k\in I} p_k e^{-np_k}. \tag{6.8}
$$

On the other hand, since $1-p \le e^{-p}$ for all $p \in [0,1]$,

$$
n^{1-\delta} \sum_{k\in I} p_k(1-p_k)^n \le n^{1-\delta} \sum_{k\in I} p_k e^{-np_k}.
$$

Furthermore, applying Parts 1 and 2 of Lemma 6.2 in the first and the third steps in the following, respectively, leads to

$$
\begin{aligned}
n^{1-\delta} \sum_{k\in I} p_k(1-p_k)^n &\ge n^{1-\delta} \sum_{k\in I} p_k \exp\left(-\frac{np_k}{1-p_k} \right) \\
&\ge n^{1-\delta} \sum_{k\in I} p_k \exp\left(-\frac{np_k}{1-\sup_{i\in I} p_i} \right) \\
&\ge n^{1-\delta} \sum_{k\in I} \exp\left(-2n(\sup_{i\in I})^2 \right) p_k e^{-np_k}.
\end{aligned}
$$

Noting the fact that

$$
\lim_{n\to\infty} \exp(-2n(\sup_{i\in I})^2) = 1
$$

uniformly by the definition of I,

$$
\lim_{n\to\infty} n^{1-\delta} \sum_{k\in I} p_k(1-p_k)^n = \lim_{n\to\infty} n^{1-\delta} \sum_{k\in I} p_k e^{-np_k},
$$

and hence, by (6.7) and (6.8), the lemma follows. □

6.4.2 Proof of Theorem 6.2

Proof: For clarity, the proof is given in two cases, respectively:

1) $p_k = ck^{-\lambda}$ for all $k \ge k_0$ for some $k_0 > 1$, and
2) $p_k \ge ck^{-\lambda}$ for all $k \ge k_0$ for some $k_0 > 1$.

Case 1: Assuming $p_k = ck^{-\lambda}$ for all $k \geq k_0$, it suffices to consider the partial series $\sum_{k \geq k_0} np_k(1-p_k)^n$. First consider

$$n^{1-\frac{1}{\lambda}} \sum_{k=k_0}^{\infty} p_k e^{-np_k} = n^{1-\frac{1}{\lambda}} \sum_{k=k_0}^{\infty} ck^{-\lambda} e^{-nck^{-\lambda}} = \sum_{k=k_0}^{\infty} f_n(k)$$

where

$$f_n(x) = n^{1-\frac{1}{\lambda}} cx^{-\lambda} e^{-ncx^{-\lambda}}.$$

Since it is easily verified that

$$f_n'(x) = -\lambda cn^{1-\frac{1}{\lambda}} x^{-(\lambda+1)}(1 - ncx^{-\lambda})e^{-ncx^{-\lambda}},$$

it can be seen that $f_n(x)$ increases over $[1, (nc)^{1/\lambda}]$ and decreases over $[(nc)^{1/\lambda}, \infty)$ for every sufficiently large n. Let $x_0 = k_0$ and $x(n) = (nc)^{1/\lambda}$. It is clear that $f_n(x_0) \to 0$ and

$$\begin{aligned} f_n(x(n)) &= n^{1-\frac{1}{\lambda}} c(nc)^{-1} e^{-nc(nc)^{-1}} \\ &= n^{1-\frac{1}{\lambda}} c(nc)^{-1} e^{-1} \\ &= \frac{1}{en^{1/\lambda}} \to 0. \end{aligned}$$

Invoking the Euler–Maclaurin lemma, with changes of variable $t = x^{-\lambda}$ and then $s = nct$,

$$\begin{aligned} n^{1-\frac{1}{\lambda}} \sum_{k=k_0}^{\infty} p_k e^{-np_k} &\sim \int_{x_0}^{\infty} n^{1-\frac{1}{\lambda}} cx^{-\lambda} e^{-ncx^{-\lambda}} dx \\ &= \frac{c}{\lambda} \int_0^{x_0^{-\lambda}} n^{1-\frac{1}{\lambda}} t^{-\frac{1}{\lambda}} e^{-nct} dt \\ &= \frac{c}{\lambda} n^{1-\frac{1}{\lambda}} \int_0^{x_0^{-\lambda}} (nct)^{-\frac{1}{\lambda}} (nc)^{-1+\frac{1}{\lambda}} e^{-nct} d(nct) \\ &= \frac{c}{\lambda} n^{1-\frac{1}{\lambda}} (nc)^{-1+\frac{1}{\lambda}} \int_0^{ncx_0^{-\lambda}} s^{-\frac{1}{\lambda}} e^{-s} ds \\ &= \frac{c^{\frac{1}{\lambda}}}{\lambda} n^0 \int_0^{ncx_0^{-\lambda}} s^{-\frac{1}{\lambda}} e^{-s} ds \\ &= \frac{c^{\frac{1}{\lambda}}}{\lambda} \int_0^{ncx_0^{-\lambda}} s^{\left(1-\frac{1}{\lambda}\right)-1} e^{-s} ds \\ &= \frac{c^{\frac{1}{\lambda}}}{\lambda} \Gamma\left(1 - \frac{1}{\lambda}\right) \left[\frac{1}{\Gamma\left(1-\frac{1}{\lambda}\right)} \int_0^{ncx_0^{-\lambda}} s^{\left(1-\frac{1}{\lambda}\right)-1} e^{-s} ds\right] \\ &\to \frac{c^{\frac{1}{\lambda}}}{\lambda} \Gamma\left(1 - \frac{1}{\lambda}\right) > 0. \end{aligned}$$

Hence by Lemma 6.3,

$$n^{1-1/\lambda}\sum_{k=1}^{\infty}p_k(1-p_k)^n \to c^{1/\lambda}\lambda^{-1}\Gamma\,(1-1/\lambda)>0,$$

and therefore $\tau_n \to \infty$.

Case 2: Assuming $p_k \geq ck^{-\lambda} =: q_k$ for all $k \geq k_0$ for some $k_0 \geq 1$, one first has

$$n^{1-\frac{1}{\lambda}}\sum_{k\geq (nc)^{\frac{1}{\lambda}}} ck^{-\lambda}e^{-nck^{-\lambda}} = n^{1-\frac{1}{\lambda}}\sum_{k\geq 1} ck^{-\lambda}e^{-nck^{-\lambda}}1\left[k\geq (nc)^{\frac{1}{\lambda}}\right].$$

Since

$$f_n(x)=n^{1-\frac{1}{\lambda}}ck^{-\lambda}e^{-nck^{-\lambda}}1\left[k\geq (nc)^{\frac{1}{\lambda}}\right]$$

satisfies the condition of Lemma 1.6 (The Euler–Maclaurin lemma) with $x(n)=(nc)^{\frac{1}{\lambda}}$ and $f_n(x(n))\to 0$, one again has

$$\begin{aligned}
&n^{1-\frac{1}{\lambda}}\sum_{k\geq[(n+1)c]^{\frac{1}{\lambda}}} ck^{-\lambda}e^{-nck^{-\lambda}}\\
&=c\int_1^{\infty} n^{1-\frac{1}{\lambda}}x^{-\lambda}e^{-ncx^{-\lambda}}1\left[x\geq [(n+1)c]^{\frac{1}{\lambda}}\right]dx\\
&=c\int_{[(n+1)c]^{\frac{1}{\lambda}}}^{\infty} n^{1-\frac{1}{\lambda}}x^{-\lambda}e^{-ncx^{-\lambda}}dx\\
&=c^{\frac{1}{\lambda}}\lambda^{-1}\Gamma\left(1-\frac{1}{\lambda}\right)\int_0^{n(n+1)c^2}\frac{1}{\Gamma\left(1-\frac{1}{\lambda}\right)}s^{\left(1-\frac{1}{\lambda}\right)-1}e^{-s}ds\\
&\to c^{\frac{1}{\lambda}}\lambda^{-1}\Gamma\left(1-\frac{1}{\lambda}\right)>0.
\end{aligned}\tag{6.9}$$

On the other hand, for sufficiently large n,

$$I^*=\left\{k:p_k\leq\frac{1}{n+1}\right\}\subseteq\{k;k\geq k_0\},$$

by Parts 1 and 2 of Lemma 6.2 at steps 2 and 4 below and (6.9) at step 7, one has

$$\begin{aligned}
n^{1-1/\lambda}\sum_{k\in I^*}p_k(1-p_k)^n &\geq n^{1-1/\lambda}\sum_{k\in I^*}q_k(1-q_k)^n\\
&\geq n^{1-1/\lambda}\sum_{k\in I^*}q_k\exp\left(-\frac{nq_k}{1-q_k}\right)
\end{aligned}$$

$$\begin{aligned}
&\geq n^{1-1/\lambda}\sum_{k\in I^*} q_k \exp\left(-\frac{nq_k}{1-\sup_{i\in I^*} q_i}\right)\\
&\geq n^{1-1/\lambda}\sum_{k\in I^*} \exp(-2n(\sup_{i\in I^*} q_i)^2)\; q_k e^{-nq_k}\\
&\geq n^{1-1/\lambda}\sum_{k\in I^*} \exp(-2/n) q_k e^{-nq_k}\\
&= \exp(-2/n) n^{1-1/\lambda}\sum_{k\in I^*} ck^{-\lambda}e^{-nck^{-\lambda}}\\
&\to c^{\frac{1}{\lambda}}\lambda^{-1}\Gamma\left(1-\frac{1}{\lambda}\right) > 0.
\end{aligned}$$

Finally,

$$\tau_n = n\sum_k p_k(1-p_k)^n \geq n^{1/\lambda}n^{1-1/\lambda}\sum_{k\in I^*} p_k(1-p_k)^n \to \infty$$

as $n \to \infty$.

□

6.4.3 Proof of Lemma 6.6

Proof: For clarity, the proof of Lemma 6.6 is given in three parts, Part 1: Preliminaries, Part 2: Part 1 of Lemma 6.6, and Part 3: Part 2 of Lemma 6.6.

Part 1: Preliminaries. Noting that the first finite terms of τ_n vanishes exponentially fast for any distribution, one may assume, without loss of generality, that $k_0 = 1$. For any given n sufficiently large, define $k^* = k^*(n)$ by

$$p_{k^*+1} < \frac{1}{n+1} \leq p_{k^*}. \tag{6.10}$$

Noting

$$\begin{aligned}
&c_0e^{-(k^*+1)} < \frac{1}{n+1} \leq c_0e^{-k^*},\\
&e^{-(k^*+1)} < \frac{1}{c_0(n+1)} \leq e^{-k^*},\\
&-(k^*+1) < -\ln(c_0(n+1)) \leq -k^*, \quad \text{and}\\
&k^*+1 > \ln(c_0(n+1)) \geq k^*,
\end{aligned}$$

k^* may be expressed as

$$k^* = \lfloor \ln(c_0(n+1)) \rfloor$$

for each n. That is to say that, although k^* is uniquely defined by any given n, each k^* may correspond to several consecutive integer values of n. For a given integer value k^*, let the said consecutive integer values of n be denoted by

$$\{n_{k^*}, n_{k^*}+1, \ldots, n_{k^*+1}-1\}, \tag{6.11}$$

specifically noting that

1) n_{k^*} is the smallest integer value of n corresponding to k^* by (6.10), that is, $k^* = \lfloor \ln(c_0(n+1)) \rfloor$;
2) n_{k^*+1} is the smallest integer value of n that satisfies $k^* + 1 = \lfloor \ln(c_0(n+1)) \rfloor$; and
3) $n_{k^*+1} - 1$ is the greatest integer value of n that shares the same value of k^* with n_{k^*}.

Since $k^* = k^*(n)$ depends on n, one may express p_{k^*} as, and define $c(n)$ by,

$$p_{k^*} = \frac{c(n)}{n}. \tag{6.12}$$

At this point, the following fact is established: for each given k^*,

$$p_{k^*} = \frac{c(n_{k^*})}{n_{k^*}} = \frac{c(n_{k^*}+1)}{n_{k^*}+1} = \cdots = \frac{c(n_{k^*+1}-1)}{n_{k^*+1}-1}. \tag{6.13}$$

There are two main consequences of the expression in (6.12). The first is that τ_n defined in (6.2) may be expressed by (6.15); and the second is that the sequence $c(n)$ perpetually oscillates between 1 and e. Both facts are demonstrated below.

Noting that the function $f_n(p) = np(1-p)^n$ increases for $p \in (0, 1/(n+1))$ and decreases for $p \in (1/(n+1), 1)$, for any n

$$\begin{aligned} f_n(p_k) &\le f_n(p_{k^*}), \quad k \le k^* \\ f_n(p_k) &< f_n(p_{k^*}), \quad k \ge k^* + 1. \end{aligned} \tag{6.14}$$

For a given n, rewrite each p_k in terms of p_{k^*}, and therefore in terms of n and $c(n)$:

$$p_{k^*+i} = e^{-i}\frac{c(n)}{n} \quad \text{and} \quad p_{k^*-j} = e^{j}\frac{c(n)}{n}$$

for all appropriate positive integers i and j. Therefore,

$$\begin{aligned} f_n(p_{k^*+i}) &= ne^{-i}\frac{c(n)}{n}\left(1 - e^{-i}\frac{c(n)}{n}\right)^n \\ &= \frac{c(n)}{e^i}\left(1 - \frac{c(n)}{ne^i}\right)^n, \\ f_n(p_{k^*-j}) &= ne^{j}\frac{c(n)}{n}\left(1 - e^{j}\frac{c(n)}{n}\right)^n \\ &= c(n)e^{j}\left(1 - \frac{c(n)e^{j}}{n}\right)^n, \end{aligned}$$

and

$$
\begin{aligned}
\tau_n &= \sum_{k\le k^*-1} f_n(p_k) + f_n(p_{k^*}) + \sum_{k\ge k^*+1} f_n(p_k) \\
&= c(n)\sum_{j=1}^{k^*-1} e^j \left(1-\frac{c(n)e^j}{n}\right)^n + c(n)\left(1-\frac{c(n)}{n}\right)^n \\
&+ c(n)\sum_{i=1}^{\infty} e^{-i}\left(1-\frac{c(n)}{ne^i}\right)^n .
\end{aligned} \tag{6.15}
$$

Next, it is to show that $c(n)$ oscillates perpetually over the interval $(n/(n+1), e)$, which approaches $[1, e)$ as n increases indefinitely. This is so because, since k^* is defined by (6.10),

$$
\frac{c(n)}{n}e^{-1} \le \frac{1}{n+1} \le \frac{c(n)}{n}
$$

or

$$
e^{-1} < \frac{n}{n+1} \le c(n) \le \frac{n}{n+1}e < e. \tag{6.16}
$$

At this point, the fact $c(n) \in [1, e)$ is established. What remains to be shown is that $c(n)$ oscillates perpetually in n. Toward that end, consider $k^*(n)$ as a mapping, which maps every positive integer value of $n \in \mathbb{N}$ to a positive integer value of $k^* \in \mathbb{N}$. The inverse of $k^*(n)$ maps every integer value $k^* \in \mathbb{N}$ to a set as in (6.11). Let

$$
\mathbb{N} = \cup\{n_{k^*}, n_{k^*}+1, \dots, n_{k^*+1}-1\} \tag{6.17}
$$

where the union is over all possible integer values of k^*. (In fact, the smallest k^* possible is $k^* = 1$ for $n_{k^*} = 1$. For this case, $p_1 = 1 - e^{-1} \approx 0.6321, p_2 = p_1 e^{-1} \approx 0.2325$, and $(1+1)^{-1} = 0.5$; therefore, $k^* = 1$ and $n_1 = 1$.)

Observe the following three facts:

1) $c(n_{k^*}) < c(n_{k^*}+1) < \cdots < c(n_{k^*+1}-1)$. This is so because of (6.13): all of them sharing the same k^* and therefore the same p_{k^*}. Furthermore, the increments of increase are all identical, namely, p_{k^*}.
2) Consider $\{n_{k*}; k^* \ge 1\}$ where n_{k*} is the smallest integer value in each partitioning set of (6.17). One has $c(n_{k^*}) = n_{k^*}p_{k^*} \to 1$. This is so because $1/n_{k^*} > p_{k^*} \ge 1/(n_{k^*}+1)$ or

$$
1 - p_{k^*} \le n_{k^*}p_{k^*} < 1, \tag{6.18}
$$

which implies that $n_{k^*}p_{k^*}$ for all sufficiently large k^* (or equivalently sufficiently large n_{k^*} or sufficiently large n),

$$
c(n_{k^*}) = n_{k^*}p_{k^*} \in (1-\varepsilon, 1) \tag{6.19}
$$

where $\varepsilon > 0$ is an arbitrarily small real value.

3) Consider $\{n_{k*+1}-1; k^* \geq 1\}$ where $n_{k*+1}-1$ is the greatest integer value in each partitioning set of (6.17). One has

$$c(n_{k^*+1}-1) = (n_{k^*+1}-1)p_{k^*} \to e.$$

This is so because

$$\begin{aligned} p_{k^*} &= p_{k^*+1}e \\ &= \frac{n_{k^*+1}-1}{n_{k^*+1}-1}p_{k^*+1}e \\ &= \frac{1}{n_{k^*+1}-1}\left(\frac{n_{k^*+1}-1}{n_{k^*+1}}\right)(n_{k^*+1}p_{k^*+1})e, \end{aligned}$$

and therefore by (6.18)

$$c(n_{k^*+1}-1) = \left(\frac{n_{k^*+1}-1}{n_{k^*+1}}\right)(n_{k^*+1}p_{k^*+1})e \to e.$$

At this point, it has been established that the range of $c(n)$ for $n \geq n_0$, where n_0 is any positive integer, covers the entire interval $[1, e)$.

Part 2: Part 1 of Lemma 6.6. Noting that $e^{-1} \leq c(n) \leq e$ (see (6.16)) and that $1-p \leq e^{-p}$ for all $p \in [0, 1]$, the desired result follows the argument below.

$$\begin{aligned} \tau_n &= c(n)\sum_{j=1}^{k^*-1} e^j \left(1-\frac{c(n)e^j}{n}\right)^n + c(n)\left(1-\frac{c(n)}{n}\right)^n \\ &\quad + c(n)\sum_{j=1}^{\infty} e^{-j}\left(1-\frac{c(n)}{ne^j}\right)^n \\ &\leq e\sum_{j=1}^{k^*-1} e^j\left(1-\frac{e^{j-1}}{n}\right)^n + e\left(1-\frac{e^{-1}}{n}\right)^n + e\sum_{j=1}^{\infty} e^{-j}\left(1-\frac{1}{ne^{j+1}}\right)^n \\ &\leq e\sum_{j=1}^{k^*-1} e^j e^{-e^{j-1}} + e\sum_{j=0}^{\infty} e^{-j}e^{-e^{-(j+1)}} \\ &\leq e^2\sum_{j=1}^{k^*-1} e^{j-1}e^{-e^{j-1}} + e^2\sum_{j=0}^{\infty} e^{-(j+1)}e^{-e^{-(j+1)}} \\ &< e^2\sum_{j=0}^{\infty} e^j e^{-e^j} + e^2\sum_{j=1}^{\infty} e^{-j}e^{-e^{-j}} := u. \end{aligned}$$

Part 3: Part 2 of Lemma 6.6. Consider, for any fixed $c > 0$,

$$\tau_n^* = c\sum_{j=1}^{k^*-1} e^j\left(1-\frac{ce^j}{n}\right)^n + c\left(1-\frac{c}{n}\right)^n + c\sum_{j=1}^{\infty} e^{-j}\left(1-\frac{c}{ne^j}\right)^n.$$

By dominated convergence theorem,

$$\tau(c) := \lim_{n\to\infty} \tau_n^* = c\sum_{j=0}^{\infty} e^j e^{-ce^j} + c\sum_{j=1}^{\infty} e^{-j} e^{-ce^{-j}},$$

and $\tau(c)$ is a nonconstant function in c on $[1, e]$.

Noting that as k^* increases (or equivalently, n_{k^*} increases or n increases),

$$1 \leftarrow c(n_{k^*}) < c(n_{k^*} + 1) < \cdots < c(n_{k^*+1} - 1) \rightarrow e$$

and that the increment between two consecutive terms $p_{k^*} \rightarrow 0$, $c(n)$ visits any arbitrarily small closed interval $[a, b] \subset [1, e]$ infinitely often, and therefore there exists for each such interval a subsequence $\{n_l; l \geq 1\}$ of $\mathbb{N}$ such that $c(n_l)$ converges, that is, $c(n_l) \rightarrow \theta$ for some $\theta \in [a, b]$. Since $\tau(c)$ is a nonconstant function on $[1, e]$, there exist two nonoverlapping closed intervals, $[a_1, b_1]$ and $[a_2, b_2]$ in $[1, e]$, satisfying

$$\max_{a_1\leq c\leq b_1} \tau(c) < \min_{a_2\leq c\leq b_2} \tau(c),$$

such that there exist two subsequences of $\mathbb{N}$, said $\{n_l; l \geq 1\}$ and $\{n_m; m \geq 1\}$, such that $c(n_l) \rightarrow \theta_1$ for some $\theta_1 \in [a_1, b_1]$ and $c(n_m) \rightarrow \theta_2$ for some $\theta_2 \in [a_2, b_2]$.

Consider the limit of τ_n along $\{n_l; l \geq 1\}$, again by dominated convergence theorem, as $n_l \rightarrow \infty$,

$$\tau_{n_l} = \left[c(n_l)\sum_{j=0}^{k^*-1} e^j \left(1 - \frac{c(n_l)e^j}{n}\right)^n + c(n_l)\sum_{j=1}^{\infty} e^{-j}\left(1 - \frac{c(n_l)}{ne^j}\right)^n\right]$$

$$\rightarrow \theta_1 \sum_{j=0}^{\infty} e^j e^{-\theta_1 e^j} + \theta_1 \sum_{j=1}^{\infty} e^{-j} e^{-\theta_1 e^{-j}} = \tau(\theta_1).$$

A similar argument gives $\lim_{n_m\to\infty}\tau_{n_m} = \tau(\theta_2)$, but $\tau(\theta_1) \neq \tau(\theta_2)$ by construction, and hence $\lim_{n\to\infty}\tau_n$ does not exist. □

6.5 Exercises

1 Show that $\zeta_n = \sum_{k\geq 1} p_k(1 - p_k)^n \rightarrow 0$ as $n \rightarrow \infty$ for any probability distribution $\{p_k\}$ on $\mathscr{X}$.

2 Prove Part 2 of Lemma 6.2, that is, for any real number $x \in (0, 1/2)$,

$$\frac{1}{1-x} < 1 + 2x.$$

3 Prove Lemma 6.5.

4 Use the $\varepsilon - N$ language to show that

a) $a_n = 1/n$ converges to 0; and
b) $b_n = (n+1)/n^2$ converges to 0.

5 Use the $\varepsilon - N$ language to show that
a) $a_n = \sum_{i=1}^{n} 3/10^i$ converges to $1/3$; and
b) $b_n = 5 + \frac{(-1)^{n+1}2}{n}$ converges to 5.

6 Show that
a) $\lim_{n\to\infty} \left(1 + \frac{1}{n}\right)^n = e$; and
b) $\lim_{n\to\infty} \left(1 - \frac{1}{n}\right)^n = 1/e$.

7 If $\{a_n\}$ is a convergent sequence, then every subsequence of that sequence converges to the same limit.

8 If $\{a_n\}$ is a sequence such that every possible subsequence extracted from that sequences converge to the same limit, then the original sequence also converges to that limit.

9 Show that
a) the sequence $\{a_n = \sin\ n\pi; n = 1, 2, \cdots\}$ does not converge; and
b) the sequence $\{a_n = \sin\ \ln n; n = 1, 2, \cdots\}$ does not converge.

10 Give a proof of the Bolzano–Weierstrass theorem: every bounded sequence has a convergent subsequence. (Hint: Let a and b be the lower and upper bounds of a bounded sequence. Consider the two intervals, $[a.(a+b)/2]$ and $[(a+b)/2, b]$.)

11 The Squeeze rule: let $\{a_n\}$, $\{b_n\}$, and $\{x_n\}$ be three sequences such that
a) $a_n \le x_n \le b_n$ for every $n \in \mathbb{N}$; and
b) $\lim_{n\to\infty} a_n = \lim_{n\to\infty} b_n = L$, for some finite constant L, then $\lim_{n\to\infty} x_n = L$. Use the $\varepsilon - N$ language to prove the above-mentioned Squeeze rule.

12 Show that every convergent sequence is bounded.

13 Let $\{a_n\}$ be a positive sequence such that $\lim_{n\to\infty} a_n = L > 1$. Prove that $\{a_n^n\} \to \infty$. (Hint: $\lim_{n\to\infty} x^n = \infty$ for any number x satisfying $x > 1$.)

14 Let

$$a_n = (-1)^n + \frac{1}{n},$$

and consider the sequence $\{a_n; n = 1, 2, \dots\}$. Find $\limsup a_n$ and $\liminf a_n$.

15 Let S be a nonempty subset of the real line R. Prove the following statements.

a) Suppose that S is bounded above, that is, there exists a constant M such that $s \leq M$ for every $s \in S$. Then $\alpha < \sup S$ if and only if there exists an $a \in S$ such that $a > \alpha$.

b) Suppose that S is bounded below, that is, there exists a constant m such that $s \geq m$ for every $s \in S$. Then $\beta > \inf S$ if and only if there exists an $b \in S$ such that $b < \beta$.

16 For every sequence $\{a_n; n = 1, 2, \cdots\}$, let $\sum_{n=1}^{\infty} a_n = \lim_{n\to\infty} \sum_{i=1}^{n} a_i$.

a) Find $\sum_{n=1}^{\infty} e^{-n}$.

b) Find $\sum_{n=10}^{\infty} \frac{1}{n(n+1)}$.

17 A series $\sum_{i=1}^{n} a_n$ is said to be convergent if

$$\sum_{i=1}^{\infty} a_n = \lim_{n\to\infty} \sum_{i=1}^{n} a_n = a$$

for some finite constant a. If such an a does not exists, then the series is said to be divergent.

a) Show that $\sum_{i=1}^{n} (-1)^n$ is divergent.

b) Show that $\sum_{i=1}^{n} n$ is divergent.

18 Show that $\sum_{n=1}^{\infty} \frac{(-1)^n}{n} = \ln 2$. (This series is known as the alternating harmonic series. Hint: Consider the Taylor expansion of $f(x) = \ln x$ at $x_0 = 1$.)

19 Show that the harmonic series, $\sum_{n=1}^{\infty} \frac{1}{n}$, diverges to infinity, that is,

$$\lim_{n\to\infty} \sum_{n=1}^{\infty} \frac{1}{n} = \infty.$$

(Hint: Consider the inequalities below.

$$\sum_{n=1}^{\infty} \frac{1}{n} = \frac{1}{1} + \frac{1}{2} + \frac{1}{3} + \frac{1}{4} + \frac{1}{5} + \frac{1}{6} + \frac{1}{7} + \frac{1}{8} + \frac{1}{9} + \cdots$$
$$> \frac{1}{1} + \frac{1}{2} + \frac{1}{4} + \frac{1}{4} + \frac{1}{8} + \frac{1}{8} + \frac{1}{8} + \frac{1}{8} + \frac{1}{16} + \cdots .$$

)

20 Show that

$$\frac{1}{n} + \ln n \leq \sum_{n=1}^{k} \frac{1}{n} \leq 1 + \ln n.$$

(Hint: For any decreasing function $f(x)$,

$$\int_a^{b+1} f(x)dx \le \sum_{i=a}^{b} f(i) \le \int_{a-1}^{b} f(x)dx,$$

where a and b are positive integers satisfying $a < b$.)

21 Find the limit of $\frac{1}{n}\sum_{k=2}^{n}\frac{1}{\ln k}$.
(Hint:

$$\frac{1}{n}\sum_{k=2}^{n}\frac{1}{\ln k} = \frac{1}{n}\sum_{k=2}^{\lfloor \ln n\rfloor}\frac{1}{\ln k} + \frac{1}{n}\sum_{k=\lfloor \ln n\rfloor+1}^{n}\frac{1}{\ln k}$$
$$< \frac{\lfloor \ln n\rfloor - 1}{n\ln 2} + \frac{n - \lfloor \ln n\rfloor}{n\ln\lfloor \ln n\rfloor}.$$

)

22 Show that the notion of dominance by Definition 6.2 is transitive, that is, if $\mathbf{p}_1$ dominates $\mathbf{p}_2$ and $\mathbf{p}_2$ dominates $\mathbf{p}_3$, then $\mathbf{p}_1$ dominates $\mathbf{p}_3$.

7

Estimation of Tail Probability

The remarkable Turing's formula suggests that certain useful knowledge of the tail of a distribution $\{p_k; k \geq 1\}$ on $\mathscr{X} = \{\ell_k; k \geq 1\}$ may be extracted from an *iid* sample. This chapter demonstrates how such information could be used to estimate parametric probability models in, and only in, the tail of the underlying distribution, possibly beyond data range.

7.1 Introduction

Consider a discrete probability distribution $\{p_k; k \geq 1\}$ for which there exists an unknown integer $k_0 > 0$ such that, for each and every $k \geq k_0$,

$$p_k = Ck^{-\lambda} \tag{7.1}$$

where $C > 0$ and $\lambda > 1$ are unknown parameters. Suppose the primary problem of interest is to estimate C and λ with an *iid* sample of size n.

There are two somewhat unusual characteristics about this problem. First, the parametric model is only imposed on the tail. For any $k < k_0$, p_k is not restricted to any particular parametric form. Second, the threshold k_0 is unknown, and therefore for any particular observed value of the underlying random element, say $X = \ell_k$, it is not known whether $k \geq k_0$. This model will be referred to as the power, or the Pareto, tail model hereafter.

In the existing literature, there exists a well-known similar problem in the continuous domain where the underlying random variable X has a continuous distribution function $F(x)$ such that $1 - F(x) = Cx^{-\lambda}$ for $x \geq x_0$ and $F(x)$ is unspecified for $x < x_0$ where x_0, $C > 0$ and $\lambda > 0$ are unknown parameters. The primary interest of this problem is also to estimate C and λ. The existing methodological approaches to this problem are largely based on extreme value theory and domains of attraction. The reference list on this problem is extensive. Some of the more commonly cited publications include Hill (1975), Haeusler and Teugels (1985), Hall and Welsh (1985), Smith (1987), Hall and Weissman (1997), and Pickands (1975). However, while many parallels can be

Statistical Implications of Turing's Formula, First Edition. Zhiyi Zhang.

drawn between this problem and the one on hand, it must be said that the two problems are fundamentally different. The extreme value theory that supports the approaches to the continuous tail problem has little relevance in the problem under current consideration due to its discrete nature. A solution to the current problem calls for an entirely different approach.

In a similar spirit, a more general parametric tail model may be entertained. Consider a discrete probability distribution $\{p_k; k \geq 1\}$ for which there exists an unknown integer $k_0 > 0$ such that, for each and every $k \geq k_0$,

$$p_k = p(k; \boldsymbol{\theta}) \tag{7.2}$$

where $\boldsymbol{\theta}$ is a vector of several parameters.

Consider a multinomial distribution with its countably infinite number of prescribed categories indexed by $\mathbb{N} = \{k; k \geq 1\}$ and its category probabilities denoted by $\{p_k; k \geq 1\}$, satisfying $0 < p_k < 1$ for all k and $\sum_{k\geq 1} p_k = 1$. This model will be referred to as the general model hereafter. Let the category counts in an *iid* sample of size n from the underlying population be denoted by $\{Y_k; k \geq 1\}$ and its observed values by $\{y_k; k \geq 1\}$. For a given sample, there are at most n nonzero y_ks.

For every integer r, $1 \leq r \leq n$, recall the following three important random variables from Chapter 1,

$$\begin{aligned}
N_r &= \sum 1[Y_k = r], \\
T_r &= \frac{r}{n - r + 1} N_r, \\
\pi_{r-1} &= \sum p_k 1[Y_k = r - 1],
\end{aligned}$$

as given in (1.1), (1.27), and (1.2).

For any given positive integer R, consider the following two random vectors of length R,

$$\begin{aligned}
\boldsymbol{T} &= (T_1, \ldots, T_R)^\tau \\
\boldsymbol{\pi} &= (\pi_0, \ldots, \pi_{R-1})^\tau.
\end{aligned}$$

In Chapter 1, it is argued that $\boldsymbol{T}$ is a reasonable estimator of $\boldsymbol{\pi}$, at least in the sense of Theorem 1.11. In this chapter, it is argued that Theorem 1.11 may support parametric tail probability models such as (7.2) in general and (7.1) in specific.

At a first glance, it would seem a bit far-fetched that a nonparametric result like Theorem 1.11 could support a parametric model in the tail, let alone a distant tail possibly beyond data range. Yet it is the very core implication of Turing's formulae. It is perhaps instructive to give a moment of thought to why such an undertaking is possible.

Consider the following disjoint random subsets of $\mathbb{N} = \{k; k = 1, 2, \dots\}$, for an *iid* sample of size n and a given positive integer R,

$$\begin{aligned}
\mathbb{K}_0 &= \{k : Y_k = 0\}\\
\mathbb{K}_1 &= \{k : Y_k = 1\}\\
&\vdots\\
\mathbb{K}_{R-1} &= \{k : Y_k = R-1\}
\end{aligned}$$

and their union

$$\mathbb{K}_{tail} = \cup_{r=0}^{R-1}\mathbb{K}_r. \tag{7.3}$$

For simplicity, assuming $p_k > 0$ for every $k \in \mathbb{N}$, $\mathbb{K}_{tail}$ in (7.3) may be thought of as a tail of the index set $\mathbb{N} = \{k; k \geq 1\}$ because the following key fact that

$$\lim_{n\to\infty} \mathrm{P}\left(\mathbb{K}_{tail} \subset \{k; k \geq k_0\}\right) = 1 \tag{7.4}$$

where k_0 is the unknown positive integer threshold in the parametric tail model in (7.1) or (7.2) (see Exercise 1).

Several summarizing remarks can now be given:

1) The summands in each π_r of $\boldsymbol{\pi}$ are subject to the event $\{Y_k = r\}$ for a fixed r. This implies that category probabilities p_k included in π_r are dynamically shifting into the tail in probability as n increases. Therefore, for a fixed R, the entire panel $\boldsymbol{\pi}$ shifts into the tail in probability. As a result, in probability, all nonzero summands in $\boldsymbol{\pi}$ will eventually have indices satisfying $k \geq k_0$. Under the power tail model, as n increases, $\boldsymbol{\pi}$ will only contain p_ks such that $p_k = Ck^{-\lambda}$, or whatever the parametric tail model may be. This is a partial reason why a consistent estimator of the tail is possible.
2) The equation in (7.4) holds regardless whether k_0 is known or not. This fact gives much flexibility to the approach and bypasses the necessity of knowing (or estimating) k_0 *a priori*.
3) By Turing's formulae in general and Theorem 1.11 in specific, $\boldsymbol{T}$ chases $\boldsymbol{\pi}$ while $\boldsymbol{\pi}$ chases the tail. In fact, $\mathbb{K}_{tail}$ of (7.3) may be thought of as a "window on tail." It is in this sense that the subsequently described estimator of the tail is said to be in Turing's perspective.

In the next several sections, the Pareto tail model in (7.1) is studied in detail. A distributional relationship between the observable $\boldsymbol{T}$ and the observable functions of C and λ, namely $\boldsymbol{\pi}$, is explored; an asymptotic distribution of $\boldsymbol{T} - \boldsymbol{\pi}$ is established; an estimator $(\hat{C}, \hat{\lambda})$ of (C, λ) based on the likelihood of the said asymptotic distribution is defined; the consistency of the estimator is established; and a large sample confidence region for (C, λ) is derived.

7.2 Estimation of Pareto Tail

Let $\mathbf{p} = \{p_k; k \geq 1\}$ be a probability distribution on $\mathscr{X}$ and $g(n, \delta) = n^{1-2\delta}$ where $\delta > 0$ is a constant. The following is a condition on $\mathbf{p}$.

Condition 7.1 *There exists a $\delta \in (0, 1/4)$ such that as $n \to \infty$,*

1) $\dfrac{g^2(n,\delta)}{n^2}\mathrm{E}(N_r) \to \dfrac{c_r}{r!} \geq 0,$

2) $\dfrac{g^2(n,\delta)}{n^2}\mathrm{E}(N_{r+1}) \to \dfrac{c_{r+1}}{(r+1)!} \geq 0,$ *and*

3) $c_r + c_{r+1} > 0$

where r is a positive integer.

Let

$$\begin{aligned}
\sigma_r^2 &= r^2\mathrm{E}(N_r) + (r+1)r\mathrm{E}(N_{r+1}),\\
\rho_r(n) &= -r(r+1)\mathrm{E}(N_{r+1})/(\sigma_r\sigma_{r+1}),\\
\rho_r &= \lim_{n\to\infty}\rho_r(n),\\
\hat{\sigma}_r^2 &= r^2N_r + (r+1)rN_{r+1}, \text{ and}\\
\hat{\rho}_r &= \hat{\rho}_r(n) = -r(r+1)N_{r+1}/\sqrt{\hat{\sigma}_r^2\hat{\sigma}_{r+1}^2}.
\end{aligned}$$

The following two lemmas are re-statements of Theorems 1.11 and 1.12, and they are given here for easy reference.

Lemma 7.1 *For any positive integer R, if Condition 7.1 holds for every r, $1 \leq r \leq R-1$, then*

$$n\left(\frac{T_1 - \pi_0}{\sigma_1}, \ldots, \frac{T_R - \pi_{R-1}}{\sigma_R}\right)^\tau \xrightarrow{L} MVN(0, \Sigma)$$

where $\Sigma = (a_{i,j})$ is a $R \times R$ covariance matrix with all the diagonal elements being $a_{r,r} = 1$ for $r = 1, \ldots, R$, the super-diagonal and the sub-diagonal elements being $a_{r,r+1} = a_{r+1,r} = \rho_r$ for $r = 1, \ldots, R-1$, and all the other off-diagonal elements being zeros.

Let $\hat{\Sigma}$ be the resulting matrix of Σ with ρ_r replaced by $\hat{\rho}_r(n)$ for all r. Let $\hat{\Sigma}^{-1}$ denote the inverse of $\hat{\Sigma}$ and $\hat{\Sigma}^{-1/2}$ denote any $R \times R$ matrix satisfying $\hat{\Sigma}^{-1} = \left(\hat{\Sigma}^{-1/2}\right)^\tau\hat{\Sigma}^{-1/2}$.

Lemma 7.2 *For any positive integer R, if Condition 7.1 holds for every r, $1 \leq r \leq R$, then*

$$n\hat{\Sigma}^{-1/2}\left(\frac{T_1 - \pi_0}{\hat{\sigma}_1}, \ldots, \frac{T_R - \pi_{R-1}}{\hat{\sigma}_R}\right)^\tau \xrightarrow{L} MVN(0, I_{R\times R}).$$

Lemma 7.3 *Under the power tail model (7.1), Condition 7.1 holds for every r,* $1 \le r \le R-1$.

Proof: Letting $\delta = (4\lambda)^{-1}$, it can be verified that

$$n^{r-2}g^2(n,\delta)\sum_{k\ge1} p_k^r e^{-np_k} \to c_r > 0$$

for every integer $r > 0$. □

See Exercise 2.

Corollary 7.1 *Under the power tail model in (7.1), the results of both Lemmas 7.1 and 7.2 hold.*

Under the power tail model in (7.1), for every $r > 0$, let every p_k in π_{r-1} be replaced by $p_k = Ck^{-\lambda}$ and denote the resulting expression as π^*_{r-1}, that is,

$$\pi^*_{r-1} = C\sum_{k\ge1} k^{-\lambda}1[Y_k = r-1].$$

Then

$$\pi_{r-1} - \pi^*_{r-1} = \sum_{k=1}^{k_0-1} 1[Y_k = r-1]\left(p_k - Ck^{-\lambda}\right),$$

and the sum is of finite terms. Since, for every k, both $E(1[Y_k = r-1])$ and $\mathrm{Var}(1[Y_k = r-1])$ converge to zero exponentially in sample size n, $E(g(n,\delta)1[Y_k = r-1])$ and $\mathrm{Var}(g(n,\delta)1[Y_k = r-1])$ converge to zero, which implies that $g(n,\delta)1[Y_k = r-1] \xrightarrow{p} 0$, which in turn implies that

$$g(n,\delta)(\pi_{r-1} - \pi^*_{r-1}) \xrightarrow{p} 0$$

for every r. This argument leads to the following two lemmas.

Lemma 7.4 *Under the power tail model in (7.1),*

$$n\left(\frac{T_1 - \pi^*_0}{\sigma_1}, \ldots, \frac{T_R - \pi^*_{R-1}}{\sigma_R}\right)^\tau \xrightarrow{L} MVN(0, \Sigma)$$

where Σ *is as in Lemma 7.1.*

Lemma 7.5 *Under the power tail model in (7.1),*

$$n\hat{\Sigma}^{-1/2}\left(\frac{T_1 - \pi^*_0}{\hat{\sigma}_1}, \ldots, \frac{T_R - \pi^*_{R-1}}{\hat{\sigma}_R}\right)^\tau \xrightarrow{L} MVN(0, I_{R\times R}) \qquad (7.5)$$

where $\hat{\Sigma}^{-1/2}$ *is as in Lemma 7.2.*

The asymptotic likelihood given by (7.5) depends on the two model parameters, λ and C, only via

$$\pi^*_{r-1} = \pi^*_{r-1}(C, \lambda) = C \sum_{k \geq 1} 1[Y_k = r-1]k^{-\lambda}, \tag{7.6}$$

$r = 1, \ldots, R$.

Definition 7.1 *If there exists a pair of values, $C = \hat{C}$ and $\lambda = \hat{\lambda}$, which maximizes the likelihood given by (7.5) in a neighborhood of $C > 0$ and $\lambda > 1$, then $(\hat{C}, \hat{\lambda})$ is said to be an asymptotic maximum likelihood estimator (AMLE) of (C, λ).*

Let $R = 2$, and denote, hereafter, $(\hat{C}, \hat{\lambda})$ as *AMLE* by Definition 7.1 but with $R = 2$. The following is a corollary of Theorem 7.5.

Corollary 7.2 *Under the model in (7.1),*

$$\hat{\Sigma}_2^{-1/2}\left(T_1 - \pi_0^*, T_2 - \pi_1^*\right)^\tau \xrightarrow{L} MVN(0, I_{2\times 2}) \tag{7.7}$$

where

$$\hat{\Sigma}_2^{-1/2} = \frac{n}{2\sqrt{1-\hat{\rho}_1^2}} \begin{pmatrix} \sqrt{1-\hat{\rho}_1} + \sqrt{1+\hat{\rho}_1} & \sqrt{1-\hat{\rho}_1} - \sqrt{1+\hat{\rho}_1} \\ \sqrt{1-\hat{\rho}_1} - \sqrt{1+\hat{\rho}_1} & \sqrt{1-\hat{\rho}_1} + \sqrt{1+\hat{\rho}_1} \end{pmatrix} \times \begin{pmatrix} 1/\hat{\sigma}_1 & 0 \\ 0 & 1/\hat{\sigma}_2 \end{pmatrix}. \tag{7.8}$$

Given a sample, an *AMLE* minimizes

$$\begin{aligned} ll_n &= ll_n(C, \lambda) \\ &:= \left(T_1 - \pi_0^*, T_2 - \pi_1^*\right) \hat{\Sigma}_2^{-1} \left(T_1 - \pi_0^*, T_2 - \pi_1^*\right)^\tau. \end{aligned}$$

The properties of *AMLE*s may be examined by the statistical behavior of an appropriately proportioned version of ll_n, namely l_n as defined in (7.9). Let

$$g^2(n, \delta)\Omega_n = \hat{\Sigma}_2^{-1}.$$

Recall that, under the power tail model,

$$g(n, \delta) = n^{1-1/(2\lambda_0)}$$

where λ_0 is the true value of λ. It can be verified that

$$\Omega_n = (\omega_{ij}(n)) \xrightarrow{p} \Omega$$

where $\Omega = (\omega_{ij})$ is some positive definite 2×2 matrix. The logarithmic likelihood given by (7.5) is negatively proportional to

$$\begin{aligned} l_n &= l_n(C, \lambda) \\ &:= n^{-1/\lambda_0} n^{2-1/\lambda_0} (ll_n(C, \lambda)/n^{2-1/\lambda_0}) \\ &= n^{2-2/\lambda_0} (T_1 - \pi_0^*, T_2 - \pi_1^*) \Omega_n (T_1 - \pi_0^*, T_2 - \pi_1^*)^\tau. \end{aligned} \tag{7.9}$$

The examination of $l_n(C, \lambda)$, a smooth random surface defined on the parameter space

$$\Theta = \{(C, \lambda); C > 0 \text{ and } \lambda > 1\},$$

is carried out separately on three subsets of Θ. Let $\Theta = \cup_{s=1}^{3} \Theta_s$ where

1) $\Theta_1 = \{(C, \lambda); \lambda \geq \lambda_0\}$
2) $\Theta_2 = \left\{(C, \lambda); \max\left(\lambda_0 - \frac{1}{2\lambda_0}, 1\right) < \lambda < \lambda_0\right\}$
3) $\Theta_3 = \left\{(C, \lambda); 1 < \lambda \leq \lambda_0 - \frac{1}{2\lambda_0}\right\}$ if $\lambda_0 - \frac{1}{2\lambda_0} > 1$, or Θ_3 is an empty set otherwise.

$\{\Theta_1, \Theta_2, \Theta_3\}$ is a partition of the parameter space along the interval $\lambda \in (1, \infty)$ with two dividing points: $\lambda_0 - 1/(2\lambda_0)$ and λ_0, provided that the first point is on the right side of 1. The necessity of examining l_n separately on these sets lies in the qualitative difference in its respective behavior on the sets. As it turns out, l_n is well behaved on Θ_1 as all of its components are convergent in some sense; whereas on Θ_2 and Θ_3, many of the components of l_n are divergent, and consequently l_n requires a different treatment on each of these sets.

This section ends with a useful lemma below.

Lemma 7.6 *Let Ω be a positive definite 2×2 matrix. For each r, $r = 1, 2$, let $a_r(n)$ and $A_r(n)$ be sequences of real values satisfying $a_r(n) \to 0$ and $A_r(n) \to -\infty$. Then*

$$\left(A_1(n), A_2(n))\Omega(A_1(n), A_2(n))\right)^\tau + \left(a_1(n), a_2(n))\Omega(A_1(n), A_2(n))\right)^\tau \to +\infty.$$

Proof: Since Ω is positive definite, $\Omega = P^\tau \Lambda P$ where P is orthogonal and Λ is diagonal matrix with two positive elements λ_1 and λ_2. Let

$$(Q_1, Q_2)^\tau = P(A_1(n), A_2(n))^\tau \quad \text{and} \quad (q_1, q_2)^\tau = P(a_1(n), a_2(n))^\tau.$$

Since $Q_1^2 + Q_2^2 = A_1^2(n) + A_2^2(n) \to +\infty$, $|Q_1| + |Q_2| \to +\infty$ and hence $\lambda_1|Q_1| + \lambda_2|Q_2| \to +\infty$. Also $P(a_1(n), a_2(n))^\tau = (q_1, q_2)^\tau \to (0, 0)^\tau$. On the other hand, for a sufficiently large n,

$$\begin{aligned} &(A_1(n), A_2(n))\Omega(A_1(n), A_2(n)) + (a_1(n), a_2(n))\Omega(A_1(n), A_2(n)) \\ &= \lambda_1 Q_1^2 + \lambda_2 Q_2^2 + \lambda_1 q_1 Q_1 + \lambda_2 q_2 Q_2 \end{aligned}$$

$$
\begin{aligned}
&\geq \lambda_1 Q_1^2 1[|Q_1| > 1] + \lambda_1 Q_1^2 1[|Q_1| \leq 1] \\
&\quad - \frac{1}{3}\lambda_1 |Q_1| 1[|Q_1| > 1] - \frac{1}{3}\lambda_1 |Q_1| 1[|Q_1| \leq 1] \\
&\quad + \lambda_2 Q_2^2 1[|Q_2| > 1] + \lambda_2 Q_2^2 1[|Q_2| \leq 1] \\
&\quad - \frac{1}{3}\lambda_2 |Q_2| 1[|Q_2| > 1] - \frac{1}{3}\lambda_2 |Q_2| 1[|Q_2| \leq 1] \\
&= \lambda_1 |Q_1| 1[|Q_1| > 1] \left(|Q_1| - \frac{1}{3}\right) + \lambda_1 |Q_1| 1[|Q_1| \leq 1] \left(|Q_1| - \frac{1}{3}\right) \\
&\quad + \lambda_2 |Q_2| 1[|Q_2| > 1] \left(|Q_2| - \frac{1}{3}\right) + \lambda_2 |Q_2| 1[|Q_2| \leq 1] \left(|Q_2| - \frac{1}{3}\right) \\
&\geq \lambda_1 |Q_1| 1[|Q_1| > 1] \frac{2}{3} - \frac{1}{3}\lambda_1 |Q_1| 1[|Q_1| \leq 1] \\
&\quad + \lambda_2 |Q_2| 1[|Q_2| > 1] \frac{2}{3} - \frac{1}{3}\lambda_2 |Q_2| 1[|Q_2| \leq 1] \\
&\geq \frac{2\lambda_1}{3} |Q_1| 1[|Q_1| > 1] - \frac{1}{3}\lambda_1 + \frac{2\lambda_2}{3} |Q_2| 1[|Q_2| > 1] - \frac{1}{3}\lambda_2 \\
&= \frac{2\lambda_1}{3} |Q_1| - \frac{2\lambda_1}{3} |Q_1| 1[|Q_1| \leq 1] - \frac{1}{3}\lambda_1 + \frac{2\lambda_2}{3} |Q_2| \\
&\quad - \frac{2\lambda_2}{3} |Q_2| 1[|Q_2| \leq 1] - \frac{1}{3}\lambda_2 \\
&\geq \frac{2}{3}(\lambda_1 |Q_1| + \lambda_2 |Q_2|) - (\lambda_1 + \lambda_2) \to +\infty.
\end{aligned}
$$

□

7.3 Statistical Properties of *AMLE*

The uniqueness and consistency of the *AMLE* of (C, λ), $(\hat{C}, \hat{\lambda})$, are respectively established in this section.

Theorem 7.1 *Under the power tail model in (7.1), the AMLE of (C, λ) uniquely exists for a sufficiently large n in a neighborhood of (C_0, λ_0), where C_0 and λ_0 are the true values of the parameters.*

The proof of Theorem 7.1 is based on the fact that the probability that the Hessian matrix of $l_n(C, \lambda)$ at the true parameter values (C_0, λ_0) is positive definite converges to 1 as $n \to \infty$. To establish the said fact, Lemma 7.7 is needed.

Toward stating Lemma 7.7, several notations are needed. Let

$$
\begin{aligned}
u_n &= \sum_{k\geq 1} 1[Y_k = 0] k^{-\lambda}, \\
v_n &= \sum_{k\geq 1} 1[Y_k = 0] (\ln k) k^{-\lambda}, \\
w_n &= \sum_{k\geq 1} 1[Y_k = 0] (\ln k)^2 k^{-\lambda},
\end{aligned}
$$

$$U_n = \sum_{k \geq 1} 1[Y_k = 1]k^{-\lambda},$$
$$V_n = \sum_{k \geq 1} 1[Y_k = 1](\ln k)k^{-\lambda}, \text{ and}$$
$$W_n = \sum_{k \geq 1} 1[Y_k = 1](\ln k)^2 k^{-\lambda}.$$

It can be verified that

$$\begin{aligned}
\frac{\partial^2 l_n}{\partial C^2} &= 2n^{2-2/\lambda_0}(u_n, U_n)\Omega_n(u_n, U_n)^\tau, \\
\frac{\partial^2 \ell_n}{\partial \lambda^2} &= -2n^{2-2/\lambda_0} C(w_n, W_n)\Omega_n(T_1 - \pi_0^*, T_2 - \pi_1^*)^\tau \\
&\quad + 2n^{2-2/\lambda_0} C^2(v_n, V_n)\Omega_n(v_n, V_n)^\tau, \\
\frac{\partial^2 \ell_n}{\partial \lambda \partial C} &= 2n^{2-2/\lambda_0}(v_n, V_n)\Omega_n(T_1 - \pi_0^*, T_2 - \pi_1^*)^\tau \\
&\quad - 2n^{2-2/\lambda_0} C(v_n, V_n)\Omega_n(u_n, U_n)^\tau.
\end{aligned} \tag{7.10}$$

Let C_0 and λ_0 be the true values of C and λ. For notation simplicity, in the statements of Lemma 7.7, let u_n, v_n, w_n, U_n, V_n, W_n be those as defined earlier but evaluated specifically at $(C, \lambda) = (C_0, \lambda_0)$.

Lemma 7.7 *Under the power tail model in (7.1), for any* $\varepsilon > 0$,

1) $n^{1-1/\lambda_0} u_n \xrightarrow{p} \frac{1}{\lambda_0 C_0^{1-1/\lambda_0}} \Gamma(1 - 1/\lambda_0)$,
2) $n^{1-1/\lambda_0} U_n \xrightarrow{p} \frac{1}{\lambda_0 C_0^{1-1/\lambda_0}} \Gamma(2 - 1/\lambda_0)$,
3) $P(n^{2-2/\lambda_0}(v_n, V_n)\Omega_n(v_n, V_n)^\tau > 0) \longrightarrow 1$,
4) $n^{1-1/\lambda_0-\varepsilon} v_n \xrightarrow{p} 0$,
5) $n^{1-1/\lambda_0-\varepsilon} V_n \xrightarrow{p} 0$,
6) $n^{1-1/\lambda_0-\varepsilon} w_n \xrightarrow{p} 0$,
7) $n^{1-1/\lambda_0-\varepsilon} W_n \xrightarrow{p} 0$,
8) $P(n^{2-2/\lambda_0}(v_n U_n - u_n V_n) > 0) \to 1$.

The proofs of the Parts 1, 3, 4, and 8 of Lemma 7.7 are given in Appendix. The proofs of Parts 2, 5, 6, and 7 are left as exercises.

Proof of Theorem 7.1: The first two expressions in (7.10) are the diagonal elements of the Hessian matrix of l_n, and the third is the off-diagonal element. The objective of establishing that the probability of the Hessian matrix being positive definite, at (C_0, λ_0), converges to 1 as $n \to \infty$ is served by showing the probabilities of the following three events each, at (C_0, λ_0), converges to 1 as $n \to \infty$:

$$(a)\ \left.\frac{\partial^2 l_n}{\partial C^2}\right|_{(C,\lambda)=(C_0,\lambda_0)} > 0,$$

$$(b)\ \left.\frac{\partial^2 l_n}{\partial \lambda^2}\right|_{(C,\lambda)=(C_0,\lambda_0)} > 0, \qquad \text{and}$$

$$(c)\ \left.\left(\frac{\partial^2 l_n}{\partial C^2}\right)\left(\frac{\partial^2 l_n}{\partial \lambda^2}\right) - \left(\frac{\partial^2 l_n}{\partial \lambda \partial C}\right)^2\right|_{(C,\lambda)=(C_0,\lambda_0)} > 0.$$

By (1) and (2) of Lemma 7.7, and the fact that Ω_n converges to a positive definite matrix in probability, (a) follows immediately.

For (b), since $n^{1-1/(2\lambda_0)}(T_1 - \pi_0^*, T_2 - \pi_1^*)^\tau$ converges to a bivariate normal vector, by (6) and (7) of Lemma 7.7, the first term of $\partial^2 l_n/\partial \lambda^2|_{(C,\lambda)=(C_0,\lambda_0)}$ is

$$\begin{aligned}
&-2n^{2-2/\lambda_0} C(w_n, W_n)\Omega_n(T_1 - \pi_0^*, T_2 - \pi_1^*)^\tau \\
&= -2Cn^{1-3/(2\lambda_0)}(w_n, W_n)\Omega_n[n^{1-1/(2\lambda_0)}(T_1 - \pi_0^*, T_2 - \pi_1^*)^\tau] \xrightarrow{p} 0.
\end{aligned}$$

Then (b) follows from (3) of Lemma 7.7.

For (c),

$$\begin{aligned}
&\left.\left(\frac{\partial^2 l_n}{\partial C^2}\right)\left(\frac{\partial^2 l_n}{\partial \lambda^2}\right) - \left(\frac{\partial^2 l_n}{\partial \lambda \partial C}\right)^2\right|_{(C,\lambda)=(C_0,\lambda_0)} \\
&= -4Cn^{4-4/\lambda_0}[(u_n, U_n)\Omega_n(u_n, U_n)^\tau]\,[(w_n, W_n)\Omega_n(T_1 - \pi_0^*, T_2 - \pi_1^*)^\tau] \\
&\quad + 4C^2 n^{4-4/\lambda_0}[(u_n, U_n)\Omega_n(u_n, U_n)^\tau]\,[(v_n, V_n)\Omega_n(v_n, V_n)^\tau] \\
&\quad - 4n^{4-4/\lambda_0}[(v_n, V_n)\Omega_n(T_1 - \pi_0^*, T_2 - \pi_1^*)^\tau]^2 \\
&\quad - 4C^2 n^{4-4/\lambda_0}[(v_n, V_n)\Omega_n(u_n, U_n)^\tau]^2 \\
&\quad + 8Cn^{4-4/\lambda_0}[(v_n, V_n)\Omega_n(T_1 - \pi_0^*, T_2 - \pi_1^*)^\tau]\,[(v_n, V_n)\Omega_n(u_n, U_n)^\tau].
\end{aligned}$$

The first, third, and fifth terms in the last expression above all converge to 0 in probability, and therefore after a few algebraic steps,

$$\begin{aligned}
&\left(\frac{\partial^2 l_n}{\partial C^2}\right)\left(\frac{\partial^2 l_n}{\partial \lambda^2}\right) - \left(\frac{\partial^2 l_n}{\partial \lambda \partial C}\right)^2 \\
&= 4C^2 n^{4-4/\lambda_0}[(u_n, U_n)\Omega_n(u_n, U_n)^\tau]\,[(v_n, V_n)\Omega_n(v_n, V_n)^\tau] \\
&\quad - 4C^2 n^{4-4/\lambda_0}[(v_n, V_n)\Omega_n(u_n, U_n)^\tau]^2 + o_p(1) \\
&= 4C^2 n^{4-4/\lambda_0}(v_n U_n - u_n V_n)^2 \det(\Omega_n) + o_p(1).
\end{aligned}$$

The desired result follows from (8) of Lemma 7.7 and the fact that Ω_n converges to a positive definite matrix in probability. □

Theorem 7.1 establishes the convexity of l_n in an infinitesimal neighborhood of (C_0, λ_0), but says nothing about l_n globally. However, the next lemma does.

Lemma 7.8 *Under the model in (7.1), if $(C, \lambda) \neq (C_0, \lambda_0)$, then there exists a constant $c > 0$ such that, as $n \to \infty$,*

$$P(l_n(\lambda, C) - l_n(C_0, \lambda_0) > c) \to 1.$$

The following theorem is a corollary of Lemma 7.8.

Theorem 7.2 *Under the power tail model in (7.1), the AMLE of (C, λ) is unique for a sufficiently large n and is consistent.*

Toward proving Lemma 7.8, the following re-expression of $l_n(C, \lambda)$ in (7.9) is needed.

$$\begin{aligned}
& l_n(C, \lambda) \\
&= n^{2-2/\lambda_0}(T_1 - \pi_0^*(C, \lambda), T_2 - \pi_1^*(C, \lambda))\Omega_n(T_1 - \pi_0^*(C, \lambda), T_2 - \pi_1^*(C, \lambda))^\tau \\
&= l_n(C_0, \lambda_0) + [n^{2-2/\lambda_0}(\pi_0^*(C_0, \lambda_0) - \pi_0^*(C, \lambda), \pi_1^*(C_0, \lambda_0) - \pi_1^*(C, \lambda))\Omega_n \\
&\quad \times (\pi_0^*(C_0, \lambda_0) - \pi_0^*(C, \lambda), \pi_1^*(C_0, \lambda_0) - \pi_1^*(C, \lambda))^\tau] \\
&\quad + \left[2n^{2-2/\lambda_0}(T_1 - \pi_0^*(C_0, \lambda_0), T_2 - \pi_1^*(C_0, \lambda_0))\Omega_n \right. \\
&\quad \left. \times (\pi_0^*(C_0, \lambda_0) - \pi_0^*(C, \lambda), \pi_1^*(C_0, \lambda_0) - \pi_1^*(C, \lambda))^\tau\right].
\end{aligned} \tag{7.11}$$

To prove Lemma 7.8, Lemma 7.9 is needed. In both proofs of Lemmas 7.9 and 7.8, it is necessary to distinguish the notation $\pi_0^*(C, \lambda)$ from that of $\pi_0^*(C_0, \lambda_0)$, and $\pi_1^*(C, \lambda)$ from that of $\pi_1^*(C_0, \lambda_0)$.

Lemma 7.9 *Under the power tail model in (7.1),*

1) *if $(C, \lambda) \in \Theta_1$ but $(C, \lambda) \neq (C_0, \lambda_0)$, then*
   a) *$n^{1-1/\lambda_0}\pi_0^*(C, \lambda) \xrightarrow{p} 0$, and*
   b) *$n^{1-1/\lambda_0}\pi_1^*(C, \lambda) \xrightarrow{p} 0$;*
2) *if $(C, \lambda) \in \Theta_2$, then*
   a) *$n^{1-1/\lambda_0-1/(2\lambda_0)}\pi_0^*(C, \lambda) \xrightarrow{p} 0$, and*
   b) *$n^{1-1/\lambda_0-1/(2\lambda_0)}\pi_1^*(C, \lambda) \xrightarrow{p} 0$;*
3) *if $(C, \lambda) \in \Theta_2 \cup \Theta_3$, then*
   a) *$n^{1-1/\lambda_0}\pi_0^*(C, \lambda) \xrightarrow{p} +\infty$, and*
   b) *$n^{1-1/\lambda_0}\pi_1^*(C, \lambda) \xrightarrow{p} +\infty$.*

The proof of Lemma 7.9 is given in Appendix.

Proof of Lemma 7.8: The proof is given for three separate cases:

1) $(C, \lambda) \in \Theta_1$ but $(C, \lambda) \neq (C_0, \lambda_0)$;
2) $(C, \lambda) \in \Theta_2$; and
3) $(C, \lambda) \in \Theta_2 \cup \Theta_3$.

For (1), it is desired to show (i) that the third term in (7.11) converges to 0 in probability, and (ii) that the second term in (7.11) converges to a positive constant in probability. Toward proving (i), first noting $\pi_0^*(C_0, \lambda_0) \geq 0$ and by the proof of Case 1 of Theorem 6.2,

$$n^{1-1/\lambda_0} \mathrm{E}(\pi_0^*(C_0, \lambda_0)) \to \frac{C^{1/\lambda_0}}{\lambda_0} \Gamma\left(1 - \frac{1}{\lambda_0}\right) > 0,$$

hence

$$n^{1-1/\lambda_0-1/(2\lambda_0)} \mathrm{E}(\pi_0^*(C_0, \lambda_0)) \to 0,$$

and therefore

$$n^{1-1/\lambda_0-1/(2\lambda_0)} \pi_0^*(C_0, \lambda_0) \xrightarrow{p} 0.$$

By (1) of Lemma 7.9,

$$n^{1-1/\lambda_0-1/(2\lambda_0)} \pi_0^*(C, \lambda) \xrightarrow{p} 0,$$

and therefore

$$n^{1-1/\lambda_0-1/(2\lambda_0)} (\pi_0^*(C_0, \lambda_0) - \pi_0^*(C, \lambda)) \xrightarrow{p} 0.$$

Similarly

$$n^{1-1/\lambda_0-1/(2\lambda_0)} (\pi_1^*(C_0, \lambda_0) - \pi_1^*(C, \lambda)) \xrightarrow{p} 0.$$

On the other hand, noting

$$n^{2-2/\lambda_0} = n^{1-1/(2\lambda_0)} n^{1-1/\lambda_0-1/(2\lambda_0)}$$

and by Theorem 1.8, each of

$$n^{1-1/(2\lambda_0)} (T_1 - \pi_0^*(C_0, \lambda_0)) \quad \text{and} \quad n^{1-1/(2\lambda_0)} (T_2 - \pi_1^*(C_0, \lambda_0))$$

converges weakly to some normal random variable, and therefore (i) follows.

Again by the proof of Case 1 of Theorem 6.2 and (1) of Lemma 7.9, the vector

$$n^{1-1/\lambda_0} (\pi_0^*(C_0, \lambda_0) - \pi_0^*(C, \lambda), \pi_1^*(C_0, \lambda_0) - \pi_1^*(C, \lambda))$$

converges to a nonzero vector in probability. Since Ω_n converges to a positive definite matrix in probability, the second term in (7.11) converges to a constant $c > 0$, and hence (ii) follows, and therefore Lemma 7.8 follows for the case of $(C, \lambda) \in \Theta_1$ but $(C, \lambda) \neq (C_0, \lambda_0)$.

For (2), it is desired to show (i) that the third term in (7.11) converges to 0 in probability, and (ii) that the second term in (7.11) converges to $+\infty$ in probability. However, (i) immediately follows from (2) of Lemma 7.9 and (ii) immediately follows from (3) of Lemma 7.9.

For (3), first by Corollary 7.2,

$$a_n = n^{1-1/\lambda_0} (T_1 - \pi_0^*(C_0, \lambda_0)) \xrightarrow{p} 0,$$
$$b_n = n^{1-1/\lambda_0} (T_2 - \pi_1^*(C_0, \lambda_0)) \xrightarrow{p} 0;$$

and second by Lemma 7.9,

$$A_n = n^{1-1/\lambda_0}(\pi_0^*(C_0, \lambda_0) - \pi_0^*(C, \lambda)) \xrightarrow{p} -\infty,$$

$$B_n = n^{1-1/\lambda_0}(\pi_1^*(C_0, \lambda_0) - \pi_1^*(C, \lambda)) \xrightarrow{p} -\infty.$$

By Lemma 7.6,

$$\begin{aligned} &l_n(C, \lambda) - l_n(C_0, \lambda_0)) \\ &\quad = (A_n, B_n)\Omega_n(A_n, B_n)^\tau + (a_n, b_n)\Omega_n(A_n, B_n)^\tau \xrightarrow{p} +\infty. \end{aligned}$$

□

Finally, by Corollary 7.2, an approximate $1 - \alpha$ confidence region for (C, λ) may be given as follows:

$$\left(\frac{T_1 - \pi_0^*(C, \lambda)}{\hat{\sigma}_1}\right)^2 - 2\hat{\rho}_1 \left(\frac{T_1 - \pi_0^*(C, \lambda)}{\hat{\sigma}_1}\right) \left(\frac{T_2 - \pi_1^*(C, \lambda)}{\hat{\sigma}_2}\right) + \left(\frac{T_2 - \pi_1^*(C, \lambda)}{\hat{\sigma}_2}\right)^2 < \frac{(1 - \hat{\rho}_1^2)}{n^2} \chi_{1-\alpha}^2(2) \tag{7.12}$$

where $\chi_{1-\alpha}^2(2)$ is the $100 \times (1 - \alpha)$ th percentile of the chi-squared distribution with 2 degrees of freedom.

Statistical inference for (C, λ) may be made by means of the confidence region given in (7.12).

7.4 Remarks

Definition 7.1 may be extended to more general tail models as described in (7.2). Under (7.2), for every $r > 0$, let every p_k in π_{r-1} be replaced by $p_k = p(k; \boldsymbol{\theta})$ and denote the resulting expression as π_{r-1}^*, that is,

$$\pi_{r-1}^* = \sum_{k \geq 1} p(k; \boldsymbol{\theta}) 1[Y_k = r - 1].$$

Then

$$\pi_{r-1} - \pi_{r-1}^* = \sum_{k=1}^{k_0 - 1} 1[Y_k = r - 1](p_k - p(k; \boldsymbol{\theta})),$$

and the sum is of finite terms. Since, for every k, both $\mathrm{E}(1[Y_k = r - 1])$ and $\mathrm{Var}(1[Y_k = r - 1])$ converge to zero exponentially in sample size n, $\mathrm{E}(g(n, \delta)1[Y_k = r - 1])$ and $\mathrm{Var}(g(n, \delta)1[Y_k = r - 1])$ converge to zero, which implies that $g(n, \delta)1[Y_k = r - 1] \xrightarrow{p} 0$, which in turn implies that

$$g(n, \delta)(\pi_{r-1}^* - \pi_{r-1}) \xrightarrow{p} 0$$

for every r. If the tail model in (7.2) supports

$$n\hat{\Sigma}^{-1/2}\left(\frac{T_1-\pi_0^*}{\hat{\sigma}_1},\ldots,\frac{T_R-\pi_{R-1}^*}{\hat{\sigma}_R}\right)^\tau \xrightarrow{L} MVN(0, I_{R\times R}) \tag{7.13}$$

where $\hat{\Sigma}^{-1/2}$ is as in Lemma 7.2, then the estimator of $\boldsymbol{\theta}$ given by the following Definition may be studied fruitfully, as suggested by $(\hat{C},\hat{\lambda})$ of (C,λ) in the Pareto tail model in (7.1).

Definition 7.2 *If there exists a vector value, $\boldsymbol{\theta}=\hat{\boldsymbol{\theta}}$, which maximizes the likelihood given by (7.13) within the space allowed by the tail model in (7.2), then $\hat{\boldsymbol{\theta}}$ is said to be an AMLE of $\boldsymbol{\theta}$.*

The theoretical issues along this path include establishing (1) weak convergence of the type similar to (7.13), (2) consistency, uniqueness, and other desirable properties of the estimator $\hat{\boldsymbol{\theta}}$, none of which is easily accomplished. However, if this path could be followed through, Turing's perspective may lend itself to support broader and finer tail probability models.

Perhaps more interestingly, the methodology discussed in this chapter may be extended to tail distributions beyond those with Pareto decay. Four cases are offered below for consideration.

Case 1. Suppose, instead of (7.1), the underlying tail model is

$$p_k = Ck^{-\lambda}(1+o(1)) \tag{7.14}$$

for all k, $k \geq k_0$, where $k_0 \geq 1$ is a positive integer, $C>0$, and $\lambda>1$ are unknown parameters, and $o(1)\to 0$ as $k\to\infty$. It can be shown that all the results established under the exact Pareto tail model (7.1) still hold. As a result, the *AMLE*, $(\hat{C},\hat{\lambda})$, may be defined as in Definition 7.1 and enjoys the same statistical properties, such as uniqueness and consistency. The proofs unfortunately become doubly tedious and are not presented in this book.

Case 2. One of the most popular thick tail models in the literature of extreme value theory is a cumulative distribution function of an underlying random variable X satisfying

$$1-F_X(x) = \mathrm{P}(X>x) = Cx^{-\lambda} \tag{7.15}$$

for all $x\geq x_0$ where x_0, $C>0$, and $\lambda>0$ are unknown parameters. Let (7.15) be referred to as the continuous Pareto tail model. The estimation of the parameters in the tail model of (7.15) may also be approached by the methodology of this chapter. Without loss of generality, suppose the support of X is $(0,\infty)$. Consider a δ-binning, where $\delta>0$ is an arbitrary but fixed positive real number, of the support of X, that is,

$$(0,\delta], (\delta, 2\delta], \ldots, ((k-1)\delta, k\delta], \ldots.$$

The probability of the kth bin is

$$\begin{aligned} p_k &= F_X(k\delta) - F_X((k-1)\delta) \\ &= C\left\{[(k-1)\delta]^{-\lambda} - (k\delta)^{-\lambda}\right\} \\ &= \frac{C}{\delta^\lambda}\left[\frac{1}{(k-1)^\lambda} - \frac{1}{k^\lambda}\right] \\ &= \frac{C}{\delta^\lambda}\frac{1}{k^\lambda}\left[\left(\frac{k}{k-1}\right)^\lambda - 1\right] \\ &= \frac{C}{\delta^\lambda}\frac{1}{k^{\lambda+1}}\left[\left(\frac{k}{k-1}\right)^\lambda k - k\right]. \end{aligned}$$

However, since

$$\lim_{k\to\infty}\left[\left(\frac{k}{k-1}\right)^\lambda k - k\right] = \lambda,$$

(see Exercise 10)

$$p_k = \frac{C\lambda}{\delta^\lambda}\frac{1}{k^{\lambda+1}}(1 - o(1)) \tag{7.16}$$

which has the form of (7.14). Consequently, by the argument in Case 1, the *AMLE* of (C, λ) of the continuous Pareto model in (7.15) may also be obtained.

Case 3. Suppose the cumulative distribution function of an underlying random variable X satisfying

$$1 - F_X(x) = \mathrm{P}(X > x) = Ce^{-\lambda x} \tag{7.17}$$

for all $x \geq x_0$ where x_0, $C > 0$, and $\lambda > 0$ are unknown parameters. Let (7.17) be referred to as the continuous exponential tail model. The estimation of the parameters in the tail model of (7.17) may also be approached by the methodology of this chapter. Consider first a transformation of X,

$$Y = e^X,$$

which entails

$$1 - F_Y(y) = Cy^{-\lambda}$$

for sufficiently large y, that is, $y \geq e^{x_0}$. By the argument of Case 2, the *AMLE* of (C, λ) of the continuous exponential model in (7.17) may also be obtained.

Case 4. Suppose the cumulative distribution function of an underlying random variable X satisfying

$$1 - F_X(x) = \mathrm{P}(X > x) = Ce^{-\lambda x^2} \tag{7.18}$$

for all $x \geq x_0$ where $x_0 \geq 0$, $C > 0$, and $\lambda > 0$ are unknown parameters. Let (7.18) be referred to as the continuous Gaussian tail model. The estimation of the parameters in the tail model of (7.18) may also be approached by the methodology of this chapter. Consider first a transformation of X,

$$Y = e^{X^2},$$

which entails

$$1 - F_Y(y) = Cy^{-\lambda}$$

for sufficiently large y, that is, $y \geq e^{x_0^2}$. By the argument of Case 2, the *AMLE* of (C, λ) of the continuous Gaussian tail model in (7.18) may also be obtained.

Cases 1–4 suggest that the methodology discussed in this chapter may be extended to models beyond the family of distributions with power decay tails. In fact, the methodology may be extended to tackle various tail models for continuous random variables, where the choice of δ, possibly of the form $\delta = \delta(n)$, in practice may prove to be an interesting issue of investigation, among many other meaningful research undertakings along this thread.

7.5 Appendix

7.5.1 Proof of Lemma 7.7

Proof: For Part 1, it suffices to show

$$n^{1-1/\lambda}\mathrm{E}(u_n) \to [1/(\lambda C^{1-1/\lambda})]\Gamma(1 - 1/\lambda), \text{ and}$$
$$n^{2-2/\lambda}\mathrm{Var}(u_n) \to 0.$$

By the argument in the proof of Lemma 6.3,

$$n^{1-1/\lambda}\mathrm{E}(u_n) \sim n^{1-1/\lambda} \sum k^{-\lambda} e^{-nCk^{-\lambda}};$$

and invoking the Euler–Maclaurin lemma,

$$\begin{aligned}
\lim\ n^{1-1/\lambda}\mathrm{E}(u_n) &= \lim\ n^{1-1/\lambda} \int_{x=1}^{\infty} x^{-\lambda} e^{-nCx^{-\lambda}} dx \\
&= \frac{1}{\lambda C^{1-1/\lambda}} \lim \int_{x=0}^{nC} y^{-1/\lambda} e^{-y} dy \\
&= \frac{1}{\lambda C^{1-1/\lambda}} \Gamma(1 - 1/\lambda).
\end{aligned}$$

$$
\begin{aligned}
\operatorname{Var}(u_n) &= \mathrm{E}(u_n^2) - [\mathrm{E}(u_n)]^2 \\
&= \left[\sum_{k\ge1}(1-Ck^{-\lambda})^n k^{-2\lambda} + \sum_{k\ne j}(1-Ck^{-\lambda}-Cj^{-\lambda})^n k^{-\lambda}j^{-\lambda}\right] \\
&\quad - \left[\sum_{k\ge1}(1-Ck^{-\lambda})^{2n} k^{-2\lambda} + \sum_{k\ne j}(1-Ck^{-\lambda})^n(1-Cj^{-\lambda})^n k^{-\lambda}j^{-\lambda}\right] \\
&\le \left[\sum_{k\ge1}(1-Ck^{-\lambda})^n k^{-2\lambda} + \sum_{k\ne j}(1-Ck^{-\lambda})^n(1-Cj^{-\lambda})^n k^{-\lambda}j^{-\lambda}\right] \\
&\quad - \left[\sum_{k\ge1}(1-Ck^{-\lambda})^{2n} k^{-2\lambda} + \sum_{k\ne j}(1-Ck^{-\lambda})^n(1-Cj^{-\lambda})^n k^{-\lambda}j^{-\lambda}\right] \\
&= \sum_{k\ge1}(1-Ck^{-\lambda})^n k^{-2\lambda} - \sum_{k\ge1}(1-Ck^{-\lambda})^{2n} k^{-2\lambda} \\
&\le \sum_{k\ge1}(1-Ck^{-\lambda})^n k^{-2\lambda}.
\end{aligned}
$$

Invoking the Euler–Maclaurin lemma whenever needed,

$$
\begin{aligned}
\lim_{n\to\infty} n^{2-2/\lambda}\operatorname{Var}(u_n) &\le \lim\ n^{2-2/\lambda}\sum_{k\ge1}(1-Ck^{-\lambda})^n k^{-2\lambda} \\
&= \lim_{n\to\infty} n^{2-2/\lambda}\int_1^\infty x^{-2\lambda}e^{-Cnx^{-\lambda}}dx \\
&= \lim_{n\to\infty}\frac{n^{-1/\lambda}}{\lambda C^{2-1/\lambda}}\Gamma(2-1/\lambda) = 0.
\end{aligned}
$$

A similar argument establishes Part 2.

For Part 3, since Ω_n converges to a positive definite matrix in probability, it suffices to show that $P(n^{1-1/\lambda}v_n > 0) \to 1$, which is clear by noting

$$
\begin{aligned}
n^{1-1/\lambda}v_n &= n^{1-1/\lambda}\sum_{k\ge2}1[Y_k=0](\ln k)k^{-\lambda} \ge n^{1-1/\lambda}\ln 2\sum_{k\ge2}1[Y_k=0]k^{-\lambda} \\
&= -n^{1-1/\lambda}(\ln 2)1[Y_1=0] + n^{1-1/\lambda}\ln 2\sum 1[Y_k=0]k^{-\lambda} \\
&= o_p(1) + n^{1-1/\lambda}(\ln 2)u_n \xrightarrow{p} \frac{\ln 2}{\lambda C^{1-1/\lambda}}\Gamma(1-1/\lambda) > 0.
\end{aligned}
$$

For Part 4, first consider the fact that a positive integer k_0 exists such that $k^\varepsilon > \ln k$ for all $k \ge k_0$.

$$
\begin{aligned}
v_n &= \sum_{k<k_0}1[Y_k=0](\ln k)k^{-\lambda} + \sum_{k\ge k_0}1[Y_k=0](\ln k)k^{-\lambda} \\
&\le \ln k_0\sum_{k<k_0}1[Y_k=0] + \sum_{k\ge k_0}1[Y_k=0]k^{-\lambda+\varepsilon}.
\end{aligned}
$$

$$\mathrm{E}\left(\ln k_0 \sum_{k<k_0} 1[Y_k=0]\right) = \ln k_0 \sum_{k<k_0} \left(1-Ck^{-\lambda}\right)^n \le k_0 \ \ln k_0\left(1-Ck_0^{-\lambda}\right)^n.$$

$$\lim_{n\to\infty} n^{1-1/\lambda-\varepsilon}\mathrm{E}\left(\ln(k_0)\sum_{k<k_0} 1[x_k=0]\right) = 0.$$

Invoking the Euler–Maclaurin lemma whenever needed,

$$\lim_{n\to\infty} \mathrm{E}\left(\sum_{k\ge k_0} 1[Y_k=0]k^{-\lambda+\varepsilon}\right) = \lim_{n\to\infty}\int_1^\infty x^{-\lambda+\varepsilon}e^{-nCx^{-\lambda}}dx$$
$$= \frac{1}{\lambda}C^{-[1-(\varepsilon+1)/\lambda)]}\lim_{n\to\infty} n^{-[1-(\varepsilon+1)/\lambda]}\int_0^{nC} y^{[1-(\varepsilon+1)/\lambda)]-1}e^{-y}dy.$$

$$\lim_{n\to\infty} n^{1-1/\lambda-\varepsilon}\mathrm{E}(v_n)$$
$$= \frac{1}{\lambda}C^{-[1-(\varepsilon+1)/\lambda)]}\lim_{n\to\infty} n^{-(1-1/\lambda)\varepsilon}\int_0^{nC} y^{[1-(\varepsilon+1)/\lambda)]-1}e^{-y}dy \to 0.$$

$$\begin{aligned}
\mathrm{Var}(v_n) &= \mathrm{E}(v_n^2) - [\mathrm{E}(v_n)]^2 \\
&= \mathrm{E}\Bigg[\sum_{k\ge1} 1[Y_k=0](\ln k)^2k^{-2\lambda} \\
&\qquad + \sum_{k\ne j} 1[Y_k=0, Y_j=0](\ln k)(\ln j)(kj)^{-\lambda}\Bigg] \\
&\qquad - \Bigg[\sum_{k\ge1}(1-Ck^{-\lambda})^{2n}(\ln k)^2k^{-2\lambda} \\
&\qquad + \sum_{k\ne j}(1-Ck^{-\lambda})^n(1-Cj^{-\lambda})^n(\ln k)(\ln j)(kj)^{-\lambda}\Bigg] \\
&= \Bigg[\sum_{k\ge1}(1-Ck^{-\lambda})^n(\ln k)^2k^{-2\lambda} \\
&\qquad + \sum_{k\ne j}(1-Ck^{-\lambda}-Cj^{-\lambda})^n(\ln k)(\ln j)(kj)^{-\lambda}\Bigg] \\
&\qquad - \Bigg[\sum_{k\ge1}(1-Ck^{-\lambda})^{2n}(\ln k)^2k^{-2\lambda} \\
&\qquad + \sum_{k\ne j}(1-Ck^{-\lambda})^n(1-Cj^{-\lambda})^n\ln(k)\ln(j)(kj)^{-\lambda}\Bigg]
\end{aligned}$$

$$\leq \sum_{k\geq 1}(1-Ck^{-\lambda})^n(\ln k)^2k^{-2\lambda}$$
$$= \sum_{k<k_0}(1-Ck^{-\lambda})^n(\ln k)^2k^{-2\lambda}$$
$$+ \sum_{k\geq k_0}(1-Ck^{-\lambda})^n(\ln k)^2k^{-2\lambda}$$
$$\leq (\ln k_0)^2k_0(1-Ck_0^{-\lambda})^n + \sum_{k\geq 1}(1-Ck^{-\lambda})^nk^{-2\lambda+2\varepsilon}.$$

$$\lim_{n\to\infty} n^{2-2/\lambda-2\varepsilon}(\ln k_0)^2k_0(1-Ck_0^{-\lambda})^n \to 0.$$

Invoking the Euler–Maclaurin lemma,

$$\lim_{n\to\infty} n^{2-2/\lambda-2\varepsilon}\sum_{k\geq 1}\left(1-Ck^{-\lambda}\right)^nk^{-2\lambda+2\varepsilon}$$
$$= \lim_{n\to\infty} n^{2-2/\lambda-2\varepsilon}\int_1^\infty x^{-2(\lambda-\varepsilon)}e^{-nCx^{-\lambda}}dx$$
$$= \lim_{n\to\infty} n^{2-2/\lambda-2\varepsilon}\left[\frac{(nC)^{-2+(2\varepsilon+1)/\lambda}}{\lambda}\int_0^{nC} y^{[2-(2\varepsilon+1)/\lambda]-1}e^{-y}dy\right]$$
$$= \frac{C^{-2+(2\varepsilon+1)/\lambda}}{\lambda}\lim_{n\to\infty} n^{-2+(2\varepsilon+1)/\lambda}n^{2-2/\lambda-2\varepsilon}$$
$$\times\left[\int_0^{nC} y^{[2-(2\varepsilon+1)/\lambda]-1}e^{-y}dy\right]$$
$$= \frac{C^{-2+(2\varepsilon+1)/\lambda}}{\lambda}\lim_{n\to\infty} n^{-1/\lambda-2\varepsilon(1-1/\lambda)}\left[\int_0^{nC} y^{[2-(2\varepsilon+1)/\lambda]-1}e^{-y}dy\right] \to 0.$$

Therefore $\mathrm{Var}(n^{1-1/\lambda-\varepsilon}v_n) \to 0$, and hence $n^{1-1/\lambda-\varepsilon}v_n \xrightarrow{p} 0$.

A similar argument establishes each of Parts 5–7.

For Part 8, it is to show that $n^{2-2/\lambda}(v_nU_n - u_nV_n)$ converges to a positive constant in probability. By the results of Parts 1 and 2 already established, it is equivalent to show that

$$n^{1-1/\lambda}\left(\frac{\Gamma(2-1/\lambda)}{\lambda C^{1-1/\lambda}}v_{n+1} - \frac{\Gamma(1-1/\lambda)}{\lambda C^{1-1/\lambda}}V_{n+1}\right)$$
$$= \frac{n^{1-1/\lambda}\Gamma(1-1/\lambda)}{\lambda C^{1-1/\lambda}}[(1-1/\lambda)v_{n+1} - V_{n+1}]$$

converges to a positive constant in probability, or that

$$n^{1-1/\lambda}[(1-1/\lambda)v_{n+1} - V_{n+1}]$$

does.

Since

$$n^{1-1/\lambda}[(1-1/\lambda)v_{n+1} - V_{n+1}]$$
$$= n^{1-1/\lambda}[(1-1/\lambda)v_n - V_{n+1}] - n^{1-1/\lambda}(1-1/\lambda)(v_n - v_{n+1}),$$

it is first to show $n^{1-1/\lambda}(v_n - v_{n+1}) \xrightarrow{p} 0$.

$$\begin{aligned}
E(v_n - v_{n+1}) &= \sum_{k\geq 1}(1-Ck^{-\lambda})^n(\ln k)k^{-\lambda} - \sum_{k\geq 1}(1-Ck^{-\lambda})^{n+1}(\ln k)k^{-\lambda} \\
&= C\sum_{k\geq 1}(1-Ck^{-\lambda})^n(\ln k)k^{-2\lambda} \\
&\leq C(\ln k_0)k_0(1-Ck_0^{-\lambda})^n + C\sum_{k\geq 1}(1-Ck^{-\lambda})^n k^{-2\lambda+\varepsilon}.
\end{aligned}$$

$$\lim_{n\to\infty} n^{1-1/\lambda} C\ln(k_0)k_0(1-Ck_0^{-\lambda})^n = 0.$$

Let $\varepsilon = \lambda/2$. Invoking the Euler–Maclaurin lemma,

$$\begin{aligned}
&\lim_{n\to\infty} n^{1-1/\lambda} C\sum_{k\geq 1}(1-Ck^{-\lambda})^n k^{-2\lambda+\varepsilon} \\
&= \lim_{n\to\infty} n^{1-1/\lambda} C\int_1^\infty x^{-2\lambda+\varepsilon}e^{-nCx^{-\lambda}}dx \\
&= \frac{C^{-1+(\varepsilon+1)/\lambda}}{\lambda}\lim_{n\to\infty} n^{-1+\varepsilon/\lambda}\int_0^{nC} y^{[2-(\varepsilon+1)-1]}e^{-y}dy \\
&= 0.
\end{aligned}$$

Therefore, $n^{1-1/\lambda}E(v_n - v_{n+1}) \to 0$. Next, it is to show

$$\text{Var}\left(n^{1-1/\lambda}(v_n - v_{n+1})\right) \to 0.$$

However, since

$$\text{Var}(v_n - v_{n+1}) \leq 2(\text{Var}(v_n) + \text{Var}(v_{n+1})),$$

it suffices to show

$$\text{Var}\left(n^{1-1/\lambda}v_n\right) \to 0.$$

In the proof of above-mentioned Part 4, it has already been shown that

$$\text{Var}(v_n) \leq \sum_{k\geq 1}\left(1-Ck^{-\lambda}\right)^n k^{-2\lambda+2\varepsilon}.$$

Therefore, letting $\varepsilon = 1/4$ and invoking the Euler–Maclaurin lemma,

$$\begin{aligned}
\lim_{n\to\infty} n^{2-2/\lambda}\text{Var}(v_n) &\leq \lim_{n\to\infty} n^{2-2/\lambda}\sum_{k\geq 1}\left(1-Ck^{-\lambda}\right)^n k^{-2\lambda+2\varepsilon} \\
&= \lim_{n\to\infty} n^{2-2/\lambda}\int_1^\infty x^{-2\lambda+2\varepsilon}e^{-nCx^{-\lambda}}dx \\
&= \frac{C^{-2+(2\varepsilon+1)/\lambda}}{\lambda}\lim_{n\to\infty} n^{-(1-2\varepsilon)/\lambda}\int_0^{nC} y^{[2-(2\varepsilon+1)/\lambda]-1}e^{-y}dy \\
&= 0.
\end{aligned}$$

Since $n^{1-1/\lambda}(v_n - v_{n+1}) \xrightarrow{p} 0$, it only remains to show that

$$n^{1-1/\lambda}[(1-1/\lambda)v_n - V_{n+1}]$$

converges to a positive constant in probability.

$$\begin{aligned}\mathrm{E}\left((1-1/\lambda)v_n - V_{n+1}\right) &= (1-1/\lambda)\sum_{k\geq 1}\left(1-Ck^{-\lambda}\right)^n(\ln k)k^{-\lambda}\\ &\quad - \sum_{k\geq 1} n\left(1-Ck^{-\lambda}\right)^n Ck^{-\lambda}(\ln k)k^{-\lambda}\\ &= \sum_{k\geq 1}\left(1-Ck^{\lambda}\right)^n(\ln k)\left[(1-1/\lambda)k^{-\lambda} - nCk^{-2\lambda}\right].\end{aligned}$$

Letting $y = nCx^{-\lambda}$ and invoking the Euler–Maclaurin lemma, after a few algebraic steps,

$$\begin{aligned}&\lim_{n\to\infty} n^{1-1/\lambda}\mathrm{E}((1-1/\lambda)v_n - V_{n+1})\\ &= \lim_{n\to\infty} n^{1-1/\lambda}\sum_{k\geq 1}\left\{\left(1-Ck^{-\lambda}\right)^n(\ln k)\left[(1-1/\lambda)k^{-\lambda} - nCk^{-2\lambda}\right]\right\}\\ &= \lim_{n\to\infty} n^{1-1/\lambda}\int_1^{\infty}(\ln x)\left[(1-1/\lambda)x^{-\lambda} - nCx^{-2\lambda}\right]e^{-nCx^{-\lambda}}dx\\ &= \frac{1}{\lambda^2 C^{1-1/\lambda}}\lim_{n\to\infty}\int_0^{nC}\ln(nC)\left[\left(1-\frac{1}{\lambda}\right)y^{-\frac{1}{\lambda}} - y^{1-\frac{1}{\lambda}}\right]e^{-y}dy\\ &\quad - \frac{1}{\lambda^2 C^{1-1/\lambda}}\lim_{n\to\infty}\int_0^{nC}(\ln y)\left[\left(1-\frac{1}{\lambda}\right)y^{-\frac{1}{\lambda}} - y^{1-\frac{1}{\lambda}}\right]e^{-y}dy.\end{aligned}$$

The first term of the last expression is zero. This is so because first,

$$\begin{aligned}&\lim_{n\to\infty}\int_0^{nC}\ln(nC)\left[\left(1-\frac{1}{\lambda}\right)y^{-\frac{1}{\lambda}} - y^{1-\frac{1}{\lambda}}\right]e^{-y}dy\\ &= \lim_{n\to\infty}\int_0^{\infty}\ln(nC)\left[\left(1-\frac{1}{\lambda}\right)y^{-\frac{1}{\lambda}} - y^{1-\frac{1}{\lambda}}\right]e^{-y}dy\\ &\quad - \lim_{n\to\infty}\int_{nC}^{\infty}\ln(nC)\left[\left(1-\frac{1}{\lambda}\right)y^{-\frac{1}{\lambda}} - y^{1-\frac{1}{\lambda}}\right]e^{-y}dy\\ &= \lim_{n\to\infty}\int_{nC}^{\infty}\ln(nC)\left[y^{1-\frac{1}{\lambda}} - \left(1-\frac{1}{\lambda}\right)y^{-\frac{1}{\lambda}}\right]e^{-y}dy,\end{aligned}$$

and second, $(1-1/\lambda)y^{-1/\lambda} - y^{1-1/\lambda} > 0$ for sufficiently large y, and third, $(1-1/\lambda)y^{-1/\lambda} - y^{1-1/\lambda}e^{-y/2}$ is decreasing for sufficiently large y, and consequently,

$$\begin{aligned}&\lim_{n\to\infty}\int_0^{nC}\ln(nC)\left[\left(1-\frac{1}{\lambda}\right)y^{-\frac{1}{\lambda}} - y^{1-\frac{1}{\lambda}}\right]e^{-y}dy\\ &\leq \lim_{n\to\infty}\ln(nC)\left[(nC)^{1-\frac{1}{\lambda}} - \left(1-\frac{1}{\lambda}\right)(nC)^{-\frac{1}{\lambda}}\right]e^{-(nC)/2}\int_{nC}^{\infty}e^{-y/2}dy\\ &\leq 2\lim_{n\to\infty}\ln(nC)\left[(nC)^{1-\frac{1}{\lambda}} - \left(1-\frac{1}{\lambda}\right)(nC)^{-\frac{1}{\lambda}}\right]e^{-(nC)/2} = 0.\end{aligned}$$

This leads to

$$\lim_{n\to\infty} n^{1-1/\lambda}\mathrm{E}((1-1/\lambda)v_n - V_{n+1})$$
$$= \frac{1}{\lambda^2 C^{1-1/\lambda}}\int_0^\infty (\ln y)\left[y^{1-\frac{1}{\lambda}} - \left(1-\frac{1}{\lambda}\right)y^{-\frac{1}{\lambda}}\right]e^{-y}dy.$$

Denoting the integrand above as

$$K(y) = (\ln y)[y^{1-1/\lambda} - (1-1/\lambda)y^{-1/\lambda}]e^{-y},$$

it is easy to see that $K(y)$ is positive on $(0, 1-1/\lambda)$, negative on $(1-1/\lambda, 1)$, and positive on $(1, \infty)$.

$$\int_0^{1-1/\lambda} K(y)dy \geq e^{-(1-1/\lambda)}\int_0^{1-1/\lambda}(\ln y)\left[y^{1-\frac{1}{\lambda}} - \left(1-\frac{1}{\lambda}\right)y^{-\frac{1}{\lambda}}\right]dy$$
$$= e^{-(1-1/\lambda)}\int_{-\ln(1-1/\lambda)}^\infty t\left[\left(1-\frac{1}{\lambda}\right)e^{-(1-1/\lambda)t} - e^{-(2-1/\lambda)t}\right]dt,$$
$$\int_{1-1/\lambda}^1 K(y)dy \geq e^{-(1-1/\lambda)}\int_{1-1/\lambda}^1 \left[y^{1-1/\lambda} - \left(1-\frac{1}{\lambda}\right)y^{-1/\lambda}\right]\ln(y)dy$$
$$= e^{-(1-1/\lambda)}\int_0^{-\ln(1-1/\lambda)} t\left[\left(1-\frac{1}{\lambda}\right)e^{-(1-1/\lambda)t} - e^{-(2-1/\lambda)t}\right]dt,$$
$$\int_0^\infty K(y)dy > \int_0^{1-1/\lambda}K(y)dy + \int_{1-1/\lambda}^1 K(y)dy$$
$$\geq e^{-(1-1/\lambda)}\int_0^\infty t\left[\left(1-\frac{1}{\lambda}\right)e^{-(1-1/\lambda)t} - e^{-(2-1/\lambda)t}\right]dt$$
$$= e^{-(1-1/\lambda)}[1/(1-1/\lambda) - 1/(2-1/\lambda)^2] > 0,$$

that is,

$$n^{1-1/\lambda}\mathrm{E}((1-1/\lambda)v_n - V_{n+1}) \to (\lambda^2 C^{1-1/\lambda})^{-1}\int_0^\infty K(y)dy > 0.$$

It still remains to show that

$$\mathrm{Var}(n^{1-1/\lambda}[(1-1/\lambda)v_n - V_{n+1}]) \to 0.$$

However, since it has been shown above that

$$\mathrm{Var}(n^{1-1/\lambda}v_n) \to 0,$$

it suffices to show that

$$\mathrm{Var}(n^{1-1/\lambda}V_{n+1}) \to 0.$$

$$
\begin{aligned}
\mathrm{Var}(V_{n+1}) &= \mathrm{E}(V_{n+1}^2) - [\mathrm{E}(V_{n+1})]^2 \\
&= \left[\sum_{k\geq 1}(n+1)Ck^{-\lambda}(1-Ck^{-\lambda})^n(\ln k)^2k^{-2\lambda}\right. \\
&\quad \left. + \sum_{k\neq j}(n+1)nCk^{-\lambda}Cj^{-\lambda}(1-Ck^{-\lambda}-Cj^{-\lambda})^{n-1}(\ln k)(\ln j)k^{-\lambda}j^{-\lambda}\right] \\
&\quad - \left[\sum_{k\geq 1}(n+1)^2C^2k^{-2\lambda}(1-Ck^{-\lambda})^{2n}(\ln k)^2k^{-2\lambda}\right. \\
&\quad + \sum_{k\neq j}(n+1)^2Ck^{-\lambda}Cj^{-\lambda}(1-Ck^{-\lambda})^n(1-Cj^{-\lambda})^n \\
&\quad \left. \times (\ln k)(\ln j)k^{-\lambda}j^{-\lambda}\right] \\
&\leq (n+1)C\sum_{k\geq 1}(1-Ck^{-\lambda})^n(\ln k)^2k^{-3\lambda} \\
&= (n+1)C\sum_{k<k_0}(1-Ck^{-\lambda})^n(\ln k)^2k^{-3\lambda} \\
&\quad + (n+1)C\sum_{k\geq k_0}(1-Ck^{-\lambda})^n(\ln k)^2k^{-3\lambda} \\
&\leq (n+1)C\sum_{k<k_0}(1-Ck^{-\lambda})^n(\ln k)^2k^{-3\lambda} \\
&\quad + (n+1)C\sum_{k\geq 1}(1-Ck^{-\lambda})^n k^{-3\lambda+2\varepsilon}.
\end{aligned}
$$

The first term in the last expression converges to zero exponentially, and therefore for $\varepsilon = 1/4$, invoking the Euler–Maclaurin lemma,

$$
\begin{aligned}
\lim_{n\to\infty}\mathrm{Var}(n^{1-1/\lambda}V_{n+1}) &\leq C\lim_{n\to\infty} n^{2-2/\lambda}(n+1)\int_1^{\infty} x^{-3\lambda+2\varepsilon}e^{-nCx^{-\lambda}}dx \\
&= \frac{C^{-2+(2\varepsilon+1)/\lambda}}{\lambda}\lim_{n\to\infty} n^{2-2/\lambda}(n+1)n^{-3+(2\varepsilon+1)/\lambda} \\
&\quad \times \int_0^{nC} y^{[3-(2\varepsilon+1)/\lambda]-1}e^{-y}dy = 0.
\end{aligned}
$$

7.5.2 Proof of Lemma 7.9

Proof of Part 1: For Part (a), first consider the function

$$f_n(x) = n^{1-1/\lambda_0}(1-C_0x^{-\lambda_0})^nCx^{-\lambda}$$

defined for $x > x_0$ where x_0 is sufficiently large a constant to ensure that $1 - C_0x^{-\lambda_0} \geq 0$. It can be easily verified that $f_n(x)$ is decreasing for all

$x \geq x(n) = [C_0(n\lambda_0 + \lambda)/\lambda]^{1/\lambda_0}$ and is increasing for $x \in [x_0, x(n)]$. For any λ satisfying $\lambda > \lambda_0 - 1$,

$$f_n(x(n)) = CC_0^{-\lambda/\lambda_0}\left(1 - \frac{\lambda}{n\lambda_0 + \lambda}\right)^n (1 + n\frac{\lambda_0}{\lambda})^{-\lambda/\lambda_0} n^{1-1/\lambda_0}$$
$$= \mathcal{O}(n^{[(\lambda_0-1)-\lambda]/\lambda_0}) \to 0.$$

Invoking the Euler–Maclaurin lemma,

$$\lim_{n\to\infty} n^{1-1/\lambda_0} \sum_{k>x_0} (1 - C_0 k^{-\lambda_0})^n C k^{-\lambda}$$
$$= \lim_{n\to\infty} n^{1-1/\lambda_0} \int_{x_0}^{\infty} (1 - C_0 x^{-\lambda_0})^n C x^{-\lambda} dx$$
$$= C \lim_{n\to\infty} n^{1-1/\lambda_0} \int_{x_0}^{\infty} e^{-nC_0 x^{-\lambda_0}} x^{-\lambda} dx$$
$$= \frac{CC_0^{(1-\lambda)/\lambda_0}}{\lambda_0} \lim_{n\to\infty} n^{(\lambda_0-\lambda)/\lambda_0} \int_0^{nC_0 x_0^{-\lambda_0}} y^{(\lambda-1)/\lambda_0 - 1} e^{-y} dy.$$

The last expression is a positive constant if and only if $\lambda = \lambda_0$ and is zero if and only if $\lambda > \lambda_0$. In fact, when $\lambda = \lambda_0$ and $C \neq C_0$, the limit is $CC_0^{(1-\lambda_0)/\lambda_0} \lambda_0^{-1}\Gamma(1 - 1/\lambda_0)$; when $\lambda = \lambda_0$ and $C = C_0$, the limit is $C_0^{1/\lambda_0}\lambda_0^{-1}\Gamma(1 - 1/\lambda_0)$. This leads to

$$n^{1-1/\lambda_0} \mathrm{E}(\pi_0^*(C, \lambda)) \to 0.$$

Next, it is to show $n^{2-2/\lambda_0}\mathrm{Var}(\pi_0^*(C, \lambda)) \to 0$. It can be verified that

$$\mathrm{Var}(\pi_0^*(C, \lambda)) \leq \sum_{k\geq 1} (1 - C_0 k^{-\lambda_0})^n C^2 k^{-2\lambda}.$$

Let

$$f_n(x) = n^{2-2/\lambda_0}(1 - C_0 x^{-\lambda_0})^2 C^2 k^{-2\lambda}$$

be defined for $x > x_0$ where x_0 is sufficiently large a constant to ensure that $1 - C_0 x^{-\lambda_0} \geq 0$. It can be verified that $f_n(x)$ is decreasing for all $x \geq x(n) = [C_0(n\lambda_0 + 2\lambda)/(2\lambda)]^{1/\lambda_0}$ and is increasing for $x \in [x_0, x(n)]$. For any λ satisfying $\lambda > \lambda_0 - 1$,

$$f_n(x(n)) = C^2 C_0^{-2\lambda/\lambda_0}\left(1 - \frac{2\lambda}{n\lambda_0 + 2\lambda}\right)^n \left(1 + n\frac{\lambda_0}{2\lambda}\right)^{-2\lambda/\lambda_0} n^{2-2/\lambda_0}$$
$$= \mathcal{O}\left(n^{2\frac{(\lambda_0-1)-\lambda}{\lambda_0}}\right) \to 0.$$

Invoking the Euler–Maclaurin lemma,

$$\lim_{n\to\infty} n^{2-2/\lambda_0} \sum_{k>x_0} (1 - C_0 k^{-\lambda_0})^n C^2 k^{-2\lambda}$$

$$= C^2 \lim_{n\to\infty} n^{2-2/\lambda_0} \int_{x_0}^{\infty} (1 - C_0 x^{-\lambda_0})^n x^{-2\lambda} dx$$

$$= C^2 \lim_{n\to\infty} n^{2-2/\lambda_0} \int_{x_0}^{\infty} e^{-nC_0 x^{-\lambda_0}} x^{-2\lambda} dx$$

$$= \frac{C^2 C_0^{(1-2\lambda)/\lambda_0}}{\lambda_0} \lim_{n\to\infty} n^{[2(\lambda_0-\lambda)-1]/\lambda_0} \int_0^{nC_0 x_0^{-\lambda_0}} y^{(2\lambda-1)/\lambda_0 - 1} e^{-y} dy.$$

The last limit is zero if and only if $\lambda > \lambda_0 - 1/2$. Part (a) immediately follows. For Part (b), first consider the function

$$f_n(x) = n^{1-1/\lambda_0}(n+1)(1 - C_0 x^{-\lambda_0})^n C_0 x^{-\lambda_0} C x^{-\lambda}$$

defined for $x > x_0$. It can be easily verified that $f_n(x)$ is decreasing for all $x \geq x(n)$ and is increasing for $x \in [x_0, x(n)]$ where

$$x(n) = \{C_0[(n+1)\lambda_0 + \lambda]/(\lambda_0 + \lambda)\}^{1/\lambda_0}.$$

For any $\lambda > \lambda_0 - 1$,

$$f_n(x(n)) = CC_0(n+1)n^{1-\frac{1}{\lambda_0}} \left[1 - \frac{\lambda_0 + \lambda}{(n+1)\lambda_0 + \lambda}\right]^n$$

$$\times \left\{ \frac{\lambda_0 + \lambda}{C_0[(n+1)\lambda_0 + \lambda]} \right\}^{1+\frac{\lambda}{\lambda_0}}$$

$$= \mathcal{O}\left(n^{1-\frac{1+\lambda}{\lambda_0}}\right) \to 0.$$

Invoking the Euler–Maclaurin lemma, for $\lambda > \lambda_0$,

$$\lim_{n\to\infty} n^{1-1/\lambda_0}(n+1) \sum_{k>x_0} (1 - C_0 k^{-\lambda_0})^n C_0 k^{-\lambda_0} C k^{-\lambda}$$

$$= CC_0 \lim_{n\to\infty} n^{1-1/\lambda_0}(n+1) \int_{x_0}^{\infty} (1 - C_0 x^{-\lambda_0})^n x^{-(\lambda_0+\lambda)} dx$$

$$= \frac{CC_0^{1-\lambda/\lambda_0}}{\lambda_0} \lim_{n\to\infty} n^{-\frac{\lambda}{\lambda_0}}(n+1) \int_0^{nC_0 x_0^{-\lambda_0}} y^{(1+\lambda/\lambda_0 - 1/\lambda_0)-1} e^{-y} dy$$

$$= \mathcal{O}\left(n^{\frac{\lambda_0-\lambda}{\lambda_0}}\right) \to 0.$$

At this point, it has been shown that $n^{1-1/\lambda_0} \mathrm{E}(\pi_1^*(C, \lambda)) \to 0$. Next, it is to show

$$n^{2-2/\lambda_0} \mathrm{Var}(\pi_1^*(C, \lambda)) \to 0.$$

Let

$$f_n(x) = n^{2-2/\lambda_0} C^2 C_0 (n+1)(1 - C_0 x^{-\lambda_0})^n x^{-(2\lambda+\lambda_0)}$$

defined for $x > x_0$. It can be easily verified that $f_n(x)$ is decreasing for all $x \geq x(n)$ and is increasing for $x \in [x_0, x(n)]$ where $x(n) = \{C_0[(n+1)\lambda_0 + 2\lambda]/(\lambda_0 + 2\lambda)\}^{1/\lambda_0}$. For any $\lambda > \lambda_0 - 1$,

$$\begin{aligned} f_n(x(n)) &= C^2 C_0 (n+1) n^{2-\frac{2}{\lambda_0}} \left[1 - \frac{\lambda_0 + 2\lambda}{(n+1)\lambda_0 + 2\lambda}\right]^n \\ &\quad \times \left\{\frac{C_0[(n+1)\lambda_0 + 2\lambda]}{\lambda_0 + 2\lambda}\right\}^{-\frac{2\lambda+\lambda_0}{\lambda_0}} \\ &= \mathcal{O}(n^{2[1-(1+\lambda)/\lambda_0]}) \to 0. \end{aligned}$$

Invoking the Euler–Maclaurin lemma, for $\lambda > \lambda_0 - 1/2$,

$$\begin{aligned} &\lim_{n\to\infty} n^{2-2/\lambda_0}(n+1) C^2 C_0 \sum_{k>x_0} (1 - C_0 k^{-\lambda_0})^n k^{-(2\lambda+\lambda_0)} \\ &= C^2 C_0 \lim_{n\to\infty} n^{2-2/\lambda_0}(n+1) \int_{x_0}^{\infty} (1 - C_0 x^{-\lambda_0})^n x^{-(\lambda_0+2\lambda)} dx \\ &= \frac{C^2 C_0^{(1-2\lambda)/\lambda_0}}{\lambda_0} \lim_{n\to\infty} n^{1-\frac{1+2\lambda}{\lambda_0}} (n+1) \int_0^{nC_0 x_0^{-\lambda_0}} y^{\frac{2\lambda+\lambda_0-1}{\lambda_0}-1} e^{-y} dy \\ &= \mathcal{O}\left(n^{2-\frac{1+2\lambda}{\lambda_0}}\right) \to 0. \end{aligned}$$

Part (b) follows. □

Proof of Part 2: In the proof of Lemma 7.9 Part 1, at each of the four times $f_n(x)$ was defined and evaluated at its corresponding $x(n)$, $\lim f_n(x) = 0$ not only for $\lambda > \lambda_0$ but also $\lambda > \lambda_0 - 1/(2\lambda_0)$. This implies that the Euler–Maclaurin lemma applies in each of the four cases when $(C, \lambda) \in \Theta_2$. In particular, it is easily verified, first

$$n^{1-1/\lambda_0-1/(2\lambda_0)} \mathrm{E}(\pi_0^*(C, \lambda)) \to 0$$

and

$$n^{2-2/\lambda_0-2/(2\lambda_0)} \mathrm{Var}(\pi_0^*(C, \lambda)) \to 0,$$

which implies

$$n^{1-1/\lambda_0-1/(2\lambda_0)} \pi_0^*(C, \lambda) \xrightarrow{p} 0,$$

second

$$n^{1-1/\lambda_0-1/(2\lambda_0)} \mathrm{E}(\pi_1^*(C, \lambda)) \to 0$$

and

$$n^{2-2/\lambda_0-2/(2\lambda_0)} \mathrm{Var}(\pi_1^*(C, \lambda)) \to 0,$$

which implies

$$n^{1-1/\lambda_0-1/(2\lambda_0)}\pi_1^*(C,\lambda) \xrightarrow{p} 0.$$

□

Proof of Part 3: First consider the case of $(C, \lambda) \in \Theta_2$. Since the Euler–Maclaurin lemma applies in all four cases of f_n, it is easily verified that

$$n^{1-1/\lambda_0}\mathrm{E}(\pi_0^*(C,\lambda)) \to +\infty$$

and

$$n^{2-2/\lambda_0}\mathrm{Var}(\pi_0^*(C,\lambda)) \to 0,$$

which in turn implies that

$$n^{1-1/\lambda_0}\pi_0^*(C,\lambda) \xrightarrow{p} +\infty.$$

Similarly,

$$n^{1-1/\lambda_0}\pi_1^*(C,\lambda) \xrightarrow{p} +\infty.$$

Second consider the case of $(C, \lambda) \in \Theta_3$. For every $(C, \lambda) \in \Theta_3$, there exists a point $(C, \lambda_1) \in \Theta_2$, where $\lambda < \lambda_1$, such that

$$\pi_0^*(C,\lambda) = \sum_{k\geq 1} 1[x_k = 0]Ck^{-\lambda} > \sum_{k\geq 1} 1[x_k = 0]Ck^{-\lambda_1} = \pi_0^*(C,\lambda_1).$$

Since $\pi_0^*(C,\lambda_1) \xrightarrow{p} +\infty$, $\pi_0^*(C,\lambda) \xrightarrow{p} +\infty$. Similarly $\pi_1^*(C,\lambda) \xrightarrow{p} +\infty$. □

7.6 Exercises

1 Assuming that the probability distribution $\{p_k; k \geq 1\}$ on an countably infinite alphabet $\mathscr{X} = \{\ell_k; k \geq 1\}$ satisfies that $p_k > 0$ for every k, $k \geq 1$, show that

$$\lim_{n\to\infty} \mathrm{P}(\mathbb{K}_{tail} \subset \{k; k \geq k_0\}) = 1$$

where $\mathbb{K}_{tail}$ is as in (7.3).

2 Provide details of the proof for Lemma 7.3.

3 In Corollary 7.2, verify that $\hat{\Sigma}_2^{-1/2}$ has the form as given in (7.8).

4 Suppose the Pareto (power decay) tail model of (7.1) is true. Let $\delta = (4\lambda_0)^{-1}$, that is,

$$g(n,\delta) = n^{1-1/(2\lambda_0)}$$

where λ_0 is the true value of the parameter λ. Let Ω_n be defined by

$$g^2(n, \delta)\Omega_n = \hat{\Sigma}_2^{-1}$$

where $\hat{\Sigma}_2^{-1}$ is as in (7.8). Show that

$$\Omega_n \xrightarrow{p} \Omega$$

where Ω is some positive definite 2×2 matrix.

5 Let $l_n(C, \lambda)$ be as in (7.9), show that
a)

$$\frac{\partial^2 l_n}{\partial C^2} = 2n^{2-2/\lambda_0}(u_n, U_n)\Omega_n(u_n, U_n)^\tau,$$

b)

$$\frac{\partial^2 \ell_n}{\partial \lambda^2} = -2n^{2-2/\lambda_0}C(w_n, W_n)\Omega_n(T_1 - \pi_0^*, T_2 - \pi_1^*)^\tau + 2n^{2-2/\lambda_0}C^2(v_n, V_n)\Omega_n(v_n, V_n)^\tau,$$

c)

$$\frac{\partial^2 \ell_n}{\partial \lambda \partial C} = 2n^{2-2/\lambda_0}(v_n, V_n)\Omega_n(T_1 - \pi_0^*, T_2 - \pi_1^*)^\tau - 2n^{2-2/\lambda_0}C(v_n, V_n)\Omega_n(u_n, U_n)^\tau.$$

6 Give a proof of Part 2 of Lemma 7.7.

7 Give a proof of Part 5 of Lemma 7.7.

8 Give a proof of Part 6 of Lemma 7.7.

9 Give a proof of Part 7 of Lemma 7.7.

10 Show that

$$\lim_{k\to\infty} \left[\left(\frac{k}{k-1} \right)^\lambda k - k \right] = \lambda$$

where $\lambda > 0$ is a constant.

References

Antos, A. and Kontoyiannis, I. (2001). Convergence properties of functional estimates for discrete distributions. *Random Structures & Algorithms*, **19**, 163–193.

Beirlant, J., Dudewicz, E.J., Györfi, L., and Meulen, E.C. (2001). Nonparametric entropy estimation: an overview. *International Journal of the Mathematical Statistics Sciences*, **6**, 17–39.

Billingsley, P. (1995). *Probability and Measure*, John Wiley & Sons, Inc., New York.

Blyth, C.R. (1959). Note on estimating information. *Annals of Mathematical Statistics*, **30**, 71–79.

Bunge, J., Willis, A., and Walsh, F. (2014). Estimating the number of species in microbial diversity studies. *Annual Review of Statistics and its Application*, **1**, 427–445.

Chao, A. (1987). Estimating the population size for capture-recapture data with unequal catchability. *Biometrics*, **43**, 783–791.

Chao, A. and Jost, L. (2012). Coverage-based rarefaction and extrapolation: standardizing samples by completeness rather than size. *Ecology*, **93**, 2533–2547.

Chao, A. and Jost, L. (2015). Estimating diversity and entropy profiles via discovery rates of new species. *Methods in Ecology and Evolution*, **6**, 873–882.

Chao, A. and Shen, T.J. (2003). Nonparametric estimation of Shannon's index of diversity when there are unseen species. *Environmental and Ecological Statistics*, **10**, 429–443.

Chao, A., Chiu, C.-H., and Jost, L. (2010). Phylogenetic diversity measures based on Hill numbers. *Philosophical Transactions of the Royal Society B: Biological Sciences*, **365**, 3599–3609.

Chao, A., Lee, S.-M., and Chen, T.-C. (1988). A generalized Good's nonparametric coverage estimator. *Chinese Journal of Mathematics*, **16(3)**, 189–199.

Cochran, W.G. (1952). The χ^2 test of goodness of fit. *Annals of Mathematical Statistics*, **25**, 315–345.

Cover, T.M. and Thomas, J.A. (2006). *Elements of Information Theory*, 2nd ed. John Wiley & Sons, Inc., Hoboken, NJ.

Statistical Implications of Turing's Formula, First Edition. Zhiyi Zhang.

David, F.N. (1950). Two combinatorial tests of whether a sample has come from a given population. *Biometrika*, **37**, 97–110.

de Haan, L. and Ferreira, A. (2006). *Extreme Value Theory: An Introduction*, Springer Science+Business Media, LLC, New York.

Emlen, J.M. (1973). *Ecology: An Evolutionary Approach*, Addison-Wesley Publishing Co., Reading, MA.

Esty, W. (1983). A normal limit law for a nonparametric estimator of the coverage of a random sample. *The Annals of Statistics*, **11**(**3**), 905–912.

Fisher, R.A. and Tippett, L.H.C. (1922). On the interpretation of chi-square from contingency tables, and the calculation of P. *Journal of the Royal Statistical Society*, **85**, 87–94.

Fisher, R.A. and Tippett, L.H.C. (1928). Limiting forms of the frequency-distribution of the largest or smallest member of a sample. *Mathematical Proceedings of the Cambridge Philosophical Society*, **24**, 180.

Fisher, R.A., Corbet, A.S., and Williams, C.B. (1943). The relationship between the number of species and the number of individuals in a random sample of an animal population. *Journal of Animal Ecology*, **12**(**1**), 42–58.

Fréchet, M. (1927). Sur la loi de probabilité de l'écart maximum. *Annales de la Société Polonaise de Mathématique*, **6**, 92.

Gini, C. (1912). Variabilità e mutabilità. Reprinted in *Memorie di metodologica statistica* (eds E. Pizetti and T. Salvemini), Libreria Eredi Virgilio Veschi, Rome (1955).

Gnedenko, B.V. (1943). Sur la distribution limite du terme maximum d'une série aléatoire. *Annals of Mathematics*, **44**, 423–453.

Gnedenko, B.V. (1948). On a local limit theorem of the theory of probability. *Uspekhi Matematicheskikh Nauk*, **3**(**25**), 187–194.

Good, I.J. (1953). The population frequencies of species and the estimation of population parameters. *Biometrika*, **40**(**3-4**), 237–264.

Grabchak, M., Marcon, E., Lang, G., and Zhang, Z. (2016). The generalized Simpson's entropy is a measure of biodiversity *<hal-01276738>*.

Haeusler, E. and Teugels, J.L. (1985). On asymptotic normality of Hill's estimator for the exponent of regular variation. *The Annals of Statistics*, **13**(**2**), 743–756.

Hall, P. and Weissman, I. (1997). On the estimation of extreme tail probabilities. *The Annals of Statistics*, **25**(**3**), 1311–1326.

Hall, P. and Welsh, A.H. (1985). Adaptive estimates of parameters of regular variation. *The Annals of Statistics*, **13**(**1**), 331–341.

Harris, B. (1975). The statistical estimation of entropy in the non-parametric case. In *Topics in Information Theory* (ed. I. Csiszar), North-Holland, Amsterdam, 323–355.

Hausser, J. and Strimmer, K. (2009). Entropy inference and the James-Stein estimator, with application to nonlinear gene association networks. *Journal of Machine Learning Research*, **10**, 1469–1484.

Heip, C.H.R., Herman, P.M.J., and Soetaert, K. (1998). Indices of diversity and evenness. *Oc eanis*, **24**(**4**), 61–87.

Hill, M.O. (1973). Diversity and evenness: a unifying notation and its consequences. *Ecology*, **54**, 427–431.

Hill, B.M. (1975). A simple general approach to inference about the tail of a distribution. *The Annals of Statistics*, **3**(**5**), 1163–1174.

Hoeffding, W. (1948). A class of statistics with asymptotically normal distributions. *The Annals of Statistics*, **19**(**3**), 293–325.

Hoeffding, W. (1963). Probability Inequalities for Sums of Bounded Random Variables. *Journal of the American Statistical Association*, **58**, 13–30.

Holste, D., Große, I., and Herzel, H. (1998). Bayes' estimators of generalized entropies. *Journal of Physics A: Mathematical and General*, **31**, 2551–2566.

Johnson, N.L. and Kotz, S. (1977). *URN Models and their Applications*, John Wiley & Sons, Inc., New York.

Jost, L. (2006). Entropy and diversity. *Oikos*, **113**, 363–375.

Kolchin, V.F., Sevastyanov, B.A., and Chistyakov, V.P. (1978). *Random Allocations*, V.H. Winston & Sons, Washington, DC.

Krebs, C.J. (1999). *Ecological Methodology*, 2nd ed. Addison-Welsey Educational Publishers, Menlo Park, CA.

Krichevsky, R.E. and Trofimov, V.K. (1981). The performance of universal encoding. *IEEE Transactions on Information Theory*, **27**, 199–207.

Kullback, S. and Leibler, R.A. (1951). On information and sufficiency. *The Annals of Mathematical Statistics*, **22**(**1**), 79–86.

Kvalseth, T.O. (1987). Entropy and correlation: some comments. *IEEE Transactions on Systems, Man, and Cybernetics*, **17**(**3**), 517–519.

Lee, A.J. (1990). *U-Statistics: Theory and Practice*, Marcel Dekker, New York.

MacArthur, R.H. (1955). Fluctuations of animal populations, and a measure of community stability. *Ecology*, **36**, 533–536.

Magurran, A.E. (2004). *Measuring Biological Diversity*, Blackwell Publishing, Ltd., Malden, MA.

Mann, H.B. and Wald, A. (1943). On stochastic limit and order relationships. *The Annals of Mathematical Statistics*, **14**(**3**), 217–226.

Marcon, E. (2014). Mesures de la Biodiversité. Technical Report, Ecologie des forêts de Guyane.

Margalef, R. (1958). Temporal succession and spatial heterogeneity in phytoplankton. In *Perspectives in Marine biology* (ed. A.A. Buzzati-Traverso), University of California Press, Berkeley, CA, 323–347.

Miller, G.A. (1955). Note on the bias of information estimates. *Information Theory in Psychology; Problems and Methods*, **II-B**, 95–100.

Miller, G.A. and Madow, W.G. (1954). On the maximum likelihood estimate of the Shannon-Weaver measure of information. Air Force Cambridge Research Center Technical Report AFCRC-TR-54-75, Operational Applications Laboratory, Air Force, Cambridge Research Center, Air Research and Development Command, New York.

Montgomery-Smith, S. and Schümann, T. (2007). *Unbiased Estimators for Entropy and Class Number. Unpublished preprint 2007*, Department of Mathematics, University of Missouri, Columbia, MO.

Mood, A.M., Graybill, F.A., and Boes, D.C. (1974). *Introduction to the Theory of Statistics*, McGraw-Hill, Inc., New York.

Nemenman, I., Shafee, F., and Bialek, W. (2002). Entropy and inference, revisited. *Advances in Neural Information Processing Systems*, vol. **14**, MIT Press, Cambridge, MA, 471–478.

Ohannessian, M.I. and Dahleh, M.A. (2012). Rare probability estimation under regularly varying heavy tails. *Journal of Machine Learning Research*, Proceedings Track, **23**, 21.1–21.24.

Paninski, L. (2003). Estimation of entropy and mutual information. *Neural Computation*, **15**, 1191–1253.

Patil, G.P. and Taillie, C. (1982). Diversity as a concept and its measurement. *Journal of the American Statistical Association*, **77**, 548–567.

Pearson, K. (1900). On a criterion that a given system of deviations from the probable in the case of a correlated system of variables is such that it can be reasonably supposed to have arisen from random sampling. *Philosophical Magazine Series 5*, **50**, 157–175. (Reprinted 1948 in Karl Pearson's Early Statistical Papers, (ed. E.S. Pearson), Cambridge University Press, Cambridge.)

Pearson, K. (1922). On the χ^2 test of goodness of fit. *Biometrika*, **14**, 186–191.

Peet, R.K. (1974). The measurements of species diversity. *Annual Review of Ecology and Systematics*, **5**, 285–307.

Pickands, J. (1975). Statistical inference using extreme order statistics. *The Annals of Statistics*, **3**(**1**), 119–131.

Purvis, A. and Hector, A. (2000). Getting the measure of biodiversity. *Nature*, **405**, 212–219.

Rényi, A. (1961). On measures of entropy and information. *Proceedings of the 4th Berkeley Symposium on Mathematical Statistics and Probabilities*, Vol. 1, 547–561.

Robbins, H.E. (1968). Estimating the total probability of the unobserved outcomes of an experiment. *The Annals of Mathematical Statistics*, **39**(**1**), 256–257.

Schürmann, T. and Grassberger, P. (1996). Entropy estimation of symbol sequences. *Chaos*, **6**, 414–427.

Serfling, R.J. (1980). *Approximation Theorems of Mathematical Statistics*, John Wiley & Sons, Inc., New York.

Shannon, C.E. (1948). A mathematical theory of communication. *The Bell System Technical Journal*, **27**, 379–423 & 623–656.

Simpson, E.H. (1949). Measurement of diversity. *Nature*, **163**, 688.

Smith, R.L. (1987). Estimating tails of probability distributions. *The Annals of Statistics*, **15**(**3**), 1174–1207.

Stanley, R.P. (1997). *Enumerative Combinatorics*, Vol. **1**, Cambridge University Press, New York.

Strehl, A. and Ghosh, J. (2002). Cluster ensembles - a knowledge reuse framework for combining multiple partitions. *Journal of Machine Learning Research*, **3**, 583–617.

Strong, S.P., Koberle, R., de Ruyter van Steveninck, R.R., and Bialek, W. (1998). Entropy and information in neural spike trains. *Physical Review Letters*, **80**, 197–200.

Taillie, C. (1979). Species equitability: a comparative approach. In —*it Ecological Diversity in Theory and Practice* (eds J.F. Grassle, G.P. Patil, W. Smith and C. Taillie), International Cooperative Publishing House, Fairland, MD, 51–61.

Tanabe, K. and Sagae, M. (1992). An exact Cholesky decomposition and the generalized inverse of the variance-covariance matrix of the multinomial distribution, with applications. *Journal of the Royal Statistical Society, Series B Methodological*, **54**(**1**), 211–219.

Tsallis, C. (1988). Possible generalization of Boltzmann-Gibbs statistics. *Journal of Statistical Physics*, **52**, 479–487.

Valiant, G. and Valiant, P. (2011). Estimating the Unseen: an $n/\log(n)$-sample estimator for entropy and support size, shown optimal via new CLTs. In *Proceedings of the 43rd Annual ACM Symposium on Theory of Computing*, STOC'11 Symposium on Theory of Computing (Co-located with FCRC 2011), San Jose, CA, USA - June 06-08, 2011 (ed. L. Fortnow and S.P. Vadhan), Association for Computing Machinery, New York, 685–694.

Vinh, N.X., Epps, J., and Bailey, J. (2010). Information theoretic measures for clusterings comparison: variants, properties, normalization and correction for chance. *Journal of Machine Learning Research*, **11**, 2837–2854.

von Mises, R. (1936). *La distribution de la plus grande de n valeurs*. Reprinted (1954) in Selected Papers 11271-294, American Mathematical Society, Providence, RI.

Vu, V.Q., Yu, B., and Kass, R.E. (2007). Coverage-adjusted entropy estimation. *Statistical Analysis of Neuronal Data*, **26**(**21**), 4039–4060.

Wang, S.C. and Dodson, P. (2006). Estimating the diversity of dinosaurs. *Proceedings of the National Academy of Sciences of the United States of America*, **103**(**37**), 13601–13605.

Yao, Y.Y. (2003). Information-theoretic measures for knowledge discovery and data mining. In: *Entropy Measures, Maximum Entropy Principle and Emerging Applications* (ed. Karmeshu), 1st edition, Springer, Berlin, 115–136.

Zahl, S. (1977). Jackknifing an index of diversity. *Ecology*, **58**, 907–913.

Zhang, Z. (2012). Entropy estimation in Turing's perspective. *Neural Computation*, **24**(**5**), 1368–1389.

Zhang, Z. (2013a). A multivariate normal law for Turing's formulae. *Sankhyā: The Indian Journal of Statistics*, **75-A**(**1**), 51–73.

Zhang, Z. (2013b). Asymptotic normality of an entropy estimator with exponentially decaying bias. *IEEE Transactions on Information Theory*, **59**(**1**), 504–508.

Zhang, Z. (2017). Domains of attraction on countable alphabets. *Bernoulli Journal* (to appear).

Zhang, Z. and Grabchak, M. (2013). Bias adjustment for a nonparametric entropy estimator. *Entropy*, **15(6)**, 1999–2011.

Zhang, Z. and Grabchak, M. (2014). Nonparametric estimation of Kullback-Leibler divergence. *Neural Computation*, **26(11)**, 2570–2593.

Zhang, Z. and Grabchak, M. (2016). Entropic representation and estimation of diversity indices. *Journal of Nonparametric Statistics*, DOI: 10.1080/10485252.2016.1190357.

Zhang, Z. and Huang, H. (2008). A sufficient normality condition for Turing's formula. *Journal of Nonparametric Statistics*, **20**(**5**), 431–446.

Zhang, Z. and Stewart, A.M. (2016). Estimation of standardized mutual information. Technical Report No. 7, University of North Carolina at Charlotte.

Zhang, C.-H. and Zhang, Z. (2009). Asymptotic normality of a nonparametric estimator of sample coverage. *The Annals of Statistics*, **37**(**5A**), 2582–2595.

Zhang, Z. and Zhang, X. (2012). A normal law for the plug-in estimator of entropy. *IEEE Transactions on Information Theory*, **58**(**5**), 2745–2747.

Zhang, Z. and Zheng, L. (2015). A mutual information estimator with exponentially decaying bias. *Statistical Applications in Genetics and Molecular Biology*, **14**(**3**), 243–252.

Zhang, Z. and Zhou, J. (2010). Re-parameterization of multinomial distribution and diversity indices. *Journal of Statistical Planning and Inference*, **140**(7), 1731–1738.

Zubkov, A.M. (1973). Limit distributions for a statistical estimate of the entropy. *Teoriya Veroyatnostei i Ee Primeneniya*, **18**(3), 643–650.

Author Index

Statistical Implications of Turing's Formula, First Edition. Zhiyi Zhang.

X

Z

Subject Index